Breakthroughs in Water Engineering, Treatment, and Hydraulics Through Nanomaterial Innovation

Tawfik A. Saleh
https://orcid.org/0000-0002-3037-5159
King Fahd University of Petroleum and Minerals, Saudi Arabia

IGI Global
Scientific Publishing
Publishing Tomorrow's Research Today

Vice President of Editorial	Melissa Wagner
Director of Acquisitions	Mikaela Felty
Director of Book Development	Jocelynn Hessler
Production Manager	Mike Brehm
Cover Design	Jose Rosado

Published in the United States of America by
IGI Global Scientific Publishing
701 East Chocolate Avenue
Hershey, PA, 17033, USA
Tel: 717-533-8845 | Fax: 717-533-7115
Website: https://www.igi-global.com E-mail: cust@igi-global.com

Library of Congress Cataloging-in-Publication Data

LCCN: 2025036623 (CIP Data Pending)
ISBN13: 9798337358864
Isbn13Softcover: 9798337358871
EISBN13: 9798337358888
British Cataloguing in Publication Data
A Cataloguing in Publication record for this book is available from the British Library.

Table of Contents

Detailed Table of Contents

The rapid growth of industrialization, urbanization, and population has led to an unprecedented increase in wastewater generation, posing serious environmental and public health challenges worldwide. Wastewater recycling has emerged as a sustainable solution to mitigate freshwater scarcity and reduce pollution loads discharged into natural water bodies. This chapter provides a comprehensive overview of physical and chemical wastewater treatment technologies with a particular focus on the integration of biomass-based materials and nanomaterials for water purification. Conventional physical and chemical methods such as coagulation–flocculation, adsorption, membrane filtration, and advanced oxidation processes are discussed alongside their limitations. Recent advances in biomass-derived adsorbents and engineered nanomaterials are highlighted, emphasizing their enhanced removal efficiency, tunable surface properties, and potential for sustainable and cost-effective wastewater recycling.

Advancing water treatment technologies has become increasingly critical as the need for sustainable wastewater management and resource recovery continues to grow. The rising complexity of water pollutants, including persistent organic compounds and emerging contaminants, has intensified the demand for purification strategies that

The growing demand for clean water and increasing pollution from domestic, industrial, and agricultural sources necessitate effective wastewater treatment strategies. This study provides a comprehensive overview of conventional treatment methods and introduces impedance and dielectric spectroscopy as complementary tools for water quality analysis. It details wastewater classification and treatment stages—pretreatment, primary, secondary, and tertiary—and highlights the valorization of industrial by-products such as fly ash and bottom ash for pollutant removal. Emphasis is placed on water reuse in agriculture, industry, and urban sectors, as well as on sludge management and its integration into eco-friendly mortars. The work also explores desalination technologies, including thermal and membrane processes, and their synergy with renewable energies. Impedance and dielectric spectroscopy offer non-destructive, frequency-dependent analyses, enhancing understanding of treatment efficiency and pollutant behavior. This dual approach promotes sustainability and resource recovery.

The growing global concern over water scarcity is closely tied to current environmental, energy, and economic challenges. Rapid technological progress, while beneficial in many domains, has contributed to the neglect of water conservation efforts, leading to critical ecological damage. T Greywater—comprising used water from bathing, dishwashing, and laundry—was collected and processed using phytoremediation in vertical and horizontal wetland setups. These systems were filled with sand and gravel substrates and supported plant growth to aid natural filtration. The treatment aimed to determine the reusability of the filtered water for secondary purposes such as bathing, irrigation, and general household use. The analysis emphasized plant adaptability and the efficiency of filtration in both horizontal and vertical flow wetlands. The microbial assessment of treated greywater revealed the presence of harmful pathogens, including E. coli, Salmonella spp., Staphylococcus aureus, Shigella spp., and Pseudomonas aeruginosa, highlighting the need for effective disinfection protocols.

The increasing contamination of aquatic ecosystems with persistent organic pollutants has intensified the search for efficient and sustainable wastewater treatment technologies. In this study, cobalt ferrite ($CoFe_2O_4$) nanoparticles and their doped counterparts—Ni@$CoFe_2O_4$ and La@Ni@$CoFe_2O_4$—were synthesized via a co-precipitation method and systematically evaluated for their photocatalytic performance in the degradation of methylene blue (MB) dye under visible light irradiation. Among the samples, La@Ni@$CoFe_2O_4$ demonstrated the highest photocatalytic activity, achieving a superior degradation rate of methylene blue due to the synergistic effects of dual doping. Additionally, the intrinsic magnetic properties of the ferrites allowed for easy recovery and reuse of the catalysts, highlighting their practical applicability in wastewater remediation. This study underscores the potential of co-doped $CoFe_2O_4$ nanostructures as efficient, magnetically separable photocatalysts for environmental applications, particularly in the treatment of dye-laden industrial effluents.

In water-limited areas like Western Tamil Nadu, where anthropogenic and environmental factors have a major impact on groundwater resources, evaluating and protecting the quality of the groundwater is essential to life support. The pressing need to use trustworthy, data driven predictive frameworks to address complicated groundwater quality issues is what spurred this study. The availability of robust and interpretable models that accurately represent seasonal and spatial variations in groundwater parameters remains a major gap despite significant efforts. To address this, the current study uses two sophisticated ensemble machine learning (ML) techniques,

Preface

Water is one of the most vital natural resources for human survival, economic development, and environmental sustainability. However, rapid population growth, industrialization, urban expansion, and climate change have placed unprecedented pressure on global freshwater resources. The increasing volume of wastewater generated from domestic, agricultural, and industrial activities has intensified the need for efficient water treatment and recycling technologies to ensure water security and environmental protection.

This book presents a comprehensive overview of water treatment and recycling technologies, with a particular emphasis on physical, chemical, biological, and advanced material-based approaches. It aims to bridge fundamental principles with recent technological advancements, highlighting both conventional treatment processes and emerging innovations such as bio-based materials, nanomaterials, and advanced oxidation systems. By integrating sustainability concepts, the book addresses current challenges related to energy consumption, cost, environmental impact, and regulatory requirements.

Nanotechnology has been developed as a consequence of significant advances in the materials science community. In the excitement surrounding these materials and technologies, however, their potential has been, frequently overhyped. The book explores these kinds of materials and forward-looking potential applications for water engineering. The book chapters are organized in a good way and make extensive use of illustrations.

The first chapters introduce an overview of the water technologies and nanomaterials for water purification. The other next chapters aim to provide the readers with information about the advanced nanomaterials and their methods of preparation with more focus on green methods such as microwave-assisted methods for the synthesis of nanomaterials and fundamentals and sources of magnetic nanoparticles. Biomaterials are good sources for the preparation of nanomaterials and their use as a support as well.

The content is structured to guide readers through the progressive stages of water treatment, from primary physical processes to advanced tertiary and recycling technologies. Special attention is given to understanding treatment mechanisms, system design considerations, performance evaluation, and the role of innovative materials in enhancing water purification efficiency. Case studies and practical examples are included to illustrate real-world applications in municipal and industrial wastewater management.

This book is intended for students, researchers, engineers, policymakers, and practitioners involved in water and environmental engineering, environmental science, and sustainability studies. It serves as both an educational resource and a reference for developing efficient, safe, and sustainable water treatment solutions. It is our hope that this book will contribute to advancing knowledge, fostering innovation, and supporting global efforts toward sustainable water management and circular water use.

ORGANIZATION OF THE BOOK

Chapter 1: Physical and Chemical Treatment Technologies for Waste-Water Recycling: An Overview of Biomass and Nanomaterials for Water Purification

This chapter provides a comprehensive overview of physical and chemical wastewater treatment technologies with a particular focus on the integration of biomass-based materials and nanomaterials for water purification. Conventional physical and chemical methods such as coagulation–flocculation, adsorption, membrane filtration, and advanced oxidation processes are discussed alongside their limitations. Recent advances in biomass-derived adsorbents and engineered nanomaterials are highlighted, emphasizing their enhanced removal efficiency, tunable surface properties, and potential for sustainable and cost-effective wastewater recycling.

Chapter 2: Nanomembrane and Electrochemical Approaches for Sustainable Water Purification and Metal Recovery: Water Treatment Methods

This chapter explores the membrane-based and electrochemical processes as systems for water purification, pollutant removal, and valuable resource extraction. Particular emphasis is placed on the role of nanomembranes and electrochemical systems in mitigating membrane fouling, enhancing selectivity, and improving overall treatment efficiency. Electrochemical techniques exhibit strong potential for the degradation of organic pollutants and the removal of heavy metal ions through controlled oxidation–reduction reactions. Various nanomembrane architectures and their functional advantages in water treatment are examined, highlighting pathways for improved contaminant rejection and recovery of strategic metals.

Chapter 3: Nanostructure-based Materials for Sorption and Photoelectrochemical Degradation of Water Pollutants: Water Purification Techniques

This chapter discusses polymer-functionalized carbon-based nanocomposites as promising materials for adsorption, and nanostructures of metal oxides for photodegradation of pollutants. The chapter will present some examples of multifunctional platforms capable of coupling PEC processes to achieve enhanced water purification. An example of the adsorption mechanisms will be discussed with highlights on the Langmuir and the Freundlich isotherm models.

Chapter 4: General Overview of Wastewater Types, Treatment Processes, and Sludge Valorization

This chapter provides a comprehensive overview of conventional treatment methods and introduces impedance and dielectric spectroscopy as complementary tools for water quality analysis. It details wastewater classification and treatment stages—pretreatment, primary, secondary, and tertiary—and highlights the valorization of industrial by-products such as fly ash and bottom ash for pollutant removal. Emphasis is placed on water reuse in agriculture, industry, and urban sectors, as well as on sludge management and its integration into eco-friendly mortars. The work also explores desalination technologies, including thermal and membrane processes, and their synergy with renewable energies. Impedance and dielectric spectroscopy offer non-destructive, frequency-dependent analyses, enhancing understanding of treatment efficiency and pollutant behavior. This dual approach promotes sustainability and resource recovery.

Chapter 5: Phytoremediation Based Wastewater Treatment Using Greywater for Sustainable Water Resource Reuse: Wastewater Treatment Using Greywater

The growing global concern over water scarcity is closely tied to current environmental, energy, and economic challenges. Rapid technological progress, while beneficial in many domains, has contributed to the neglect of water conservation efforts, leading to critical ecological damage. T Greywater—comprising used water from bathing, dishwashing, and laundry—was collected and processed using phytoremediation in vertical and horizontal wetland setups. These systems were filled with sand and gravel substrates and supported plant growth to aid natural filtration. The treatment aimed to determine the reusability of the filtered water for secondary purposes such as bathing, irrigation, and general household use. The analysis emphasized plant adaptability and the efficiency of filtration in both horizontal and vertical flow wetlands. The microbial assessment of treated greywater revealed the presence of harmful pathogens, including E. coli, Salmonella spp., Staphylococcus aureus, Shigella spp., and Pseudomonas aeruginosa, highlighting the need for effective disinfection protocols.

Chapter 6: Tuning Cobalt Ferrite Nanostructures Through Ni and La Doping for Efficient Visible-Light-Driven Wastewater Purification

The increasing contamination of aquatic ecosystems with persistent organic pollutants has intensified the search for efficient and sustainable wastewater treatment technologies. In this study, cobalt ferrite (CoFe2O4) nanoparticles and their doped counterparts—Ni@CoFe2O4 and La@Ni@CoFe2O4—were synthesized via a co-precipitation method and systematically evaluated for their photocatalytic performance in the degradation of methylene blue (MB) dye under visible light irradiation. Among the samples, La@Ni@CoFe2O4 demonstrated the highest photocatalytic activity, achieving a superior degradation rate of methylene blue due to the synergistic effects of dual doping. Additionally, the intrinsic magnetic properties of the ferrites allowed for easy recovery and reuse of the catalysts, highlighting their practical applicability in wastewater remediation. This study underscores the potential of co-doped CoFe2O4 nanostructures as efficient, magnetically separable photocatalysts for environmental applications, particularly in the treatment of dye-laden industrial effluents.

Chapter 7: Data-Driven Prediction of Groundwater Contamination in the Western Tamil Nadu

In water-limited areas like Western Tamil Nadu, where anthropogenic and environmental factors have a major impact on groundwater resources, evaluating and protecting the quality of the groundwater is essential to life support. The pressing need to use trustworthy, data driven predictive frameworks to address complicated groundwater quality issues is what spurred this study. The availability of robust and interpretable models that accurately represent seasonal and spatial variations in groundwater parameters remains a major gap despite significant efforts. To address this, the current study uses two sophisticated ensemble machine learning (ML) techniques,

Chapter 8: Remote Sensing and Artificial Intelligence Driving Liquid Intelligence for Water Sustainability Financing

This chapter explores how innovative financing models, including blended finance, microcredit, and public–private partnerships, can be strengthened through technological integration. Remote sensing provides accurate data on water resources, quality, and scarcity, while artificial intelligence enables predictive analytics, smart infrastructure management, and optimized investment decisions. Together, these tools enhance transparency, efficiency, and resilience in financing frameworks. The study argues that combining financial innovation with data-driven intelligence not only advances SDG 6 but also fosters equity, health, and long-term economic sustainability.

We welcome suggestions from readers towards improvements that can be incorporated in future editions of this book.

We hope that you enjoy the book!

Tawfik A. Saleh

Acknowledgement

I would like to acknowledge the support provided by King Fahd University of Petroleum & Minerals. I would like to express my gratitude to the many people who saw me through this book; to all those who provided support, read, wrote, offered comments, allowed me to quote their remarks, and assisted in the editing, proofreading, and design. Above all, I want to thank my parents, relatives, wife, children, and the rest of my family, who supported and encouraged me in spite of all the time it took me away from them. Last but not least: I beg forgiveness of all those who have been with me over the course of the months and whose names I have failed to mention.

Chapter 1
Physical and Chemical Treatment Technologies for Waste-Water Recycling:
An Overview of Biomass and Nanomaterials for Water Purification

Tawfik Abdo Saleh
https://orcid.org/0000-0002-3037-5159
King Fahd University of Petroleum and Minerals, Saudi Arabia

ABSTRACT

The rapid growth of industrialization, urbanization, and population has led to an unprecedented increase in wastewater generation, posing serious environmental and public health challenges worldwide. Wastewater recycling has emerged as a sustainable solution to mitigate freshwater scarcity and reduce pollution loads discharged into natural water bodies. This chapter provides a comprehensive overview of physical and chemical wastewater treatment technologies with a particular focus on the integration of biomass-based materials and nanomaterials for water purification. Conventional physical and chemical methods such as coagulation–flocculation, adsorption, membrane filtration, and advanced oxidation processes are discussed alongside their limitations. Recent advances in biomass-derived adsorbents and engineered nanomaterials are highlighted, emphasizing their enhanced removal efficiency, tunable surface properties, and potential for sustainable and cost-effective wastewater recycling.

DOI: 10.4018/979-8-3373-5886-4.ch001

1. OVERVIEW

Wastewater recycling is a critical component of sustainable water management strategies, particularly in regions experiencing water stress and limited freshwater resources. Traditional wastewater treatment systems rely heavily on physical and chemical processes to remove suspended solids, organic matter, nutrients, and toxic contaminants. While these methods are widely implemented and technologically mature, they often suffer from high operational costs, secondary pollution, and limited efficiency in removing emerging contaminants such as pharmaceuticals, dyes, heavy metals, and persistent organic pollutants.

In recent years, the use of biomass-derived materials and nanomaterials has gained significant attention as complementary or alternative solutions for water purification. Biomass materials, including agricultural waste, biochar, and natural polymers, offer advantages such as low cost, renewability, and environmental compatibility. Nanomaterials, on the other hand, exhibit unique physicochemical properties—such as high surface area, enhanced reactivity, and tunable surface chemistry—that significantly improve contaminant removal efficiency.

This overview explores the synergy between conventional physical and chemical treatment technologies and emerging biomass- and nanomaterial-based approaches. Emphasis is placed on adsorption, membrane-based processes, and advanced oxidation techniques, where these materials have shown promising performance. The integration of sustainable materials into wastewater recycling systems represents a key step toward achieving efficient, environmentally friendly, and economically viable water treatment solutions.

2. WATER Scarcity And Water Treatment And Recycling Technologies

2.1. Water Scarcity

Water scarcity has become one of the most pressing global challenges of the twenty-first century, driven by population growth, climate change, industrial expansion, and increasing agricultural demand. As freshwater resources continue to decline, wastewater recycling has emerged as an essential strategy for sustainable water supply and environmental protection. Effective wastewater treatment is necessary not only to enable water reuse but also to prevent the discharge of harmful contaminants into ecosystems. Physical and chemical treatment technologies form the backbone of most conventional wastewater treatment systems. Physical processes such as sedimentation, filtration, and membrane separation are primarily used to remove suspended solids and particulate matter, while chemical processes—including coagulation–flocculation, chemical precipitation, adsorption, and advanced oxidation—target dissolved pollutants and toxic compounds. Despite their widespread application, these technologies face limitations related to energy consumption, chemical usage, sludge generation, and reduced efficiency against complex or low-concentration contaminants. To address these challenges, researchers have increasingly explored the incorporation of biomass-based materials and nanomaterials into wastewater treatment processes. Biomass materials derived from natural and waste sources offer sustainable alternatives for pollutant removal through adsorption and ion exchange mechanisms. Meanwhile, nanomaterials such as metal oxides, carbon-based nanostructures, and nanocomposites provide superior removal capabilities due to their high surface area, catalytic activity, and selective interactions with contaminants.

2.2. Water Treatment and Recycling Technologies

The water treatment and recycling technologies can be classified into primary, secondary, and tertiary treatment stages, reflecting the progressive improvement in water quality, Figure 1.

Primary technologies focus on the physical removal of suspended and settleable solids. Processes such as screening, filtration, sedimentation, flotation, and coagulation reduce turbidity and particulate matter, serving as essential pretreatment steps that protect downstream units.

Secondary technologies are mainly biological processes, including aerobic and anaerobic treatments, which target biodegradable organic matter. These processes significantly reduce biochemical oxygen demand (BOD) and chemical oxygen demand (COD) through microbial degradation.

Tertiary technologies represent advanced physical and chemical treatments designed to remove dissolved salts, nutrients, toxic compounds, and emerging pollutants. Techniques such as reverse osmosis, ion exchange, adsorption, oxidation, electrodialysis, and crystallization enable high-quality water recovery suitable for reuse and recycling.

Figure 1. Classification of water recycling technologies

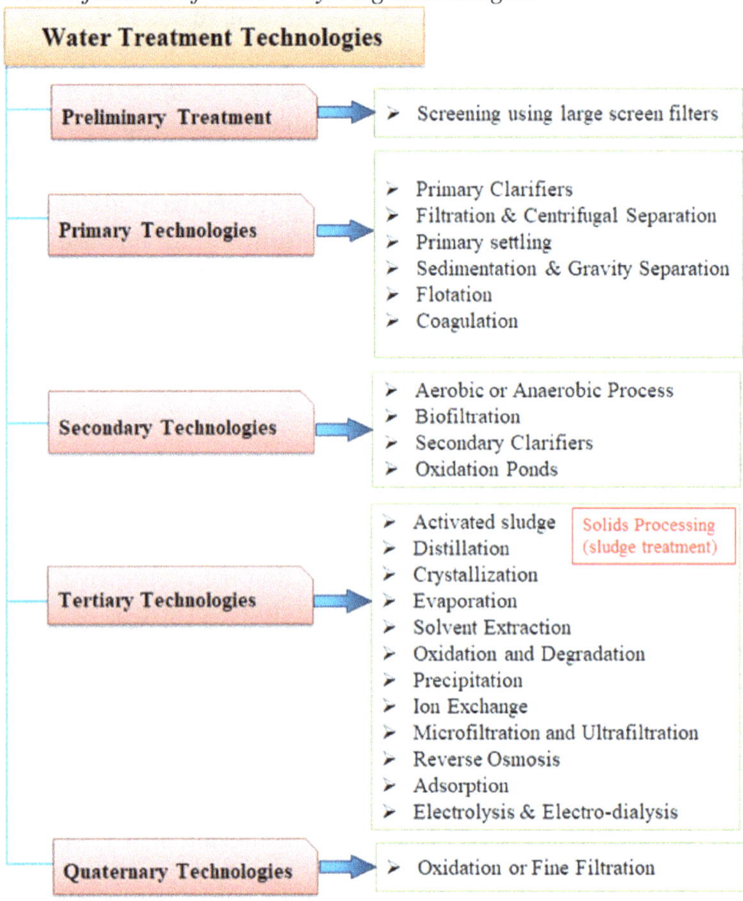

3. PHYSICAL WASTEWATER TREATMENT TECHNOLOGIES

Physical wastewater treatment technologies are primarily designed to remove suspended solids, particulate matter, and insoluble contaminants without altering their chemical structure. These processes are commonly employed as preliminary or primary treatment stages and play a crucial role in reducing pollutant load before advanced treatment.

3.1 Screening and Sedimentation

Screening is the first physical barrier in wastewater treatment, where large debris such as plastics, rags, and organic residues are removed using coarse and fine screens. Following screening, sedimentation allows heavier suspended particles to settle under gravity in sedimentation tanks. This process effectively reduces turbidity and total suspended solids (TSS), improving the efficiency of downstream treatment units.

3.2 Filtration

Filtration involves the passage of wastewater through porous media such as sand, gravel, or synthetic filters to remove fine suspended particles. Rapid and slow sand filtration systems are widely used in municipal and industrial wastewater treatment. Filtration enhances water clarity and reduces microbial load, making it an essential step prior to disinfection or advanced purification processes.

3.3 Membrane Separation Processes

Membrane technologies, including microfiltration (MF), ultrafiltration (UF), nanofiltration (NF), and reverse osmosis (RO), are advanced physical treatment methods that separate contaminants based on size and molecular weight. These processes are highly effective in removing pathogens, organic compounds, salts, and heavy metals. However, membrane fouling and high operational costs remain significant challenges.

3.4 Flotation Techniques

Dissolved air flotation (DAF) is a physical separation process used to remove oils, greases, and low-density suspended solids. Fine air bubbles attach to particles, causing them to float to the surface, where they are skimmed off. Flotation is particularly effective for industrial wastewater containing fats, oils, and grease (FOG), as well as algal-laden effluents.

4. CHEMICAL WASTEWATER TREATMENT TECHNOLOGIES

Chemical treatment technologies involve the addition of chemical agents to wastewater to transform, neutralize, or remove dissolved pollutants. These processes are essential for targeting contaminants that cannot be effectively removed by physical methods alone.

4.1 Coagulation and Flocculation

Coagulation–flocculation is one of the most widely applied chemical treatment processes. Coagulants such as aluminum sulfate, ferric chloride, and polyaluminum chloride destabilize colloidal particles, allowing them to aggregate into larger flocs. These flocs are subsequently removed by sedimentation or filtration. The process is highly effective for removing turbidity, organic matter, and some heavy metals.

4.2 Chemical Precipitation

Chemical precipitation is used to remove dissolved inorganic contaminants, particularly heavy metals and phosphates, by converting them into insoluble forms. Reagents such as lime, sulfides, and carbonates are added to induce precipitation, followed by solid–liquid separation. This method is commonly applied in industrial wastewater treatment and nutrient removal.

4.3 Advanced Oxidation Processes (AOPs)

Advanced oxidation processes rely on the generation of highly reactive species, particularly hydroxyl radicals (•OH), to degrade persistent organic pollutants. Technologies such as ozonation, UV/H_2O_2, Fenton, and photo-Fenton processes are effective in mineralizing pharmaceuticals, dyes, and endocrine-disrupting compounds. Despite their high efficiency, AOPs can be energy-intensive and require careful control of operating conditions.

5.4 Ion Exchange

Ion exchange involves the reversible exchange of ions between wastewater and a solid resin. This process is particularly effective for removing nitrates, ammonium, and heavy metals. Synthetic resins and natural ion exchangers are widely used, although regeneration and disposal of spent resins pose environmental and economic concerns.

5. BIO-BASED WASTEWATER TREATMENT TECHNOLOGIES

Bio-based wastewater treatment technologies utilize biological materials or biologically derived substances to remove contaminants in an environmentally friendly and sustainable manner. These approaches align closely with circular economy principles and green chemistry concepts.

5.1 Biomass-Derived Adsorbents

Agricultural residues, forestry waste, algae, and microbial biomass have been extensively investigated as low-cost adsorbents for wastewater treatment. Materials such as rice husk, coconut shell, sawdust, and biochar exhibit high affinity for dyes, heavy metals, and organic pollutants due to their functional surface groups. Their abundance and renewability make them attractive alternatives to conventional adsorbents.

5.2 Biochar and Activated Carbon from Biomass

Biochar produced through the pyrolysis of biomass has gained significant attention for water purification applications. Its porous structure, high surface area, and surface functionality enable effective adsorption of organic and inorganic contaminants. Biomass-derived activated carbon further enhances adsorption performance and is increasingly used in advanced wastewater treatment systems.

5.3 Biopolymers and Natural Coagulants

Natural polymers such as chitosan, alginate, cellulose, and starch derivatives are used as eco-friendly coagulants and flocculants. These biopolymers offer advantages over synthetic chemicals, including biodegradability, low toxicity, and reduced sludge production. Chitosan, in particular, has demonstrated high efficiency in turbidity and heavy metal removal.

5.4 Algae- and Microorganism-Based Systems

Algae and microorganisms play a vital role in bio-based wastewater treatment through nutrient uptake, biosorption, and biodegradation. Algal systems are effective in removing nitrogen, phosphorus, and organic matter while simultaneously producing valuable biomass. These systems are increasingly integrated with physical and chemical processes to enhance overall treatment efficiency.

6. NANOMATERIALS IN ENHANCING WATER PURIFICATION AND RECYCLING

Nanomaterials have emerged as powerful tools in advanced wastewater treatment and recycling due to their unique physicochemical properties, including high surface-to-volume ratios, enhanced reactivity, tunable surface chemistry, and catalytic activity. These properties enable superior removal of a wide range of contaminants, including heavy metals, dyes, pharmaceuticals, pathogens, and emerging pollutants that are often resistant to conventional treatment methods.

6.1 Types of Nanomaterials Used in Water Treatment

Commonly investigated nanomaterials include metal and metal oxide nanoparticles (e.g., TiO_2, ZnO, Fe_3O_4), carbon-based nanomaterials (e.g., carbon nanotubes, graphene oxide, fullerenes), and nanocomposites. Metal oxide nanoparticles are widely used as photocatalysts and adsorbents, while carbon-based nanomaterials offer exceptional adsorption capacity and mechanical stability. Nanocomposites combine multiple functionalities, such as magnetic separation and catalytic degradation, enhancing process efficiency and reusability.

Nanomaterials can be classified as:

- Zero dimension (0D); for example Q-dot, Powder, carbon nanosphere
- One dimension (1D); for example wire/nanotube, carbon nanofiber, carbon nanotube, titania nanorods, core-shell
- Two dimensions (2D); for example, thin films, multilayers, graphene, flower-like nanosheets,
- Three dimensions (3D); crumpled graphene, hybrid nanostructures

Carbon-based materials and nanomaterials are among the most common materials used in water purification. Figure 2 depicts different examples of carbon structures. Graphene, as an example of this class, is a single carbon layer of the graphite structure. It is a two-dimensional carbon allotrope made up of a conjugated system of sp^2 carbons arranged in a honeycomb structure. It is similar in nature to polycyclic aromatic hydrocarbons of quasi-infinite size. Some carbon structures can be used to form 3D nanostructured materials. For example, carbon nanotubes are used as a spacer between graphene nanosheets, Figure 3.

Figure 2. Examples of carbon nanostructures

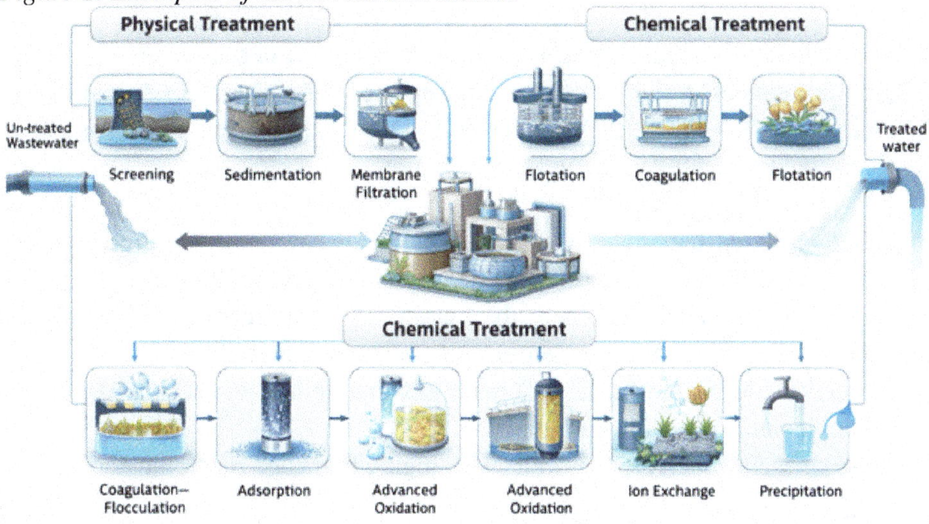

Figure 3. Carbon nanotubes used as a spacer between graphene nanosheets

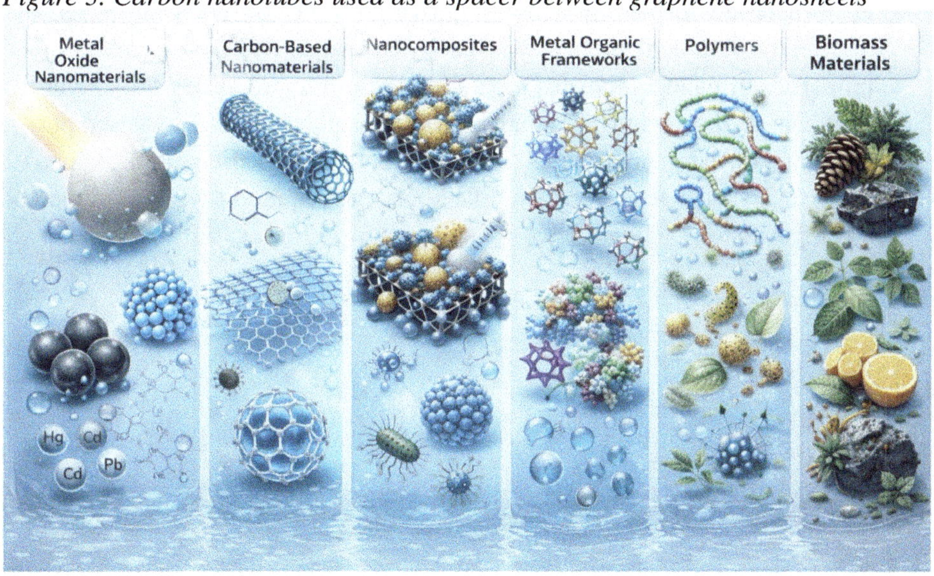

6.2 Classification of Nanomaterials Based on Their Morphology

The shape and structures of the nanomaterials play a major role in their efficiency. Figure 4 introduces examples of various shapes of nanomaterials (Hummers and Offeman 1958; Saleh et al., 2016; Saleh & Al-Daous, 2016). Nanomaterials can be categorized based on their chemical composition as:

- Carbon-based nanomaterials which consist of carbon with different structures, including nanotubes, fullerenes, graphene, and a nanosphere.
- Metal-based materials, where the main components of these materials are metals. Examples are silver, gold, and copper nanoparticles.
- Metal oxide nanomaterials are materials that consist of metal and oxygen, like titania, silica, and alumina.
- Dendrimers which are materials with highly branched macromolecules with dimensions in the nanoscale. The surface of dendrimers has numerous chains that can be modified toward specific functions. Another type is called nanocomposites, which consists of more than one type of nanomaterial.
- Nanocomposites are described as multiphase materials where at least one of the phases has at least one dimension in the nanoscale. They are special classes of materials formed by combining two phases of different materials, and they tend to retain the properties of each material used in their formation. The properties of nanocomposites are a function of the properties of the constituent phases, their relative amounts, and the geometry of the dispersed phase, the shape of the nanoparticles, size, distribution, and orientation. A matrix, filler, resin, and a reinforcement, like fibers, are required for the formation of a composite.

Figure 4. Illustration of various shapes of nanomaterials

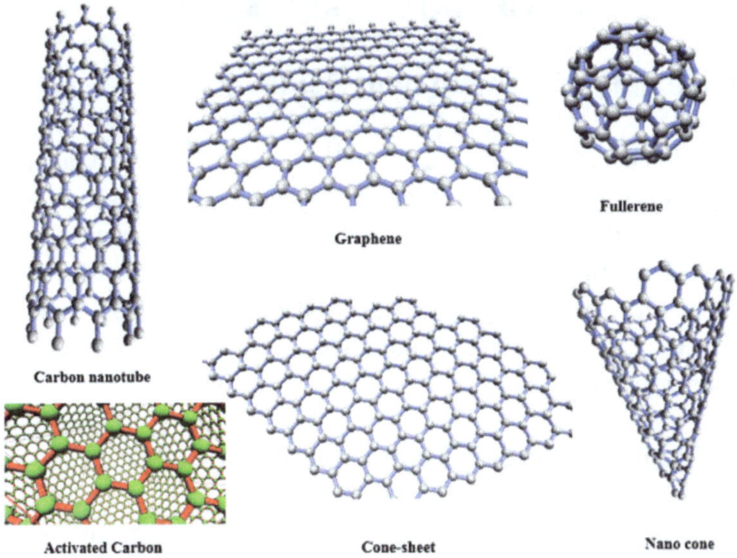

6.3 Nanomaterials in Adsorption and Membrane Technologies

Nanomaterials significantly improve adsorption processes by providing abundant active sites for pollutant binding. Functionalized nanoparticles and graphene-based materials exhibit strong interactions with contaminants through electrostatic attraction, complexation, and π–π interactions. In membrane technologies, nanomaterials are incorporated into polymeric membranes to improve permeability, selectivity, hydrophilicity, and antifouling properties, thereby extending membrane lifespan and reducing operational costs.

6.4 Nanocatalysts and Advanced Oxidation

Nanomaterials play a critical role in advanced oxidation processes by acting as catalysts for the generation of reactive oxygen species. Photocatalytic nanomaterials such as TiO_2 and doped metal oxides enable the degradation of persistent organic pollutants under UV or visible light. These processes contribute to pollutant mineralization, making treated wastewater suitable for reuse applications.

7. UNDERSTANDING THE MECHANISMS, BENEFITS, AND CHALLENGES OF WATER TREATMENT

7.1 Mechanisms of Contaminant Removal

Water treatment technologies rely on multiple mechanisms to remove pollutants, including physical separation, chemical transformation, adsorption, ion exchange, biodegradation, and photocatalysis. Adsorption mechanisms involve surface interactions between contaminants and treatment materials, while chemical processes alter pollutant structure through oxidation, reduction, or precipitation. Biological and bio-based systems utilize metabolic pathways and biosorption to eliminate organic matter and nutrients.

7.2 Benefits of Advanced Treatment Technologies

The integration of physical, chemical, bio-based, and nanomaterial-assisted technologies offers numerous benefits, such as enhanced contaminant removal efficiency, reduced treatment time, improved water quality, and increased potential for water reuse. These advanced systems enable the removal of trace contaminants and emerging pollutants, contributing to improved environmental protection and public health. Furthermore, the use of sustainable materials supports circular economy and resource recovery strategies.

7.3 Challenges and Limitations

Despite their advantages, advanced wastewater treatment technologies face several challenges. High capital and operational costs, energy consumption, material stability, and secondary pollution remain significant concerns. For nanomaterials in particular, issues related to environmental fate, toxicity, aggregation, and recovery must be carefully addressed. Additionally, scaling up laboratory-scale technologies to industrial applications requires robust process optimization and regulatory compliance.

8. Wastewater Recycling Methods

Wastewater recycling involves the treatment and reuse of wastewater for various applications, thereby reducing freshwater demand and environmental discharge. Recycling methods vary depending on water quality requirements and intended reuse.

8.1 On-Site and Decentralized Recycling Systems

On-site wastewater recycling systems are commonly used in residential buildings, commercial facilities, and industrial plants. These systems treat wastewater locally using compact physical, chemical, and biological units, enabling water reuse for non-potable applications such as toilet flushing, cooling systems, and landscape irrigation.

8.2 Centralized Recycling and Reclamation Plants

Centralized wastewater treatment and reclamation plants employ multi-stage treatment processes to produce high-quality reclaimed water. Advanced treatment trains often include membrane filtration, disinfection, and advanced oxidation to meet stringent reuse standards. Reclaimed water from centralized systems is used in agriculture, industrial processes, groundwater recharge, and, in some cases, indirect potable reuse.

8.3 Industrial Wastewater Recycling and Zero Liquid Discharge (ZLD)

Industrial wastewater recycling focuses on recovering water and valuable resources from effluents generated by manufacturing processes. Zero liquid discharge systems combine evaporation, crystallization, and advanced membrane technologies to eliminate liquid waste discharge entirely. Although energy-intensive, ZLD systems support sustainable industrial water management and regulatory compliance.

9. SELECTION OF WATER TREATMENT TECHNOLOGIES

The type of required wastewater treatment determines the selection of the treatment technologies. The water treatment methods can be summarized as shown in Figure 5. The highly polluted water, having color, solid wastes are first treated with primary and secondary processes, followed by the tertiary water treatment technologies. If the biological oxygen demand is negligible, then the secondary process is not required. If water is colorless without any solids and polluted due to inorganic, organic, and biological pollutants, then only tertiary water treatment is required. In general, groundwater is polluted by toxic metal ions and anions, and only tertiary water technologies are required for its treatment. Surface water is being contaminated due to inorganic, organic, and biological pollutants require secondary and tertiary treatment methods. Wastewater is highly polluted, and it may be colored with solid wastes containing inorganic, organic, and biological pollutants, which requires a good hyphenation of primary, secondary, and tertiary treatment technologies. The choice of the tertiary water treatment technologies depends on the types of pollutants present in water, and the optimum selection can be done by considering the required quality of water, type of pollutants, and their sources.

Figure 5. A general scheme of the most common water treatment methods

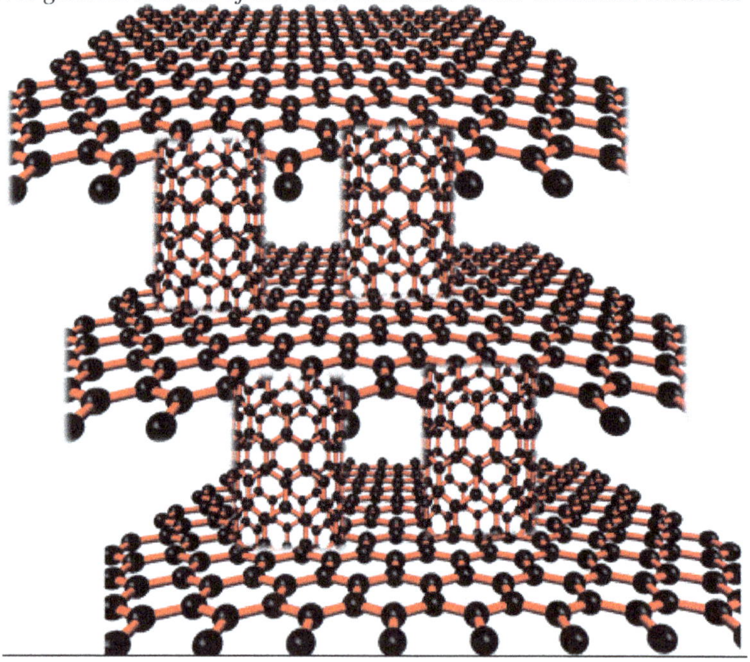

10. WASTEWATER MANAGEMENT

Wastewater is generated in industry, water purification, and recycling technologies. Therefore, waste management is very important from the environmental and health points of view because, generally, waste products are toxic. The waste generated may enter water resources and contaminate them again. A large amount of solid waste is generated in screening, filtration, centrifugal separation, sedimentation and gravity separation, coagulation, flotation, aerobic process, anaerobic process, evaporation, and precipitation processes. Micro- and ultra-filtration and reverse osmosis produce water containing a high concentration of pollutants. On the other hand, some gases are produced in electrolysis and anaerobic processes that are dangerous to the environment. In adsorption, the production of exhaustive adsorbents is quite high. Many methods have been developed and used for the management of waste products generated during water treatment processes. The most important waste management methods are recycling of waste products into useful products such as fertilizers, fillers, building materials, etc. Some toxic waste products used to be burned. Suggestions have also been given to dig the waste products underground in airtight iron or plastic containers. Nowadays, fly ash, red mud, sand, etc., are being used as useful and economic adsorbents for the removal of pollutants from water. The large quantities of these exhausted adsorbents produced have been used successfully as fillers and building materials.

REFERENCES

Alazab, A. A., & Saleh, T. A. (2023). Underwater superoleophobic cellulose/acrylamide-modified magnetic polyurethane foam for efficient oil/water separation. *Materials Chemistry and Physics*, *302*, 127609.

Alharbi, SE, Zarewa, MU, Saleh, TA, 2025., Integrated adsorption–electrocoagulation for dyes removal using solid petcoke-derived activated carbon coupled with predictive machine learning models, Diamond and Related Materials, 159, Part B, 112825

Alkenani, A., & Saleh, T. A. (2022). Synthesis of amine-modified graphene integrated membrane as protocols for simultaneous rejection of hydrocarbons, metal ions, and salts from water, Journal of Molecular Liquids, 367. *Part B*, *120291*. Advance online publication. DOI: 10.1016/j.molliq.2022.120291

Fadillah, G., Alarifi, N. T. S., Suryawan, I., & Saleh, T. A. (2024). Advances in designed reactors for water treatment process: A review highlighting the designs and performance. *Journal of Water Process Engineering*, *63*, 105417. DOI: 10.1016/j.jwpe.2024.105417

Hummers, W. S.Jr, & Offeman, R. E. (1958). Preparation of graphitic oxide. *Journal of the American Chemical Society*, *80*(6), 1339. DOI: 10.1021/ja01539a017

Hummers, W., & Offeman, R. (1958). Preparation of graphitic oxide. *Journal of the American Chemical Society*, *80*, 1339.

Iijima, S. (1991). Helical microtubules of graphitic carbon. *Nature*, *354*, 56–58.

Ito, A., Sato, T., & Goto, T. (2014). Transparent anatase and rutile TiO2 films grown by laser chemical vapor deposition. *Thin Solid Films*, *551*, 37–41.

Krishnan, D., Raidongia, R., Shao, J., & Huang, J. (2014). Graphene Oxide Assisted Hydrothermal Carbonization of Carbon Hydrates. *ACS Nano*, *8*, 449–457.

Li, X., Qin, Y., Picraux, S. T., & Guo, Z.-X. (2011). Noncovalent assembly of carbon nanotube-inorganic hybrids. *Journal of Materials Chemistry*, *21*, 7527–7547.

Park, S., & Ruoff, R. S. (2009). Chemical methods for the production of graphenes. *Nature Nanotechnology*, *4*, 217–224.

Saleh, T. A. (2024). Materials, nanomaterials, nanocomposites, and methods used for the treatment and removal of hazardous pollutants from wastewater: Treatment technologies for water recycling and sustainability. *Nano-Structures & Nano-Objects*, *39*, 101231.

Saleh, T. A., & Al-Daous, M. A. (2016). *Graphene science handbook: Fabrication methods*. CRC Press., DOI: 10.1201/b19606-39

Saleh, T. A., Muhammad, A. M., Tawabini, B., & Ali, S. A. (2016). Aminomethylphosphonate chelating ligand and octadecyl alkyl chain in a resin for simultaneous removal of Co(II) ions and organic contaminants. *Journal of Chemical & Engineering Data*, *61*(9), 3377–3385. DOI: 10.1021/acs.jced.6b00475

Chapter 2
Nanomembrane and Electrochemical Approaches for Sustainable Water Purification and Metal Recovery:
Water Treatment Methods

Maha Aqeel Alaqeel
Imam Abdulrahman bin Faisal University, Dammam, Saudi Arabia

Marwah Ahmed Alabdulrahem
Imam Abdulrahman bin Faisal University, Dammam, Saudi Arabia

Tawfik Abdo Saleh
https://orcid.org/0000-0002-3037-5159
King Fahd University of Petroleum and Minerals, Saudi Arabia

ABSTRACT

Advancing water treatment technologies has become increasingly critical as the need for sustainable wastewater management and resource recovery continues to grow. The rising complexity of water pollutants, including persistent organic compounds and emerging contaminants, has intensified the demand for purification strategies that surpass the capabilities of conventional methods. This chapter explores the membrane-based and electrochemical processes as systems for water purification,

DOI: 10.4018/979-8-3373-5886-4.ch002

pollutant removal, and valuable resource extraction. Particular emphasis is placed on the role of nanomembranes and electrochemical systems in mitigating membrane fouling, enhancing selectivity, and improving overall treatment efficiency. Electrochemical techniques exhibit strong potential for the degradation of organic pollutants and the removal of heavy metal ions through controlled oxidation–reduction reactions. Various nanomembrane architectures and their functional advantages in water treatment are examined, highlighting pathways for improved contaminant rejection and recovery of strategic metals.

1. INTRODUCTION

1.1 Water Pollution

1.1.1 Water Pollution Definition and Classification

Pollution refers to the introduction of harmful substances, gases, or solid or liquid phases at levels that negatively affect the environment. Pollution is classified based on the affected environment into air, soil, and water pollution. Any change in physical or chemical properties of water that influences the quality of surface or groundwater is called water pollution. A wide range of organic and inorganic pollutants with high toxicity and low ability of biodegradation ability are released into wastewater. Therefore, effective and advanced water treatment technologies are required to protect aquatic environments and maintain a safe water supply for human and ecological health (Ariho et al., 2024; Nurafifa et al., 2023; Cai et al., 2023).

Research reported different classifications of water pollutants by nature, source, and effect. Contaminants can be classified based on the source and nature of pollutants into three categories: chemicals, such as heavy metals; physical, like suspended sediments; and biological, such as pathogens. Physical pollutants are substances that can change the physical nature of the water, such as suspended solids, turbidity, color, and temperature of water, whereas chemical contaminants include organic and inorganic substances, pesticides, and nutrients like nitrogen and phosphorus, which are able to change water composition. Lastly, living organisms such as bacteria, viruses, fungi, and algae blooms or their by-products that degrade water quality are biological pollutants (Dubey et al., 2022). Pollutants can also be categorized as biodegradable or non-biodegradable based on their biodegradability. As some pollutants are biodegradable, breaking down by bacteria or fungus in natural processes such as organic wastes, others are non-biodegradable, such as aluminum, plastics, persistent organic pollutants (POPs), and glass, and accumulate and persist for long periods in the environment (Ariho et al., 2024).

Pollutants can be divided into two categories: organic and inorganic. Organic pollutants are a group of contaminants that contain carbon-hydrogen bonds and are often biodegradable, such as oxygen-demanding wastes, plant nutrients, sewage, synthetic organic compounds, and oil. Dissolved oxygen in water is a significant key for the survival of aquatic plants and animals, with optimum concentrations ranging between 4 and 6 ppm. In this regard, deoxygenation below that level indicates the presence of pollution since organic materials break down by bacteria, which consume oxygen during the decomposition process, leading to reduced oxygen levels (Dubey et al., 2022). Inorganic pollutants, which might comprise fine metal particles, metal-organic complexes, organometallic compounds, mineral acids, or inorganic salts, can undergo various types of interactions, such as redox-oxidation, acid-base reactions, and colloid formation, which are affected by chemical and biological processes in water. As a result, the level of toxicity for metals in aquatic environments increased, affecting the growth of plants and other aquatic organisms (Dubey et al., 2022).

1.1.2 Water Pollution Causes and Negative Impact

Water pollution is caused by both natural causes and human activities. Climate change and natural disasters are considered one of the major natural causes of water pollution, whereas human activities mainly stem from industrial uses and farming activities. Naturally, climate change includes extreme weather changes such as heavy rainfall and higher temperatures, varying the chemical and physical characteristics of water bodies by changing pH levels and oxygen concentration. For example, during intense rainfall, surface runoff contaminated with dissolved harmful chemicals and gases flows into rivers and lakes, affecting water quality, while in dry environments, higher temperatures and low water flow lead to concentrated harmful substances and high oxygen levels. In addition, floods are one of the natural disasters that cause rivers to overflow with different pollutants into pure and freshwater sources. These natural processes contribute to affecting water composition as these processes increase the level of contamination, threatening sustainable development and safe living conditions (Ariho et al., 2024; Nurafifa et al., 2023; Cai et al., 2023).

Water contamination by pollutants also arises from anthropogenic activities, including sediment runoff, urbanization, industrial processes, agriculture, domestic sewage, and waste disposal, that introduce harmful substances into water, known as emerging pollutants. The presence of emerging pollutants in ground and surface water is increasing from excessive use of detergents, fertilizers, industrial chemicals, pesticides, pharmaceuticals, and packaging materials. These pollutants destroy ecosystems and threaten aquatic life, creating serious risks to human health. In other words, these pollutants affect marine life by reducing oxygen levels, varying pH, and introducing toxic substances that harm fish, plants, and microorganisms (Ariho et al., 2024; Nurafifa et al., 2023; Cai et al., 2023).

Since contaminants leach into lakes, groundwater, and rivers, it is crucial to distinguish the types and sources of pollution, as it helps to target prevention and treatment. More than 80% of untreated wastewater from human activities is released into rivers and oceans, spreading pollution that contributes to over 50 diseases. Pollution in water supplies transmits infectious diseases such as cholera, typhoid, and hepatitis. In addition, chemical pollutants can cause poisoning, cancers, and neurological disorders. In consequence, the level of water pollution is assessed and monitored through physical, chemical, and biological processes and indicators (Ariho et al., 2024; Nurafifa et al., 2023; Cai et al., 2023; Norén et al., 2025).

One of the most serious chemicals produced mainly from human activities such as industry and agriculture, is persistent organic pollutants (POPs). These highly resistant degradation and toxic groups of chemicals include substances like pesticides, chlorinated and brominated compounds, polycyclic aromatic hydrocarbons (PAHs), and polychlorinated biphenyls (PCBs). Persistent organic pollutants can travel long distances through air and water. Additionally, it causes serious health problems such as hormonal disruption, cancer, and immune disorders by accumulating in the fatty tissues of living organisms. To reduce their harmful effects, treatments of contaminated sites and changing to safer chemical alternatives to protect ecosystems and public health are proposed solutions (Nurafifa et al., 2023).

Another significant reason for water pollution linked to industrial, agricultural, and urban activities is industrial discharges. It is considered one of the major contributors since factories release untreated or partially treated effluents containing toxic chemicals and heavy metals such as mercury, lead, cadmium, and chromium into water bodies, where they affect the quality of water and change its composition (Asif et al., 2022). Also, excess nutrients such as nitrogen and phosphorus lead to harmful algal blooms (HABs) that reduce the level of oxygen and create dead zones where aquatic life cannot survive (Ariho et al., 2024; Nurafifa et al., 2023; Norén et al., 2025). Furthermore, agricultural runoff is a significant contributing factor to the overuse of pesticides and fertilizers, which introduce nitrates, phosphates, and persistent organic compounds into rivers and groundwater as these harmful chemical pollutants bioaccumulate (Ariho et al., 2024).

In addition, oil spills and discharge of oily wastewater from accidental releases during transportation, leaks from underwater pipelines, and offshore drilling affect and damage the water composition. Since leakage or spillage of crude oil results in the release of aliphatic or aromatic hydrocarbons and other toxic substances that coat marine organisms and block oxygen exchange. In addition, spilled oil undergoes physical and chemical changes such as spreading, dispersion, and biodegradation.

The rate of these changes, influenced by wind, the presence of organisms, light exposure, and temperature, threatens marine life (Asif et al., 2022; Valjarević, 2025).

Plastic pollution is also one of the most serious threats to marine ecosystems. These plastics break down into microplastics smaller than 5 mm into micro- and nanoplastics that are easily ingested by marine organisms and fish and lead to direct effects on human health. Floating plastics also affect marine transport and water quality. It is estimated that about 8–10 million metric tons of plastic waste enter the oceans annually. The proposed solution to reduce the impact of these pollutants is to use recyclable and biodegradable materials and products to reduce the amount of plastic waste entering the marine environment and the ocean. However, due to their high global spread, stability, and persistence, plastics remain a long-term pollutant in aquatic environments. Plastic pollution emerged in aquatic environments because of many sources, including the degradation of single-use plastics and the improper disposal of solid and electronic waste. In water systems, due to their extremely small size and high surface reactivity allow them to interact with other contaminants and travel long distances. Research reported that concentrations of these particles are up to 1000 particles per liter in freshwater and up to 10000 particles per cubic meter in aquatic environments. Overall, collectively, all these sources contribute to a significant and global water issue affecting sustainable development for both human health and ecological balance, making it one of the most global environmental challenges of the current century (Norén et al., 2025; Kumar et al., 2025; Fouda-Mbanga et al., 2025).

1.2 Nanoparticles

1.2.1 Nanoparticles Definition and Characteristic

Nanoparticles are tiny particles with at least one dimension roughly between 1 and 100 nanometers (nm). Over the past two decades, nanotechnology has grown very quickly and gained a lot of attention in different fields such as medicine, industrial fields, and environmental management. Nanoparticles (NPs) exhibit unique optical, thermal, electronic, mechanical, magnetic, and catalytic properties that greatly differ from their bulk form due to their nanoscale size. At this scale, materials begin to show exceptional size and surface effects related to surface and quantum effects. As nanoparticles become smaller, their surface area-to-volume ratio increases, meaning a larger proportion of their atoms are on the surface and available for interaction. This results in enhancement of their properties, such as reactivity, strength, and electrical and magnetic properties, making them more effective in special applications. This allows us to improve important features such as surface-to-volume ratios, catalytic, magnetic, or other behaviors (Abdullah et al., 2025; Bayda et al., 2019; Sun, 2019; Joudeh & Linke, 2022).

Magnetically at room temperature, materials including elements such as cobalt, nickel, or iron are magnetic materials, while other metals such as gold, silver, or palladium are non-magnetic; however, at the nanoscale size, magnetic behavior appears due to the unequal distribution of electrons affected critically by size and shape. For instance, bulk FeAl is non-magnetic, becoming magnetic as nanoparticles. In optical and electronic terms, semiconductor and metallic nanoparticles show significant emission and absorption properties, related to localized surface plasmon resonance and quantum confinement, that are influenced by the surrounding environment, besides the size and shape of the particles. These properties are improved by a high degree of crystallinity and a smaller size. For example, Ag and Au nanoparticles display color changes with a variety of particle sizes and media, making them very effective in sensing and imaging applications. Catalytically, since nanomaterials possess high surface reactivity and selectivity, yielding higher catalytic activity (Joudeh & Linke, 2022).

In terms of mechanical characteristics, they exhibit their distinct behavior at the nanoscale because of strong forces between nanoparticles, such as electrostatic and van der Waals forces, which improve flexibility, strength, and elasticity. In thermal characteristics, the extremely small size improves the efficiency of heat transfer due to increased Brownian motion effects, which is the random movement of microscopic particles suspended in a fluid, by providing a higher number of electrons for heat transfer and conduction of electrons. Moreover, their thermoelectric performance and thermal conductivity are enhanced because of their high surface area; however, their melting points are lower than bulk materials due to surface energy effects. Overall, their performance depends on several factors, including size, surface effect, shape, composition, interparticle spacing, and support materials, making them valuable in environmental and industrial applications, including removing chemicals, heavy metals, and microbes (Joudeh & Linke, 2022).

1.2.2 Nanoparticles History, Application, and Negative Impact on the Environment

The origins of nanotechnology trace back to Richard Feynman's 1959 lecture, "There's Plenty of Room at the Bottom," where he first introduced the idea of nanoparticles. Research and experiments in nanotechnology continued in the following decades to develop the concept and its application. An example of experimental nanoscience from the early 1990s is Sumio Iijima's discovery of carbon nanotubes introduced to design and building of materials with specific shapes and properties at the nanoscale, making a major step beyond the synthesis of fullerenes, resulting in controlled engineering of carbon structures at the nanoscale (Bayda et al., 2019; Drexler, 1981; Iijima, 1991).

Due to their flexible characteristics, these nanoscales give rise to fascinating chemical phenomena that make nanoparticles promising in a long list of applications for catalysis, medicine, and industrial fields, including agriculture, food technology, and environmental remediation. The chemistry of nanoparticles arises from their extremely small size and their interactions at the atomic and molecular levels. Nanoscience is a branch of science focused on the study of material properties at the nanoscale since it combines principles from physics, chemistry, biology, and materials science, allowing scientists to change and control at the level of individual atoms and molecules. Therefore, understanding their chemical behavior allows scientists to use these exceptional properties in technology and industry for specific purposes to meet human and industrial needs (Altammar, 2023; Aroor et al., 2025; Astruc, 2020; Sun, 2019).

Determining the possible use of nanoparticles is also greatly aided by their structure and chemical makeup. Because metallic nanoparticles like gold and silver exhibit surface plasmon resonance, they can be used in sensing, imaging, and biosensing technologies to detect trace concentrations. For instance, metal oxide nanoparticles like zinc oxide, titanium dioxide, and copper oxide show strong photocatalytic activity as valence electrons are excited to the conduction band, forming reactive oxygen species that break down environmental pollutants, while gold nanoparticles can detect low concentrations of mercury in water. Additionally, binding to chemical groups or molecules to improve stability, dispersion, and compatibility with host materials, functionalization, and surface modification play a significant role in predicting their performance (Altammar, 2023). Nanoparticles are very effective in the biomedical field because of their large surface-to-volume ratio, which improves their surface reactivity, catalytic efficiency, and interaction with biological systems. They can be used in biomedical and pharmaceutical applications, such as drug delivery, contrast agents, and therapeutic carriers, due to their biocompatibility, bioactivity, antimicrobial, and anti-inflammatory properties. Because of its capacity to enhance bioabsorption and regulate the release of therapeutic agents through surface modification with biomolecules or dyes, it can also be utilized in advanced medical treatment of diseases. (Abdullah et al., 2025; Jiang et al., 2023; Azzaoui et al., 2022; Alam et al., 2025).

Another nano application is in the agriculture sector, as ensuring sustainable food production for the growing global population is a major challenge for modern agriculture. Sustainable agriculture seeks to meet the increasing food needs of the growing global population while reducing environmental negative impact to protect natural resources and maintain long-term productivity. However, traditional farming practices mostly depend on chemical fertilizers, pesticides, and excessive usage of water, which can cause serious environmental and health concerns, including soil and water contamination. It is estimated that over 2.5 million tons of pesticides are applied annually, with more than 90% lost to the environment. To address these issues, safer and more efficient pesticide delivery methods are required. Nanotechnology presents a promising solution by enabling controlled and targeted delivery of chemicals, reducing waste, toxicity, and costs. The production and use of nanoparticles is growing in popularity in agricultural applications because some of them are nontoxic and biodegradable. For instance, nano fertilizers and nano pesticides exhibit excellent ability to carry and release nutrients or active agents in a controlled manner, improving nutrient efficiency and lowering required doses of chemicals. Many factors influencing this process include nanoparticle concentration and the combination of different particles (Jiang et al., 2023; Azzaoui et al., 2022; Alam et al., 2025).

Furthermore, taking advantage of large surface areas and active sites with small size and high mobility allows nanoparticles to move easily through soil and groundwater, making them effective for direct treatment of contaminated sites of soil and water. Therefore, nanoparticles have been widely used for environmental remediation of contaminated land, sediments, and subsurface water due to their high surface area and reactivity towards contaminant removal. Research reported that nano remediation technology application for cleaning up soil and groundwater is a fast, cost-effective, and environmentally friendly method for cleaning polluted sites (Alam et al., 2025; Reihanifar et al., 2024; Alazaiza et al., 2021). Also, a study reported that the initial field-scale application of nanoparticles for soil and groundwater remediation was documented two decades ago. It reported that the activity of nanoparticles in the injected soil was up to 56 days, with travel distance in groundwater exceeding 20 meters. In this process, different types of nanoparticles were used, such as metal oxides, nanoscale zeolites, nanometals, carbon nanotubes, and titanium dioxide, helping to restore polluted environments more efficiently than traditional methods (Alazaiza et al., 2021).

Additionally, one of the significant applications for nanotechnology in the air is pollution monitoring by using nanomaterials as adsorbents, catalysts, filters, and sensors. These materials can capture pollutants through adsorption mechanisms that allow photocatalytic degradation of pollutants. Additionally, released gases can be filtered by nanostructured membranes with extremely fine pores, and hazardous gases like nitrogen dioxide, sulfur dioxide, and hydrogen sulfide can be detected by nano sensors. In specific applications, carbon nanotubes are developed as nano-adsorbents for harmful gases like carbon dioxide and methane due to their outstanding ability to functionalize active chemical groups with high surface area and pore volume. For example, modified carbon nanotubes with amine groups achieved significant CO_2 adsorption, while zinc oxide nanostructures are used to remove hydrogen sulfide at high temperatures, demonstrating the great ability of nanomaterials designed for air pollution remediation (Saleem et al., 2022).

In contrast, the use of nanomaterials may increase concerns about their potential negative impacts on environmental and human health since the high reactivity of nanoparticles results in their reacting with living organisms and microbes in different ways (Saleem et al., 2022; Del Prado-Audelo et al., 2021). These particles might undergo various chemical changes, such as dissolution, oxidation, reduction, and interactions with inorganic compounds, which can change their structure and chemical form. These redox reactions are significant for nanomaterials made of active redox metals such as silver and iron. The rate of these reactions depends on different environmental factors, including the presence of electron donors and acceptors, oxygen levels, light exposure, and organic matter. Although these reactions cause toxic effects, as they may release ions or reactive oxygen species, this study reported that nanoparticles are less toxic than dissolved metal ions (Feculak et al., 2025).

Even though the high surface-to-volume ratio of nanomaterials provides an advantage for pollutant adsorption, it also contributes to challenges such as flocculation, forming larger clusters either with other nanoparticles of the same type, called homoaggregation, or with different environmental particles, named heteroaggregation (Feculak et al., 2025). Therefore, achieving a balance between physical stability and effective surface activity is essential to enhance pollutant interactions while maintaining nanoparticle stability. For this purpose, non-ionic surfactants can be used to reduce flocculation; functionalizing nanoparticles with groups such as –COOH and $–NH_2$ can both improve pollutant removal and prevent aggregation under specific pH conditions (Del Prado-Audelo et al., 2021).

1.2.3 Nanoparticles Classification

The classification of nanoparticles might be based on source, chemical composition, or dimensions. Nanoparticles can be classified according to their chemical composition, which directly determines their physical, chemical, biological, and functional properties by predicting optical, electrical, catalytic, magnetic, and biological behavior. They are classified into organic, inorganic, carbon-based, and hybrid or composite categories. Firstly, organic nanoparticles are composed of organic molecules such as proteins, lipids, polymers, or carbohydrates. These types of nanomaterials are typically biodegradable, biocompatible, and nontoxic, making them ideal for biomedical applications such as targeted drug delivery and cancer therapy. However, they are sensitive to heat and light. In general, surface morphology, carrying capacity, composition, and stability are major parameters that determine the potential field of application of organic nanomaterials (Joudeh & Linke, 2022; El Kalliny et al., 2023).

Inorganic nanoparticles are composed mainly of non-carbon elements such as metal, metal oxide, ceramics, and semiconductor nanoparticles. The major classes of inorganic nanomaterials are metal nanoparticles, which might be monometallic, bimetallic, or polymetallic; metal oxide nanoparticles; semiconductors; and ceramic nanoparticles. The most used metals are gold (Au), silver (Ag), iron (Fe), and aluminum (Al), while the most used metal oxides are TiO_2, ZnO, and Fe_3O_4. Since it has unique optical, magnetic, and catalytic activities related to strong surface plasmon resonance, this type is used in sensing, catalysis, and photothermal therapy. Thirdly, carbon-based nanomaterials, made entirely of carbon atoms, such as fullerenes, carbon nanotubes, graphene, and carbon quantum dots, as shown in Figure 1 (Aydin et al., 2024). This group displays exceptional electrical and mechanical properties by combining nanoscale properties with the strength and conductivity of sp^2-hybridized carbon, making them suitable for energy storage, bioimaging, and environmental sensing. For example, carbon quantum dots displayed significant properties such as biocompatibility, high water solubility, and low toxicity (Ameer et al., 2025). Finally, hybrid or composite nanoparticles combine two or more materials, such as polymer-coated metal particles, to enhance multifunctional performance. Overall, material-based classification is important to predict nanoparticle stability, toxicity, reactivity, and potential applications (Joudeh & Linke, 2022; El-Kalliny et al., 2023; Ameer et al., 2025; Galal Magdy et al., 2024; Ana Flavia Burlec et al., 2023).

Figure 1. Carbon-based nanoparticles

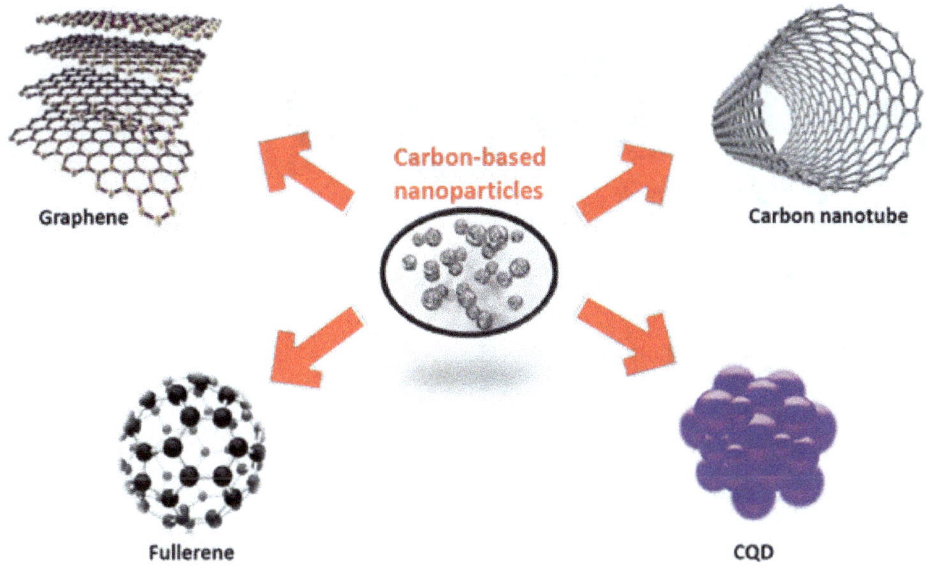

(Aydin et al., 2024)

Another major categorization is according to their dimensional characteristics, describing how many of their external dimensions fall within the nanoscale range of 1–100 nm. In this regard, nanoparticles are typically classified, as proposed by Pokropivny and Skorokhod, as zero-dimensional (0D), one-dimensional (1D), two-dimensional (2D), or three-dimensional (3D). Zero-dimensional nanomaterials, such as quantum dots and nanospheres, have all three dimensions within the nanoscale, showing strong quantum confinement effects, where electrons are restricted in all directions, resulting in unique optical, magnetic, and electronic properties that lead to good usage for solar cells and lasers, while 1D nanomaterials, such as nanowires, nanorods, and nanotubes, are nanoscale in two dimensions; however, they are extended in one dimension, making them valuable for electrical conduction and mechanical enhancement. Two-dimensional nanomaterials, including graphene, nanosheets, nano thin films, and nanoplates, are nanoscale in thickness but have large lateral dimensions, resulting in exceptional surface areas and significant electrical and thermal conductivity. Lastly, three-dimensional nanomaterials might be networks or assemblies of nanostructures such as nanofibers, nanotubes, or porous nanocomposites where nanoscale features exist throughout the material Figure 2 (Joudeh & Linke, 2022; El-Kalliny et al., 2023; Alam et al., 2025).

Figure 2. Dimensionality-based classification of nanomaterials

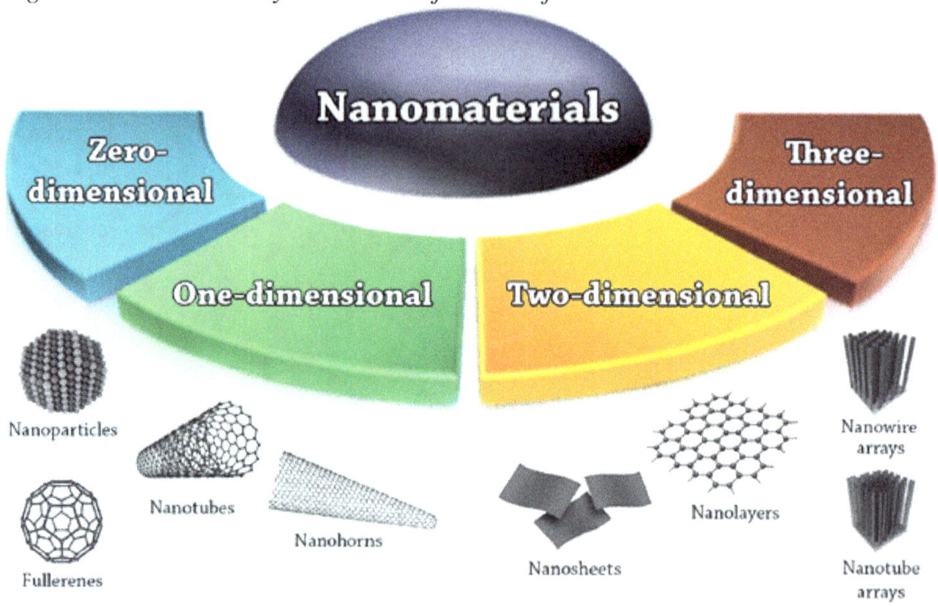

(Joudeh & Linke, 2022)

Quantum influences become more noticeable as the radius of a material approaches the exciton Bohr radius, which is the separation distance between the electron and hole, changing properties like electron affinity and magnetism significantly compared to bulk materials. Thus, dimension-based classification is very important since it helps to explain their size-dependent physical and chemical behaviors, which are essential for the design of advanced specific applications (Joudeh & Linke, 2022; El-Kalliny et al., 2023). Furthermore, nanoparticles may be classified in terms of morphology of structure into tubes, spheres, cubes, or rods, and core-shell, or layered arrangements. The characteristics and reactivity of nanomaterials are influenced by the distinct surface area and aspect ratio that each shape offers. In this regard, internal features are important; whether they are single or polycrystalline features or layered or core-shell constructions, full detailed structure information is helpful and key in designing and determining suitable applications (Joudeh & Linke, 2022).

Nanoparticles can also be classified based on their origin or source, which is important for understanding their environmental and biological fate and impacts. They are generally divided into three categories: natural, incidental, and engineered nanoparticles. Natural nanoparticles are produced in nature from biological, geological, or atmospheric processes, while incidental nanoparticles are formed as unintentional by-products of human activities, such as combustion and emissions of industrial processes. Engineered nanoparticles are mostly designed and synthesized for specific applications in medicine, electronics, catalysis, and energy storage. Additionally, nanoparticles produced from green synthesis methods are considered a subcategory of engineered nanoparticles. This source-based classification helps determine nanoparticle exposure and environmental behavior (Joudeh & Linke, 2022; El-Kalliny et al., 2023).

2. WATER TREATMENT

2.1 Wastewater Treatment Technologies

Reliable treatment methods are required to produce high-quality drinking water. In the past, traditional treatments mainly depended on coagulation, flocculation, sedimentation, and filtration, while in recent decades the primary methods used by municipal and industrial facilities were clarification, granular media filtration, and chlorination. Recently, the interest in advanced water treatment technologies has been rising, especially in removing emerging contaminants. The technologies that consistently separate or eliminate a variety of contaminants, such as nutrients, salts, suspended solids, organics, metals, and micro-pollutants, are necessary to increase the supply of safe drinking water. The development of these methods to operate more successfully and efficiently has been the focus of recent research, including membrane filtration and electrochemical processes like electrocoagulation and electroextraction. Electrochemical techniques apply electric current or fields to destabilize contaminants, causing precipitation or moving ions across selective barriers, whereas membranes separate contaminants by selective permeability based on size exclusion or charge. However, these methods have many limitations when applied separately. In electrochemical processes, electrode consumption and energy requirements are major problems in this method, while fouling and concentrating management are in the membrane. Consequently, it is critical to optimize these technologies for resource recovery, fouling control, and energy efficiency (Razali et al., 2023).

2.2. Nanomembrane Filtration

2.2.1 Membrane-Based Technology

Membrane technology has a long scientific history that dates to the 18th century; however, major industrial development began in the 1960s, specifically when Loeb and Sourirajan produced the first high-performance reverse osmosis (RO) membrane made from asymmetric cellulose acetate, marking the start of significant development. This membrane illustrates excellent separation efficiency with high-rate flow, demonstrating the valuable technique, membrane system, as an industrial separation tool (Aydin et al., 2024). Membrane filtration has significant advantages among other separation techniques, since it has high efficiency, low energy consumption, room temperature procedures, and a simple operating process (Aydin et al., 2024)

Membrane technology utilizes a variety of material types that might be porous, ceramic, metallic, or even liquid-based with selective carriers for specific industrial and environmental applications. High selectivity and flux, good mechanical strength, resistance to chemicals and heat, and minimal fouling are all necessary for a good performance membrane. Depending on contaminants' chemical interactions, size, or charge, membranes act as selective barriers allowing them to purify water and block pollutants, as membranes might be permeable or semi-permeable (Aydin et al., 2024; Othman et al., 2021). Membrane-based water treatments are strongly related to the characteristics of the materials used membrane. The first generation of membranes that are applied in water remediation are polymeric membranes, followed by ceramics and other inorganics that are employed due to their strong chemical and thermal stability (Yusof et al., 2023). Most fabricated organic polymers are made from cellulose or non-cellulose materials, which are not good for hard environments. According to a study, inorganic ceramic membranes are better than polymeric organic membranes for treating highly contaminated wastewater, such as oily and highly turbid water. As a result, ceramic membranes are preferred for highly contaminated wastewater due to their exceptional chemical stability, mechanical durability, and capacity to function at high temperatures (Othman et al., 2021).

Membranes are categorized depending on the size of their pores as microfiltration (MF), ultrafiltration (UF), nanofiltration (NF), and reverse osmosis (RO). MF and UF generally use larger pore sizes that operate at low pressures of around 10 to 30 psi to eliminate microorganisms, colloids, and suspended solids. Microfiltration primarily has pores that typically range in size from 50 to 500 nm, which apply size exclusion to remove particles, while ultrafiltration membranes have smaller pores less than 50 nm and separate macromolecules or colloids according to surface interactions and size (Aydin et al., 2024; Othman et al., 2021). Because of the great interception effect of ultrafiltration membranes with varying molecular weights of organics, such as humic acids (HAs), proteins, polysaccharides of nucleic acids, and bacteria, this type of membrane can be applied in wastewater purification and environmental applications. Nano and ultra membranes provide a high recovery rate, quick filtration speed, and good retention performance with minimum expense or introduction of new impurity by different types of mechanisms like electrostatic repulsion, hydrophobic adsorption, or size exclusion (Othman et al., 2021; Zhang et al., 2023; Liang et al., 2021).

Nanofiltration and reverse osmosis are high-pressure from 75 to 250 psi procedures that eliminate smaller solutes such as salt, multivalent ions, and organic molecules (Othman et al., 2021). Nanomembranes have pore sizes around 2 nm or smaller and rely on a combination of sieving and diffusion. Consequently, nanomembranes are useful for partial desalination and reduce the level of hardness (Aydin et al., 2024; Chaturvedi & Dave, 2019). The nanofiltration procedure is used in separating numerous bacteria, organic pollutants, viruses, insecticides, calcium, and magnesium salts from water with lower energy usage. Reverse osmosis membranes, which are nonporous with pores of about 0.3–0.6 nm, apply a diffusion mechanism to remove nearly all dissolved solids, inorganic ions, and small organic pollutants with dense membranes when pressure exceeds the natural osmotic pressure (Aydin et al., 2024; Othman et al., 2021; Liang et al., 2021; Chaturvedi & Dave, 2019). NF and RO are widely used in the production of high-quality drinking water by desalination of advanced wastewater (Aydin et al., 2024; Liang et al., 2021). The permeability of membrane filtration types is listed in Figure 3.

Figure 3. The permeability of different membrane filtration types

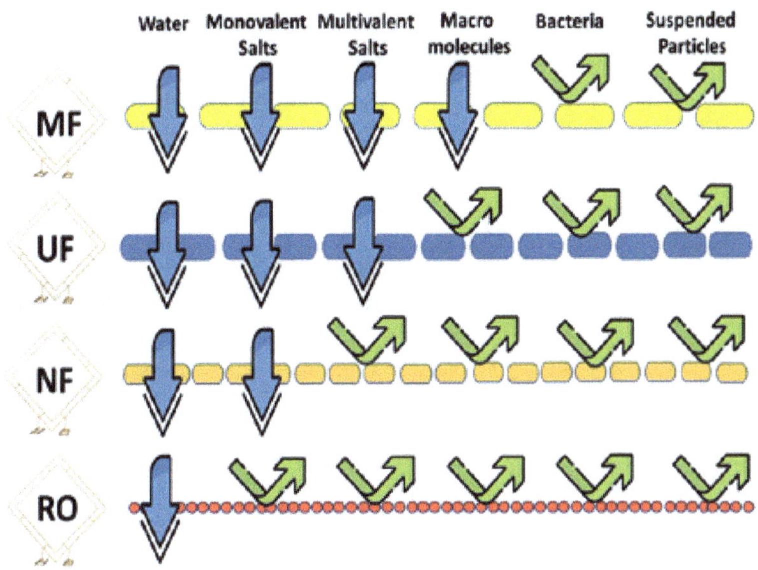

(Aydin et al., 2024)

2.2.2 Nanostructured Membranes, Functionalization of Nanoparticles for Selective Surface Modification

Due to the growth of nanotechnology in recent decades, a new generation of highly developed membrane materials has developed, known as nanomembranes. Nanomaterials mostly consist of nanofibers, nanoparticles, or two-dimensional layer materials. Nanomaterials can be differentiated as per their dimensions, ranging from 1 to 100 nm, Figure 4. Such materials have remarkable physical and chemical characteristics, with permeation properties related to significant stability and large surface area (Aydin et al., 2024). The addition of nanostructures to the membrane increased water flux due to their hydrophilic nature, resulting in reduced energy consumption and enhanced membrane performance with more economical and effective purification (Aydin et al., 2024; Othman et al., 2021). Compared to previous typical membrane designs, these materials proved a great enhancement in selectivity and durability by increasing water permeability. These developments have made membrane technology more reliable and effective than many traditional membrane-based processes in water treatment methods (Aydin et al., 2024). Commonly, microbes, organic compounds, metallic ions, and salts are the most common contaminants in wastewater with a range of sizes between 1 and 10 nm that can be filtered out by nanomembranes (Aydin et al., 2024). The application of advanced Membranes made of nanomaterials is receiving great attention because of their potential to address clean water shortages.

Figure 4. Nanomaterial-based membrane

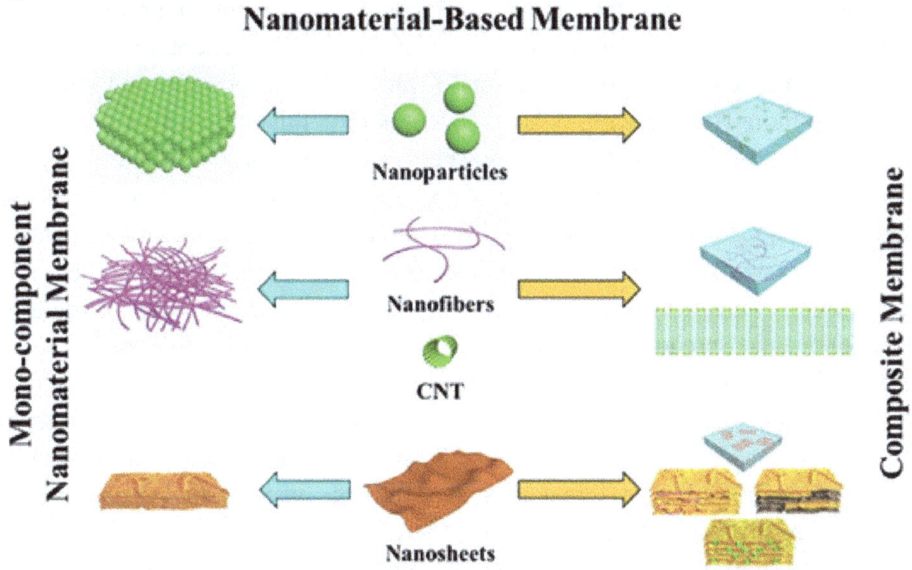

(Aydin et al., 2024)

Nanomembrane design strongly relies on the structure of nanomaterials, as the addition of nanomaterials in the membrane provides further advantage by combining a continuous phase of the polymer with distributed nanoparticles. Nanomaterials are purposefully positioned within the membrane matrix. Therefore, choosing the right dispersed and continuous phases is crucial since it directly influences properties such as pore formation, transport pathways, and overall membrane morphology that controls permeability and mechanical stability compared to typical polymeric membranes (Chaturvedi & Dave, 2019). In addition, there are many factors that affect the efficiency of the designed membrane, including temperature, configuration, characteristics of the feedwater, flow rate, pH, and membrane composition (Othman et al., 2021).

Identification of targeted contaminants is key to selecting the appropriate nanomembrane, as it is important to understand the specific pollutants to be removed. Choosing a nanomembrane for water treatment to be able to detect specific pollutants relies on matching the physicochemical characteristics of the membrane with the type of contaminants that are present. Membrane chemistry and structure, performance standards, economic and environmental considerations, and contaminant characteristics are very important to consider to make an efficient selection. Functionalizing nanoparticles by adding specific biomolecules or functional groups to their surface to modify their physical and chemical characteristics to employ them for targeted uses. Common functionalization methods might be by chemicals, covalent bonding, physical adsorption, and electrostatic force, each of which has its own benefits and drawbacks. While chemical functionalization and covalent bonding provide strong and stable linkages, the uniform density, distribution, and orientation of surface groups are difficult to control. In terms of physical adsorption, despite that it is easier and more flexible with its noncovalent nature, this might cause desorption, whereas electrostatic forces allow binding between charged biomolecular species and pairing of complementary charges; however, their stability is affected by external conditions. Because nanoparticles without modification have negative effects in biological systems, surface functionalization is considered a critical strategy that enhances biocompatibility in various real applications, such as analytical, imaging, therapeutic, and diagnostic systems, resulting in achieving selective surface modification (Aydin et al., 2024).

Based on a specific expected effect, a range of ligands can be purposefully attached to nanoparticles on platforms enabling either single-function or multimodal performance. For instance, in aqueous environments, water-soluble polymers like polyethylene glycol (PEG) can be applied to enhance significant features, including solubility, dispersion, and biocompatibility, whereas bulky or hydrophobic ligands may be introduced to prevent aggregation of nanoparticle cores (Aydin et al., 2024). A study investigating surface engineering demonstrated that modified polyvinylidene fluoride (PVDF) membranes with polyacrylic acid (PAA) and titanium dioxide nanoparticles showed superior enhanced antifouling behavior by self-assembly or grafting-from polymerization. In the grafting method, better dispersion of TiO_2 efficiency is achieved, resulting in membranes that significantly improve antifouling performance. Additionally, long-term stability of nanoparticles is confirmed during cleaning and operation related to the covalent incorporation of nanoparticles on the PAA surface (Aydin et al., 2024).

Moreover, research reported that overcoming conductivity limitations in polybenzimidazole (PBI) membrane can be achieved with functionalization of nanostructure methods using sulfonated silica nanoparticles, polymer-modified SiO_2, and polyvinylimidazole to enhance proton conductivity and mechanical stability during acid doping and MEA (membrane electrode assembly) fabrication. These desired modifications aim to operate steadily and efficiently in specific circumstances, optimizing membrane performance. In conclusion, compared to regular membrane performance, nanocomposite membranes display greater efficiency in removing contaminants with higher water flow affected by specific modifying properties of surfaces (Aydin et al., 2024). Table 1 represents a comparison highlighting the benefits and drawbacks as well as potential applications of different nanomaterials in wastewater remediation.

2.2.3 Nanomembrane Classification

Nanomembranes used for water purification are classified into four categories: organic polymeric membranes, inorganic membranes, biomimetic membranes that mimic natural biological structures, and hybrid or composite membranes, as shown in Figure 5. The performance of the nanomembrane is strongly influenced by the material used and characteristics like hydrophobicity and surface charge. For example, organic polymeric membranes: most commercial membranes are made from polymers such as cellulose derivatives, polyethersulfone (PES), polypropylene, polyvinylidene fluoride (PVDF), and polysulfone (PSF) (Othman et al., 2021). Some research confirmed that PES is used in numerous applications due to mechanical, thermal, and chemical stability (Othman et al., 2021; Zahid et al., 2018). However, the presence of aromatic ether and sulfone structure reduces the efficiency of separation by making it raise the possibility of being susceptible to fouling. To address these limitations, mixed matrix membranes (MMMs), which combine polymer matrices with inorganic nanofillers, are used and prepared by different methods such as graft polymerization, coating, interfacial polymerization, and blending. A variety of inorganic materials can be used, such as sulfides, CNTs, graphene oxide, nitrides, and TiO_2 nanoparticles or nanotubes to improve MMMs' filtration efficiency, including mechanical strength, hydrophilicity, and antifouling behavior (Othman et al., 2021). Recently, green and sustainable nanomaterials production using non-toxic methods has been used to reduce environmental impact while maintaining high purification efficiency (Othman et al., 2021).

Figure 5. Polymer nanocomposite membrane representation

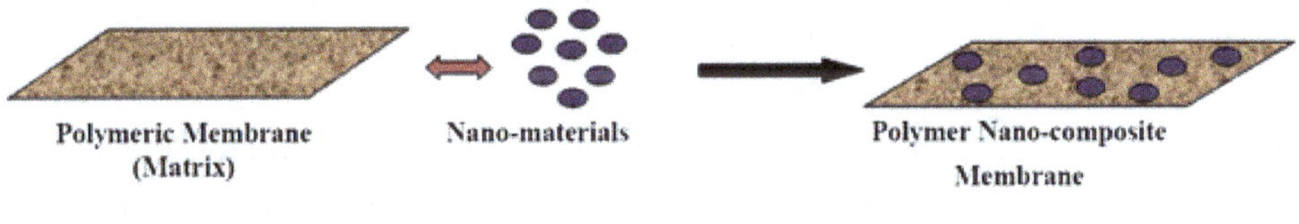

(Zahid et al., 2018)

There are several specialized subtypes within these categories: carbon-based nanomembranes, such as fullerenes, carbon nanotubes (CNTs), and graphene; electrospun nanofiber membranes; mixed-matrix or nanocomposite membranes that comprise nanoparticles, and two-dimensional inorganic membranes like graphene oxide (Divya Bajpai Tripathy & Gupta, 2023). CNTs are nanomaterials that are formed into tube shapes as multi-walled carbon nanotubes (MWCNT) or single-walled carbon nanotubes, Figure 6. For groundwater remediation, MWCNTs are oxidized with nitric acid to achieve higher levels of sorption capacity. The study reported that this type exhibits a greater capacity three to four times in removing heavy metals like Pb (II), Cu (II), and Cd (II) compared to the two sorbents that are commonly used in water purification are granular and powdered activated carbon (Chaturvedi & Dave, 2019).

Figure 6. An illustration of a carbon nanotube and a carbon nanotube with multiple walls

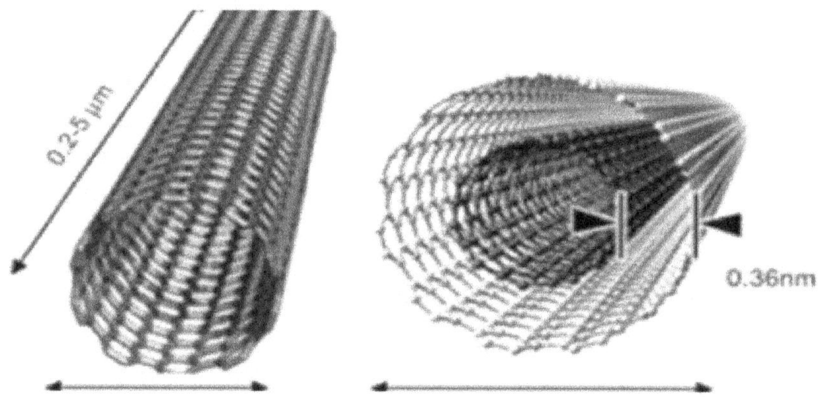

(Chaturvedi & Dave, 2019)

Additionally, a study reported that carbon nanomembranes (CNMs), two-dimensional ultrathin carbon sheets, have significant selectivity in regulating molecular exchange by organizing molecular exchange across all material types, including gas, liquid, and solid. Molecules are arranged on a solid surface and cross-linked to produce CNMs to create a film in two dimensions, followed by separation of the resultant nanomembrane (Divya Bajpai Tripathy & Gupta, 2023). The nature of thin-film CNMs, clearly presented in Figure 7 displays a CNM crossing a hexagon in a gold mesh image captured by He ion microscopy. These materials have structural and chemical properties due to their tunable nature and strong physical and chemical stability. Features such as nanopore distribution, tunable surface, and thickness are obtained from the parent self-assembled monolayer (SAM). Also, the simple fabrication methods, besides all these advantages, make CNMs have more applications in filtration and sensing (Divya Bajpai Tripathy & Gupta, 2023). The variations in membrane structure affect selective separation characteristics and transport behavior. Thus, each type of nanomembrane is appropriate for a particular purification application as listed in Table 1.

Figure 7. The CNM image is positioned on a hexagonal supporting group

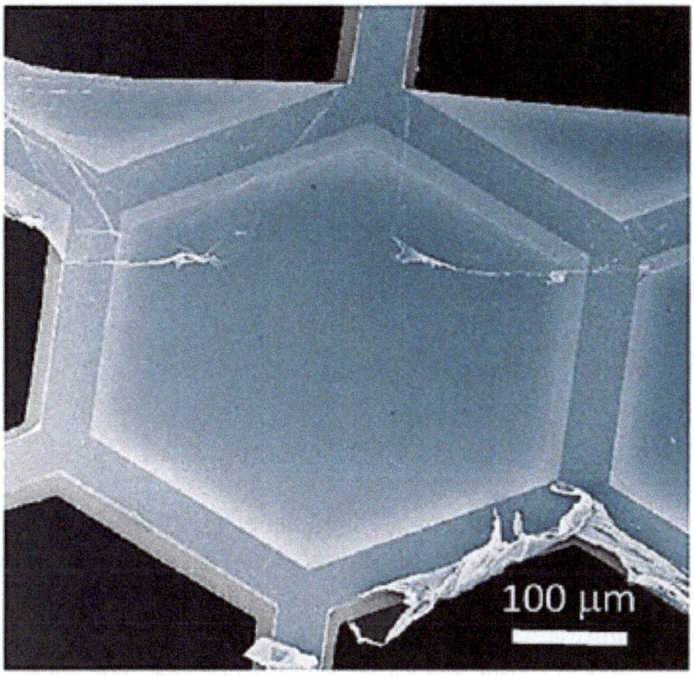

(Divya Bajpai Tripathy & Gupta, 2023)

Graphene is well known as a high-performance conductive membrane because of its exceptional mechanical strength, electrical characteristics, and high ability to create ultrathin separation layers. It has distinctive inherent antibacterial properties, helping to inhibit bacterial growth, decrease the chance of membrane biofouling, and improve microbial removal efficiency during the process. Despite its high conductivity, Water permeability is significantly increased by the addition of hydrophilic polymers. These modifications resulted in raising the water transport pathway, which can be achieved using non-toxic and low-cost hydrophilic polymers (Divya Bajpai Tripathy & Gupta, 2023). Graphene oxide membranes are considered hydrophilic due to the presence of oxygen-rich functional groups such as –OH, –COOH, and epoxy that improve water permeability with moderate electrical conductivity. In the other hand, reduced graphene oxide membranes contain fewer oxygen groups and more non-oxidized graphitic regions, creating smoother water pathways. Consequently, this type allows water to pass more quickly; however, poor hydrophilicity reduces the water flow through the membrane, limiting its performance. The Formula structure of graphene oxide and reduced graphene oxide are presented in Figure 8 a and b (Qi et al., 2023).

Figure 8. Represents structural formula in a) graphene oxide (b) reduced graphene oxide

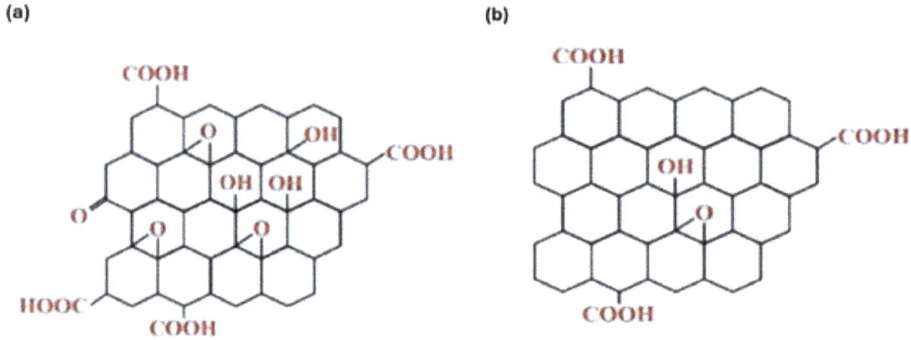

(Qi et al., 2023)

In terms of hybrid membranes, different nanocomposite membranes in generally used and well known as conventional mixed-matrix, nanocomposite, thin-film nanocomposite, surface-located nanocomposite, and thin-film composite with nanocomposite substrate (Othman et al., 2021). Blended nanocomposite membranes and thin-film nanocomposite (TFC) membranes are common types of polymer nanocomposite membranes shown in Figure 9. In blended nanocomposite membranes, nanoparticles are directly mixed into the polymer casting solution before membrane formation to create membranes with nanoparticles evenly trapped in the polymer matrix. On the other hand, the TFC type is formed by creating a selective nanostructured thin layer through the addition of nanoparticles onto the surface of the membrane either through pressure-induced deposition or self-assembly, such as dip-coating (Zahid et al., 2018). Hybrid inorganic–organic membrane materials have been applied more recently; the incompatibility of inorganic fillers and organic polymers may result in serious challenges like nanoparticle agglomeration, which can be addressed by using dispersion agents or surface treatment with nanoparticles. Utilizing nanofillers such as silica, graphene oxide, TiO_2, and carbon nanotubes greatly improves hydrophilicity, thermal stability, and mechanical strength and selectivity. To ensure long-term membrane performance and under high-pressure filtration conditions, functionalization of the surface is required to prevent nanoparticles from leaching (Zahid et al., 2018)

Figure 9. (a) Membranes made of blended polymer nanocomposite; (b) membrane made of thin-film polymer nanocomposite

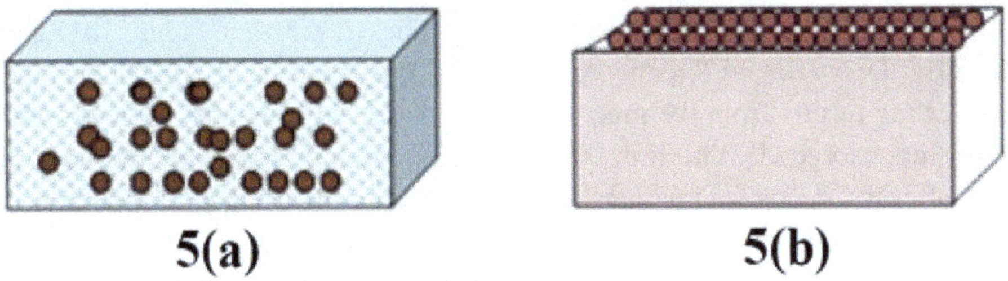

(Zahid et al., 2018)

Another classification of nanomembrane based on configuration is commonly hollow-fiber or flat-sheet. Membrane sheets and spacers are wrapped around a central permeate tube, whereas hollow-fiber membranes, commonly applied in microfiltration and ultrafiltration for drinking water treatment, consist of bundles of fine capillary fibers at both ends and are placed in a cylindrical module. Appropriate configuration selection is crucial for effective water treatment since designing affects features like hydrodynamics, fouling behavior, and energy consumption (Othman et al., 2021). The technique of phase inversion is utilized in the production of polymer nanocomposite membranes in either hollow fiber or flat sheet configurations by dispersing nanomaterials in polymer solution before the process (Zahid et al., 2018).

In addition, membrane morphology can be categorized as either asymmetric or symmetric. While symmetric membranes have uniform pore configurations throughout their cross-section, asymmetric membranes are composed of a thin and dense layer supported by a porous substrate that lowers the resistance to transport. The selectivity of the membrane is influenced by pore size and molecular weight cut-off (MWCO), especially for ultrafiltration membranes, since asymmetrical membranes can be made from one or more materials. Intrinsic performance is determined by pore structure and fouling, which is a secondary layer that forms on the membrane surface, affecting the change in the resultant efficiency and rejection characteristic (Othman et al., 2021).

2.2.4 Nanomembrane Challenging

Nanomaterials have outstanding physicochemical attributes with a great ability to remove pollutants, demonstrating potential applications in water remediation. Despite their recent widespread use, there are some restrictions on their large-scale commercial use in water purification, such as cost-effectiveness, correct dispersion of nanoparticles, recovery, retention, intrinsic toxicity, decrease in activity, and availability of reuse. Using nanomembranes in water treatments faces significant challenges; the most critical one is fouling and scaling. The accumulation of suspended solids, biological organisms, and natural organic or inorganic matter causes fouling, which might be completely or partially blocked shown in Figure 10 (Zahid et al., 2018; Divya Bajpai Tripathy & Gupta, 2023; Ullah et al., 2020). On the other hand, scaling results from the precipitation of dissolved salts after their solubility limits are exceeded. With time, these affect separation efficiency by blocking water pathways, decreasing permeate flux, shortening the lifespan of membranes, and increasing transmembrane pressure (Divya Bajpai Tripathy & Gupta, 2023). Nanoparticle dispersion over a given substrate is necessary for better performance; however, its performance might be affected by a reduction in effective surface area resulting from NP aggregation, particularly when it is used for wastewater with a high pollutant concentration (Puri et al., 2021). The performance of the membrane is influenced by the type of membrane, including pore size and material properties, the quality of the feed water, and the operating environment, like pH, flow velocity, and temperature (Zahid et al., 2018; Divya Bajpai Tripathy & Gupta, 2023). Mechanisms such as standard blocking, intermediate blocking, complete pore blockage, cake layer formation, and gel-layer formation are progressed through fouling advances, each of which results in decreased flux and increased consumption of energy. Although methods like antifouling coating such as such as nano silver, copper, gold, graphene oxide, carbon nanotubes, TiO_2, and zinc-based nanomaterials, and chemical cleaning reduce the effect of fouling and scaling, these procedures are expensive, varying their resultant efficacy based on the type of contaminants and membrane system (Divya Bajpai Tripathy & Gupta, 2023; Puri et al., 2021). Since the reusage of the nanomembrane lowers the efficiency of filtration and operation, membrane fouling is considered a serious issue (Puri et al., 2021). Additionally, it is important to ensure minimal fouling by automatically backwashing from two to four times per hour before following the sequence of filtration in membrane large-scale filtration systems (Othman et al., 2021).

Figure 10. Blocking of membrane pores due to foulant accumulation

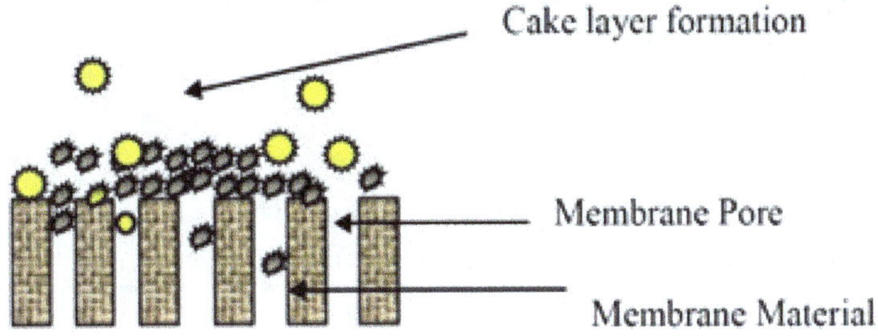

(Zahid et al., 2018)

The cost-effectiveness is another major challenge related to the use of raw materials, like advanced polymers, ceramics, and functional nanoparticles, as complex fabrication methods are required to design nanostructures. To address this, employing alternative fabrication that uses inexpensive materials graphene and other two-dimensional nanomaterials, enhances the membrane performance and lowers production costs. Applying this technique on a large scale for water treatment requires addressing technical and environmental issues (Divya Bajpai Tripathy & Gupta, 2023). Also, nanomembranes, particularly polymeric or nano-composite membranes, frequently face trade-off phenomena. Means that increasing permeability typically decreases selectivity and rejection of contaminants, and vice versa (Divya Bajpai Tripathy & Gupta, 2023). Moreover, under harsh conditions, some organic polymeric membranes are mechanically brittle or chemically sensitive. For example, disinfection by additions of chlorine may result in degradation and shorten membrane lifespan because of exposure to oxidizing agents or extremely high pH (Divya Bajpai Tripathy & Gupta, 2023). Furthermore, the membrane on a large-scale water purification system is supplied with hardware and software to guarantee continuous operation. Despite that, production of membranes is becoming economic; however, other expenses like cost of spare parts, periodic maintenance, and electricity are required for all these facilities, increasing their costs and expenses (Othman et al., 2021).

The possibility of accidental complexes or secondary contaminants being released during the treatment procedures and the decrease in NP activity that emerges over time are concerns. These problems raise the cost and increase the risk of environmental emission of nanomaterials functionalized with nanoparticles because they may eventually accumulate and contaminate the surrounding area. The study reported that some metallic nanoparticles, like Nano-Ag, ZnO, and TiO_2, are restricted in their durability and availability. Furthermore, TiO_2 nanoparticles are required to be activated by ultraviolet before starting the procedure with low biodegradability, which raises concerns about possible long-term effects for both terrestrial and aquatic species. Specifically, the partial solubility of some nanomaterials increases the difficulties in removing such materials from treated water, expanding the obstacles faced to get high quality water (Puri et al., 2021).

Table 1. Comparison of benefits, drawbacks, and various uses for nanomaterials in the wastewater purification process (Puri et al., 2021)

Different types of nanomaterials	Advantages	Disadvantages	Application
Metal-based nanoadsorbents	Highly reactive, faster kinetics, mostly water-insoluble, easier separation	Cost-effective, higher aggregation tendency, sometimes sludge production, Intrinsic toxicity of some metals	Antibiotic removal, Heavy metal ions, and dye removal capability
Carbon-based nanoadsorbents	Tunable surface functional groups, higher thermal and mechanical stability	Expensive regeneration costs, Loss of performance after the regeneration process	Heavy metal removal, noxious waste removal
Nanocomposite membranes	High fouling resistance, hydrophilicity, thermal and mechanical stability, and better water permeability	Discharge of nanoparticles, membrane pore blocking	Good sorption and filtration capability, Dye and odor removal from water, disinfection of bacteria
Polymer-based nanoadsorbents	Higher chemical functionality, mechanical stability, and better adsorption rate	Extraction of binary or ternary complexes, higher maintenance cost	Heavy metal ions removal, dye removal

2.2.5 Nanomembrane Application in Water Treatment:

2.2.5.1 Removal of Organic Pollutants

Pollutants might include both polar and nonpolar functional groups, resulting in membrane fouling. This indicates the importance of removing these contaminants effectively and efficiently. Natural organic matter (NOM) is found mostly in ground and surface water. The presence of these can negatively affect the quality of water, like taste, odor, or color (Othman et al., 2021; Puri et al., 2021). Also, it can help to carry different hazardous pollutants, such as pesticides and heavy metals, forming stable organometallic complexes that increase their mobility and toxicity. Disinfection processes like chlorination produce harmful by-products such as trihalomethanes (THMs) and adsorbable organic halides (AOX) when organic molecules react with disinfection chemicals. Reverse osmosis, nano, and ultra membrane effectively removed organic and hazardous by-products, which are presented in a variety of sizes;

the efficiency of removal depends on the membrane pore size. Nano and ultrafiltration are particularly used in removing such pollutants with significant efficiency, with the advantage of not producing any secondary hazardous material. Furthermore, these filters are effective in removing other contaminants such as phthalates, pollutants derived from plastic, up to 99.9% (Othman et al., 2021)

A study conducted in 2019 examined the ability of a polysulfone nanocomposite membrane combined with gold nanoparticles (CT-GNPs/PSF) and reported that the membrane showed exceptional removal of humic acid (HA) with self-cleaning properties. The nanocomposite membrane with a hydrophilic surface having negative charges improves many features compared to native polysulfone, such as fouling resistance, permeation of water, flux recovery, and degradation of HA. Moreover, because of their chemical durability, small pore sizes, and increased water permeability, thin-film nanocomposite (TFN) membranes are also equally effective. Additionally, many organic contaminants can be removed using developed functionalized multi-walled carbon nanotubes due to enhanced hydrophilicity and porosity. The potential of reusing such membranes is very high, as they can be regenerated through alkaline treatments (Puri et al., 2021).

2.2.5.2 Removing Inorganic Pollutants

Polymeric membrane performance in removing heavy metals is enhanced by carbon-based nanomaterials such as carbon nanotubes (CNTs), graphene, or graphene oxide. The presence of such materials as graphene oxide containing functional groups like COOH or OH provides pollutant binding capacity with remarkable mechanical, electrical, and thermal properties from graphene itself. In this regard, modified polymeric membranes with graphene oxide enhance heavy-metal rejection. For instance, polyethersulfone (PES) membranes modified with graphene oxide can remove up to 65% Cd, 70% Cu, and 85% Zn, whereas carboxylated graphene oxide composites reach 90% removal of Cd and Ni. All these confirmed that graphene oxide is a successful nanofiller for water purification as it has an adjustable oxidation state and hydrophilic functional groups, resulting in improving membrane stability and structure (Iqbal et al., 2024)

Additionally, a study confirmed the enhancement in metal ion selectivity, fouling resistance, and membrane strength achieved by modification of the cellulose acetate membrane with nano chitosan particles (CSNPs). Converting chitosan, derived from deacetylating chitin found in crustacean shells, into nanoparticles using ionic gelation and other techniques to increase its surface area. The incorporation of such materials improves membrane performance, resulting in high adsorption capacity, improved by the presence of functional groups, like amine and hydroxyl, that make strong selective interactions with metal ions. This nanomembrane is successfully improving the removal of two widespread pollutants, Fe (III) and Mn (II), considered toxic even at low concentrations, from agricultural and industrial wastewater, offering a scalable and efficient way to reduce heavy metal pollution (Morsy et al., 2025).

Reverse osmosis is mainly used to purify seawater and brackish groundwater; however, removing suspended solids, ions, organic matter, and microorganisms is essential by using a nanomembrane as a pre-treatment of water to avoid membrane fouling or blocking. Since these nanomembranes offer a great, affordable substitute for water softening instead of ion exchange, they help in reducing the level of hardness based on the type of membrane and operation conditions. The study reported that a combination of nanofiltration with Integrated systems, like biological oxidation, enhances the efficiency of treatments and membrane lifespan (Othman et al., 2021). According to another study, heavy metals such as arsenic, copper, iron, lead, and chromium can be effectively reduced by hybrid membrane systems by a combination of both conventional treatments with pressure-driven membrane processes, offering economic and environmental benefits. For instance, reducing the high concentration of iron in water and combining microfiltration with air oxidation are useful strategies that lead to good-quality water. Arsenic can also be removed using ultrafiltration, reverse osmosis, and hybrid coagulation-membrane systems, reaching 99% arsenic reduction, whereas nanofiltration through the membrane surface's electrostatic repulsion and size exclusion mechanisms removes up to 97% (Othman et al., 2021).

In recent years, metal-based nanoparticles have received great attention for their possible uses in the technology of membranes in industrial applications, with a great increase in their production, which have unique physical and chemical properties that enable them to be used. Since permeability and selectivity between membranes are improved by specific additions of metal nanoparticles, different metallic nanoparticles are mostly synthesized with specific desired modifications in sizes and shapes that can be achieved using various physical and chemical techniques, like Al, Ag, Au, Ni, Fe, and Cu. For example, it has been demonstrated that adding copper nanoparticles to polymeric membranes increases the removal of heavy metals, while platinum nanoparticles have been applied to enhance the surface of membranes and capacity to separate gases (Aydin et al., 2024). In another study, flexible sodium titanate nanobelt membranes showed a great capacity to adsorb radioactive ions Sr^{2+} and Cs^{+} through ion-exchange processes followed by oil extraction after rutile phase transition (Puri et al., 2021). Comparing the characteristics of membranes made of nanomaterials are presented in Table 2 (Aydin et al., 2024).

Table 2. Comparing nanomaterial-based membranes' properties (Aydin et al., 2024)

Material	Pore size control	Mechanical strength	Ion selectivity	applicability	Low cost	Toxicity
Polymeric membranes	X	√	√	√	√	√
Block copolymers	√	X	√	√	X	X
Metal-based membranes	√	X	√	√	√	X
Carbon-based membranes	√	√	X	X	X	X

2.2.5.3 Removing Surfactants and Dyes From Wastewater

Another research reported that wastewater contaminated with surfactants can be treated by membrane-based technologies like ultrafiltration, microfiltration, and nanofiltration, which are emerging in industrial applications. The study reported that ceramic ultrafiltration membranes can remove both anionic and non-ionic surfactants moderately. Additionally, at high pressure, polymeric nanofiltration membranes can remove up to 99.7% of the chemical oxygen demand (COD) from laundry wastewater [M11]. The research aims to enhance the removal of cationic surfactants from water by a combination of membrane filtration and ion-exchange procedures. This method used polymeric ultrafiltration and nanofiltration to purify a solution containing cetrimonium bromide (CTAB) and benzalkonium chloride (BAC), followed by batch ion exchange with strongly acidic resins. While experiments were conducted at surfactant concentrations below, close to, and above the critical micelle concentration (CMC), the permeate obtained was sent to ion exchange units. CMC refers to the surfactant concentration that causes micelles to form, and any addition of surfactants to the system will also cause the formation of micelles. Results confirmed that neither membrane filtration nor ion exchange itself was completely successful or efficient. However, using the combination of ultrafiltration and ion exchange reached complete removal of CTAB and BAC at concentrations below and near the CMC. In the case of high concentrations higher than 1,000 mg/L, ultrafiltration is combined with C150H resin to successfully eliminate almost 90% of BAC, whereas under similar conditions, CTAB removal reached 99.6% (Klimonda & Kowalska, 2021).

The versatility of nanomembranes in dye and contaminant remediation has been shown in numerous studies. A study conducted on modified microporous polyvinylidene fluoride (PVDF) membranes containing zwitterionic poly (sulfobetaine methyl methacrylate) (PSBMA) and halloysite nanotubes (PSBMA/HNTs/PVDF) reported the successful removal of heavy metal ions and soluble dyes from wastewater by offering many active sites based on simple filtration-adsorption that effectively purifies viscous oils with a significant level of antifouling resistance. Electro-sprayed hierarchical $Mg(OH)_2$ nanostructures on membrane capsules have great efficiency towards basic dyes, especially flower-like $Mg(OH)_2$ nanostructures, which showed the highest adsorption efficiency among other examined morphologies (Puri et al., 2021).

2.3 Electrochemical Techniques Categories and Advantages

Electrochemical technologies have become efficient and effective methods for removing contaminants from wastewater. These methods, which use electrosorption and electrokinetic effects to treat water, are becoming increasingly popular (Alkhadra et al., 2022). Electrical energy is used to separate, convert, or coagulate ions or metals from wastewater, producing repulsion or attraction forces causing the movement of solute ions and contaminants with a great level of precision (Alkhadra et al., 2022; Fadillah et al., 2025). Depending on the strength of the electric field, the electrophoretic velocity and mobility of target species will be. Electrophoretic mobility is influenced by the size and charge of the involved species (Fadillah et al., 2025). These methods have important uses in the environment, as they are considered economic, with low operating expenses, great selectivity, and efficiency for water purification (Youssra El Hamdouni et al., 2022). Even though the ohmic resistance, concentration polarization, and overpotentials of Faradaic activation cause loss of energy of systems, the amount of loss relies on the quantity of removed contaminated ions, not the total volume of treated water. Thus, these techniques are particularly ideal for treating brackish water desalination and low-salinity applications (Alkhadra et al., 2022; Dube et al., 2025). Selective binding to toxic monitored species from complex water matrices can be enhanced by tuning surface chemistry and immobilizing functional groups at the surfaces of electrodes at ambient temperature and pressure, depending only on the electric potential applied as an energy source (Alkhadra et al., 2022).

Recent studies highlight the successful electrochemical processes with significant efficiency in modern water treatments for a variety of pollutants, such as heavy metals, pathogens, salts, pharmaceuticals, dyes, and textiles, considered sustainable, powerful, and energy-efficient tools (Alkhadra et al., 2022). These processes depend on reactions that take place at the electrode surface, which might be transformation, oxidation, reduction, or aggregation caused by an applied electric field (Alkhadra et al., 2022). Over traditional chemical treatment, electrochemical technologies offer advantages like less production of sludge, reduced chemical input, simpler automation, and the ability to generate reactive species such as hydroxyl radicals, coagulant ions, and hydrogen peroxide (Dube et al., 2025). In this regard, waste handling expense and the impact on the environment are reduced, which is documented in numerous studies on both municipal and industrial wastewater treatments by electrochemical processes (Dube et al., 2025). Although these techniques demonstrated good sustainability, which can be greater by combining them with renewable energy sources, they might produce toxic byproducts such as trihalomethanes, which are mostly connected to disinfection processes based on chlorine, particularly in saline water treatment (Dube et al., 2025). Electrochemical methods offer alternative ways with less energy consumption than other technologies like conventional reverse osmosis (RO), as seen in Figure 11, which demonstrates the energy usage of traditional reverse osmosis with great energy consumption compared to electrochemical water purification. Most contaminant types can be removed by the electrochemical process due to effective oxidation and reduction reactions in breaking down complex compounds (Dube et al., 2025). According to several reported studies, the cost of electrochemical procedures ranged from moderate to high due to initial setup and the considerable expense of ongoing maintenance for electrodes to avoid fouling (Dube et al., 2025).

Figure 11. Comparing the energy usage of conventional and electrochemical reverse osmosis

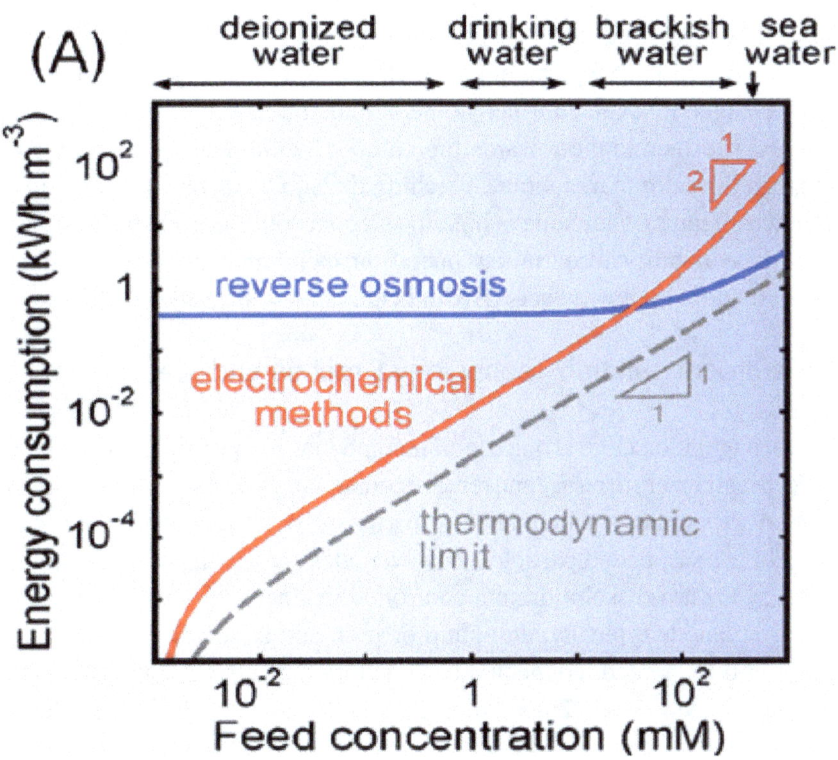

(Dube et al., 2025)

Based on the main mechanism controlling the removal of pollutants, electrochemical water-treatment methods can be categorized into three main groups used to improve water quality. The first one is electrochemical separation methods, which rely on ionic migration caused by the presence of an electric field to exclude impurities without chemically transforming them. These technologies are particularly useful for desalination and the removal of dissolved ions, such as electrodialysis (ED), reverse electrodialysis (RED), and electrodeionization (EDI) (Dube et al., 2025). The second category includes electrochemical conversion processes to transform contaminants chemically by either oxidation or reduction reactions. Such methods can be used to degrade pharmaceuticals, dyes, persistent organic pollutants, or toxic inorganic pollutants, like electrochemical oxidation (EO), electrochemical reduction (ER). The variation in degradation efficiency is related strongly to oxidizing agents produced in the process, and the ability of pollutants to degrade, while electrochemical reduction and oxidation depend on electrode material with applied current strength (Dube et al., 2025; Ashish Kumar Das & Chen, 2024). Electrochemical aggregation and removal procedures are the third type that remove impurities by producing gas bubbles or coagulants, such as electrocoagulation (EC) and electroprecipitation. Heavy metals, turbidity, and suspended solids can all be effectively removed using these techniques (Al-Marri et al., 2023). A driving force, like pressure, concentration gradients, or electrical potential, is necessary for them to operate. For instance, in electrodialysis (ED) and reverse electrodialysis (RED), a differently concentrated stream is generated by applying an electrical potential to separate ions by ion-exchange membranes. Because of induced direct current (DC), positively charged ions are confined at the anionic membrane's surface when they travel across the membrane towards the cathode, while negatively charged ions pass through the anionic membrane, reaching the anode, resulting in two different concentrated streams, where one is high in salt concentration, and the other is less. To ensure dependability in continuous operation, membranes must be manufactured at a low cost and with few defects (Aydin et al., 2024; Othman et al., 2021).

2.3.1 Electrocoagulation Technology Process and Mechanism

Electrocoagulation (EC) is one useful method in water purification that employs micro-electrical current to purify and remove contaminants. Both electro- and chemical coagulation are similar in principle of formation of larger aggregation of pollutants by destabilizing suspended particles; however, electrocoagulation generates metal coagulant by electro-dissolution in a controlled manner. Typical EC contains two electrodes, an anode typically aluminum or iron, and a cathode, immersed in the water that will be treated, connected to an external electrical source, forming an electrolytic cell (Mao et al., 2023; Patel et al., 2024; Dube et al., 2025). Aluminum anodes release Al^{3+}, whereas iron anodes produce Fe^{2+} and Fe^{3+} ions. Fe^{3+} has a higher adsorption capacity and is a stronger coagulant relative to Fe^{2+}. As a result, iron anodes used at specific pH ranges encourage oxidation from Fe^{2+} to efficient Fe^{3+} ions. Despite the high expense of aluminum electrodes, the performance of aluminum electrodes is greater due to their exceptional coagulation effectiveness than iron. Also, for desired specific pollutant removal, aluminum anodes can be coupled with other cathodic materials like stainless steel or iron. The EC method combines the benefits of electrochemical reactions, flotation, and coagulation into one system (Mao et al., 2023).

During the process and under applied current, metal ions, which mostly are aluminum or iron, are released from the sacrificial anode by the electrolytic oxidation process, the reaction represented in Equation (1). Hydroxyl ions are produced simultaneously at the cathode (Equation 2), resulting from water hydrolysis to react with metal ions, forming monomeric and polymeric metal hydroxide species (Equations 3, 4, and 5) that start the process of contaminant capturing by neutralizing the surface charges for pollutants like solid particles, oil, and dye pollutants (Dube et al., 2025; Mao et al., 2023; Patel et al., 2024). Neutralized species become unstable in the water and start to aggregate, making large flocs with higher weight, forming sweep coagulation that may rise to the water's surface or sink to the bottom. The resulting flocs may be raised to the surface by the generated hydrogen gas bubbles at the cathode's surface, which may adhere to them. These flocs can bind with organic materials, inorganic pollutants, and suspended solids to be separated from water. In this regard, this combination of floating and sedimentation successfully eliminates pollutants from water (Dube et al., 2025; Mao et al., 2023; Patel et al., 2024). The electrocoagulation process schematic diagram is shown in Figure 12 (Sadaf et al., 2024).

Figure 12. Diagram showing the electrocoagulation process

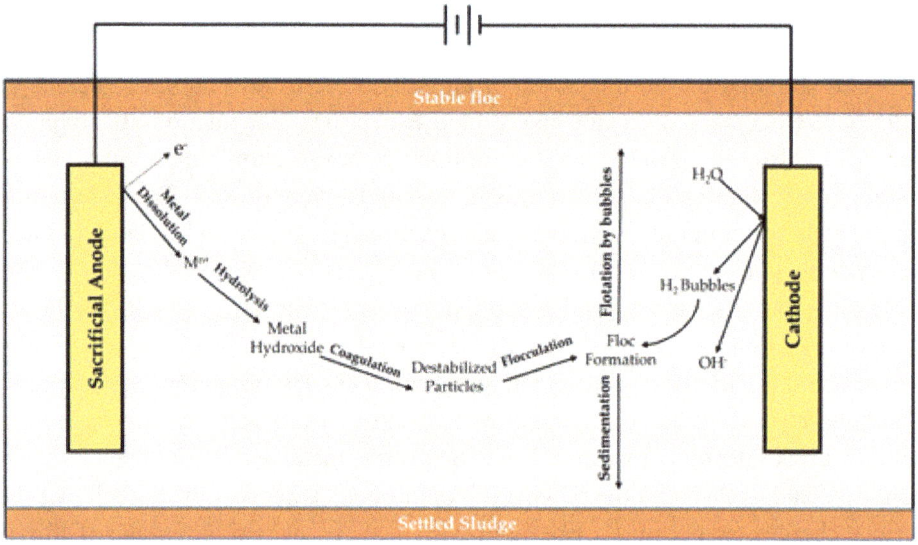

(Sadaf et al., 2024)

$$M\ ne^{-1} + M^{+n} \quad (1)$$

$$2e^{-1} + 2H_2O\ H_2 + OH^- \quad (2)$$

$$4Fe^{+2}_{(aq)} + 10\ H_2O + O_{2(aq)}\ 4Fe(OH)_{3\ (s)} + 8H^+ \quad (3)$$

$$Fe^{+2}_{(aq)} + 2OH^-\ Fe(OH)_{2\ (s)} \quad (4)$$

$$Al{+}3_{(aq)} + nH2O\ Al(OH)_n^{3-n}{}_{(s)} + nH^+ \quad (5)$$

Electrocoagulation, first applied in 1906 by Dieterich, aimed to treat ship bilge and sewage water, where the basic mechanism employs the metal anodes' in-situ oxidation to generate charged species that capture and collect pollutants. After neutralization of their negative charge, the separation and removal of these contaminants becomes easier (Dube et al., 2025). This method is used widely in industry to remove a variety of pollutants, such as heavy metals, and break down organic compounds like dyes. Research and studies in this technology are continuous to enhance it, as the promising wastewater way provides reasonably simple, affordable, and on-site processes easier (Dube et al., 2025). Despite these advantages and benefits in EC, this process has many operational drawbacks, like passivation of electrodes, residual metal ions that need to be treated, unequal anode dissolution, and the need for regular electrode replacement to maintain treatment easier (Dube et al., 2025). Mass of dissolved metal during electrocoagulation is linked to other factors like operation time, molecular weight of the anode material, number of electrons involved in the electrochemical reaction, and applied current by Faraday's law, presented in Equation (6). In the equation, the mass of dissolved anode (g) is represented by mA, t is the operation time in seconds, Mw represents the molecular weight of the anode in g/mol, I is the current (A), F is Faraday's constant (96,487 C/mol), and z represents the number of electrons engaged in the process (Mao et al., 2023).

$$mA = \frac{I \times t \times Mw}{z \times F} \tag{6}$$

The relationship states that the mass of dissolved metals is directly proportional to the applied current and time of procedure; however, the experimental observation of anode dissolution in practical applications frequently differs from the calculated value obtained by applying Faraday's law in real-world applications. Along with the expected dissolution of metal in the anode, disparity indicates that the possibility of other electrochemical reactions may take place, which is mentioned in some cases that under alkaline conditions and at high enough anodic potentials, oxygen evolution can occur at the surface of the anode, as shown in Equation (7). Generating oxygen gas, protons, and electrons can affect the efficiency of released metal ions and overall performance (Mao et al., 2023; Patel et al., 2024).

$$2H_2O\ O_2 + 4H + 4e^{-1} \tag{7}$$

One of the studies aimed to enhance effectiveness for wastewater treatment and applied a continuous electrocoagulation–flotation (ECF) system designed specifically for municipal wastewater treatment, reporting significant removal of contaminants using both iron and aluminum anodes at an optimal current density of 120 A/m^2. According to the study, the aluminum anode exhibited super faradaic behavior, resulting in higher residual concentrations in the effluents and cost of operation compared to an anode made of iron, meaning that the rate at which the aluminum anode dissolved exceeded the theoretically calculated value from Faraday's law. This is explained by more electrode material being consumed, increasing the quantities of residual aluminum to be treated, making the cost higher. Also, the coagulant generation becomes more difficult, making treatment efficacy less predictable. This system successfully has good repeatability and identical outcomes, about 98% of total phosphorus, 60% of biological oxygen demand (BOD), 94% of suspended solids, 40% of total nitrogen, and 90% of total chemical oxygen demand from treated water; however, resulting performance depends on the weather easier (Dube et al., 2025).

2.3.2 Parameters Influencing the Electrocoagulation Process (EC):

2.3.2.1 Effects of Electrode Materials, Size, Shape, and Configuration

According to numerous studies, electrode material is one of the most important parameters affecting the electrocoagulation process, as, depending on the material, the energy efficiency, subsequent reaction with contaminants, and mechanism used for removing pollutants and electro-generated coagulants vary. The most popular sacrificial anodes used in operations are aluminum and iron because of their dependability and accessibility, which generate monomeric and polymeric hydroxo-complexes that form flocs after neutralizing surface charge (Patel et al., 2024). Regarding the studies, aluminum-based electrodes are mostly used to eliminate heavy metals, whereas for degradable organic compound removal and wastewater purification, iron electrodes are frequently used easier (Dube et al., 2025). In some cases, heavy metals were effectively eliminated by the Fe electrode with less required energy and expense; however, it corrodes more quickly than Al (Sadaf et al., 2024).

Other anode materials that have been examined include Pt, Pb, Cu, Zn, and Mg. In terms of Cu and Zn anodes, their performance was similar, except that the copper electrodes may produce secondary pollutants. Although the platinum electrode has high stability and conductivity, it may form fine coagulating particles that reduce its efficiency, while nitrite and ammonia are removed with high efficiency using the magnesium electrode. Lead is considered an economical option, but it corrodes quickly (Sadaf et al., 2024).

Moreover, the size of electrodes effectively influences total contaminant removal, as it affects spatially distributed current density throughout the reactor. Similarly, the geometry of electrodes and shapes also affects the dynamics of the generated coagulants' dispersion significantly. Therefore, in electrocoagulation (EC) systems designed for wastewater treatment, the choice of electrode material, geometry, and size should be optimized. The creation of high-efficiency EC that recovers the resource is crucial, which can be applied by treating a variety of wastewater, aiming to reduce energy consumption and operating costs (Patel et al., 2024).

Another factor affecting the extraction of pollutants by electrocoagulation is the arrangement of the electrodes in the cell. Arranged electrodes might be one of three arrangements: monopolar–parallel (MP–P), bipolar–series (BP–S), or monopolar–series (MP–S), as shown in Figure 13. In MP-P arrangements, electrodes are connected in parallel to the power source and relatively to other electrodes, while in the BP-S system, the internal electrodes are electrically isolated from the source of power and from each other, making a series pathway. In the MP-S arrangement, intermediate electrodes are connected in series (Mao et al., 2023; Patel et al., 2024). Current is distributed across linked electrodes based on their electrical resistance in parallel configurations that result in significant energy savings, while in series arrangements, a larger potential difference is needed and required. In reported cases, the BP–S configuration is applied for improving pollutant removal efficiencies, whereas monopolar systems operate with lower expenses. Generally, the efficiency and cost of a system are significantly impacted by both the choice of material and the linking strategy of electrodes (Mao et al., 2023).

Figure 13. Different electrode arrangements in an electrochemical cell

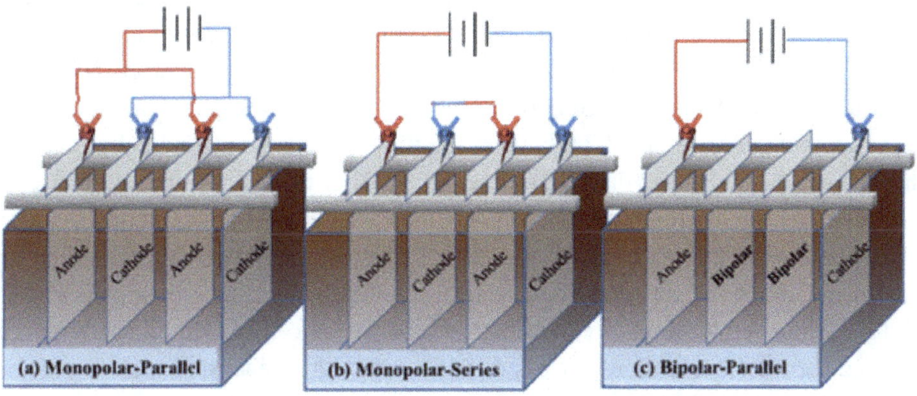

(Patel et al., 2024)

2.3.2.2 Effects of Current Intensity, Density, and pH:

In electrocoagulation (EC), both applied current density and intensity are important operational factors in a system. Current density is a distributed current over the effective surface area of the electrode calculated from the submerged part of the electrode in the treated water based on its geometric measurements (length × width) and expressed in A/m^2 or mA/cm^2, while the total applied current in the EC system, expressed in A or mA, is current density (Patel et al., 2024). The rate of coagulation generation, size of produced bubble, and mass transfer at the surface of the electrode are affected by current density, which is reported in a range from 0.01 to 880 A/m^2, which impacts total process performance. Increasing current density enhances the efficiency of pollutant removal by raising the number of released coagulants and forming smaller bubbles that more effectively stick to contaminants. However, applying excessive current densities increases electrode passivation, wastes energy, causes secondary reactions, and reduces the Faradaic efficiency of electrodes with significant differences in anode dissolution from expected values, affecting overall electrocoagulation efficiency (Mao et al., 2023)

A study conducted examines the effect of increasing current intensity in removing chemical oxygen demand (COD) and Total Dissolved Solids (TDS). The results shown in Figures 14a and b indicate that, up to 60 minutes, increasing the current intensity improves the biochemical oxygen demand (COD) capture process, with the percentages reaching 40%, 45%, and 47.5% at 1 A, 1.5 A, and 2 A, respectively. The same pattern was obtained for TDS, suggesting larger floc formation by increasing the intensity. Nevertheless, after 60 minutes, the amount of removal is reduced related to restabilization of metal hydroxides in longer treatment durations with little enhancement at 120 minutes (Patel et al., 2024).

Another crucial factor is the pH value, because the initial pH gives mineral species with different degrees of stability, solubility, hydrolysis pathways, and colloidal instability behavior under different media, whether neutral, acidic, or alkaline, which affects the formation of the resulting masses and the efficiency of the processing (Mao et al., 2023; Patel et al., 2024). According to published studies, the formation of metal hydroxyl complexes, whether polymeric or monomeric, is controlled by the aggregation efficiency, which is affected by the pH of the water, as practical experiments have reported that the ideal conditions for the best water treatments are in the range of semi-neutral to slightly alkaline, whereas acidic or highly alkaline environments can affect the balance by forming soluble compounds that reduce the flocculation process or encourage other reactions such as oxygen release (Mao et al., 2023).

Figure 14. Effect of current intensity on the removal of TDS at pH 9, EG = 2 cm, and stirrer speed of 200 rpm

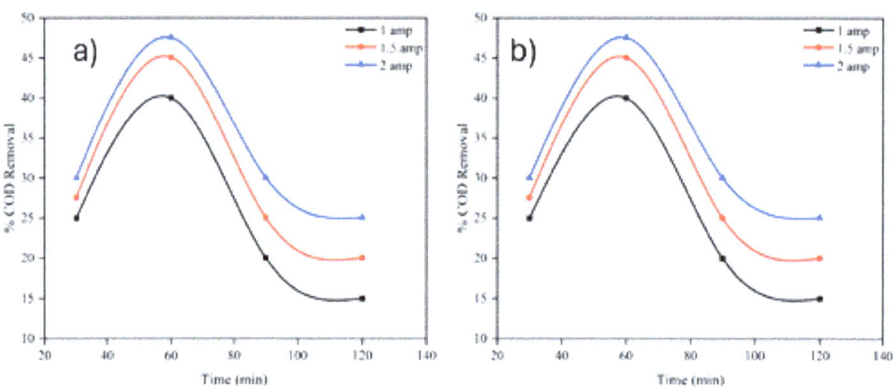

(Patel et al., 2024)

In the case of an iron electrode, the metal undergoes a series of changes: from Fe^{2+} generation, followed by oxidation into Fe^{3+}, and the consequent production of insoluble hydroxides, while in an Al electrode, the resulting hydroxides of amphoteric aluminum contribute to neutralize pH (Mao et al., 2023). In the case of an iron electrode, the metal undergoes a series of changes: from Fe^{2+} generation, followed by oxidation into Fe^{3+}, and the consequent production of insoluble hydroxides, while in an Al electrode, the resulting hydroxides of amphoteric aluminum contribute to neutralize pH. Figure 15 showed that at low pH Al^{3+} is formed, solid $Al(OH)_3$ in natural medium, and $Al(OH)_4^-$ in alkaline conditions, while increasing pH, Fe^{2+} is oxidized to Fe^{3+} and then to insoluble hydroxides like $Fe(OH)_2$ and FeOOH at high values of pH (Mao et al., 2023).

The study result showed that COD removal of debrominated wastewater was highest at a level of pH=9 with 40% by 60 minutes, and increasing treatment time resulted in raising the formation of metal hydroxide and competition of adsorption sites and decreased overall removal, dropping to 15% after 120 minutes. In TDS removal, the efficiency of the process remains consistent across the range of 23.25% to 26.49% pH levels, probably related to efficient settling, which was supported by stable floc density, Figure 16 (Mao et al., 2023). Controlling pH is important for achieving an efficient and economical electrolysis process because by adjusting the pH, the researcher can control and enhance the instability of contaminants and their precipitation (Patel et al., 2024).

Figure 15. pH diagrams at 25 °C of aluminum and iron in water, the dashed lines represent the electrochemical stability range of water (Mao et al., 2023)

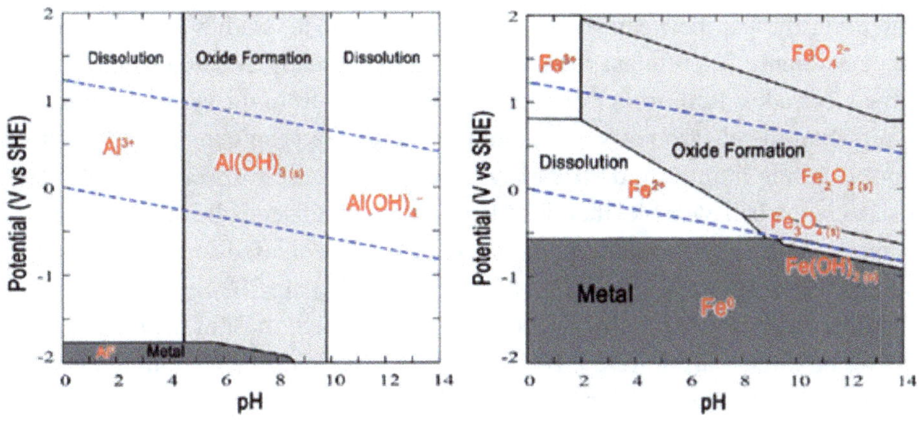

Figure 16. Impact of pH on COD and TDS reduction at I = 1 A, electrode spacing = 2 cm, and stirrer speed 200 rpm (Patel et al., 2024)

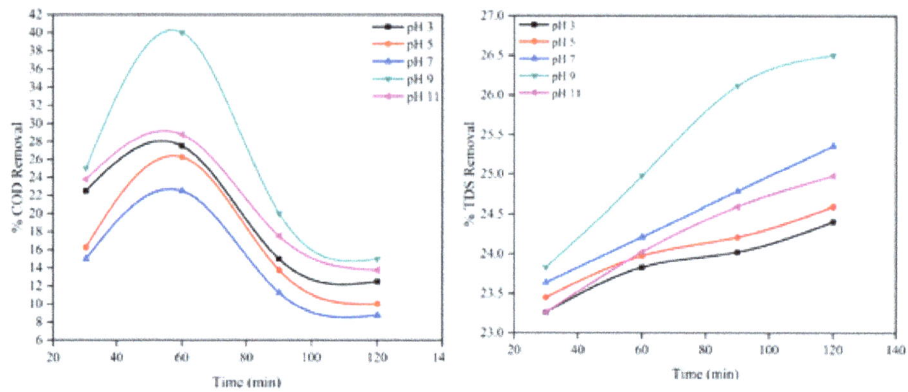

2.3.2.3 Effect of Separation Distances Between Electrodes And Power Source

EC processes are affected by other factors such as the separation distance between electrodes, initial concentration of pollutants, time, and type of power sources. Regarding the gap between the cathode and the anode, studies have shown that the performance of pollutant removal in general is greatly affected by the spacing of the electrodes. Removal of dissolved solids (DSS) from the system is enhanced by reducing the electrode spacing, like in chemical oxygen demand (COD) treatment, resulting in improved efficiency from 45% at 2.5 cm spacing to 50% at 1.5 cm spacing, a behavior like that observed in total dissolved solids (TDS) removal. Generally, these results highlight the importance of decreased separation distance, as larger electrode gaps reduce EC efficiency (Patel et al., 2024).

Additionally, the type of power source is also crucial; most of the EC processes depend on direct current (DC). However, long duration time leading to electrode passivation affects the efficiency of the process, which can be solved by switching to using pulsed current (AC) or polarity-reversal current (PRC). Fouling resulting from oxide layers produced at the surface of the electrode that obstruct the flow of current is reduced, encouraging the stable release of metal ions. Also, another solution that can be applied is the addition of a sufficient amount of chloride ions to dissolve the passive layer. PRC is relatively easier to apply, reversing the direction of current and reducing the possibilities of electrode passivation, while AC generates smoother electrode surfaces (Mao et al., 2023).

Numerous studies confirmed the outstanding improvement obtained by AC or APC power sources, like electrode longevity, energy consumption, and removing pollutants when applied in metals, dyes, and bilge water treatment. This, by increasing hydrogen evolution and reducing passive layers, leads to longer-lasting performance and reduces the costs, as shown in Figure 17 (Sadaf et al., 2024).

Electrolysis is an environmentally friendly technology that does not require the addition of chemicals and produces minimal secondary pollutants while maintaining high pollutant removal efficiency. Therefore, despite the cost of used electricity, developing industrial applications is crucial (Sadaf et al., 2024). Variations in electrode material, dimensions, and configuration are affecting the efficiency of the removal process (Patel et al., 2024). Applied current density and intensity, pH values, type of power source, and separation distance between electrodes should be optimized in operation, aiming to achieve low power consumption and high effectiveness of removal (Sadaf et al., 2024). Collectively, the a need for electrode design for specific target removal and achieving desired treatment goals. Table 3 provides a summary of the different configurations and operational conditions that were applied in different studies aiming to enhance the efficiency of treatment and energy-efficient and cost-effective (Patel et al., 2024).

Figure 17. Comparing the production of hydrogen and passivation in bilge water using AC and DC modes

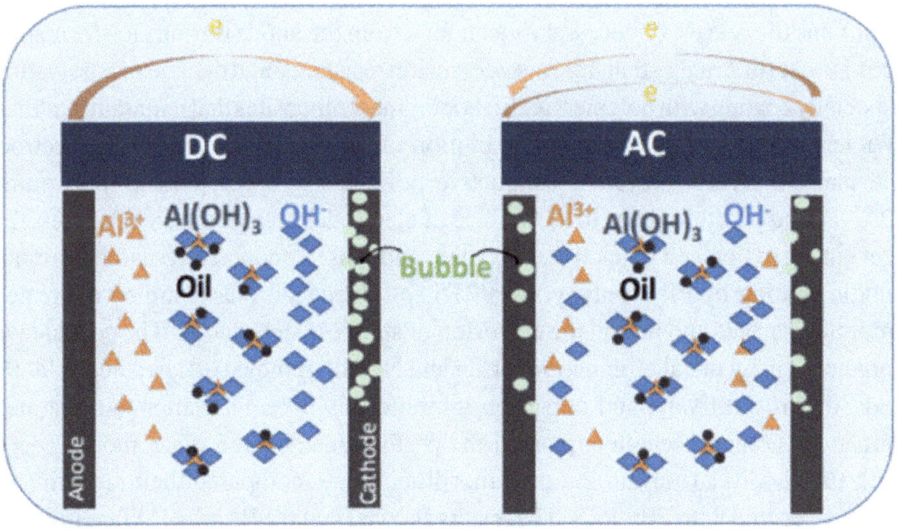

(Sadaf et al., 2024)

Table 3. Electrocoagulation process of actual wastewater from industry in batch mode (Patel et al., 2024; (Sadaf et al., 2024)

Anode	Cathode	Electrode Gap	Size (L × W)/ Area (electrode)	Current Intensity/ voltage	pH	Time	Pollutant	Removal %
Al	Al	1 cm	7 × 4 cm	5 Volts	7	10 min	COD	98.09%
Fe	Fe	3 cm	17 × 3 cm	0.2 A	7.3	120 min	COD	38%
Al	Al	1.5 cm	56.7 cm^2	216.4 A/m^2	6	30 min	Turbidity	97.43%
Al	Al	0.5 cm	10 × 10 cm	75 A/m^2	7.7	120 min	Calcium	88%
							COD	50%
Al	Fe	1.5 cm	13 × 8 cm	120 A/m^2	7	60 min	Phenol	64.7%
							COD	96.8%
Fe	Fe	40 cm	-	13 mA/cm^2	7	30 min	Cr^{3+}	100%
Al	Al	5 mm	-	1.2 mA/cm^2	3	60 min	Turbidity	100%
Fe	Fe	5 mm	-		4			
Al	Al	5 mm	-	0.6 mA/cm^2	3	60 min	$PO4^{-3}$	100%
							$SO4^{-2}$	79%

2.3.3 Surface Modification of Electrodes

Surface modification of electrodes, like coating, creates a protective layer that maintains the release of necessary metal ions from the anode during floc formation. This is a useful strategy that improves corrosion resistance and decreases passivation, especially coating with polymers and nanopolymer composites that maintain treatment efficiency, preventing excessive dissolution of the anode and extending electrode lifespan. Applying a layer of conductive polymers like polypyrrole, polyaniline (PANI), and polythiophene to the surface of electrodes creates physical barriers that prevent corrosion and block water, oxygen, and ions. Conducted research examines anodic coating by PPy (polypyrrole)/PTS (p-toluene sulfonate) aimed to create a smoother surface and inhibit the corrosion of stainless steel anode. The coated layer enhances heavy metal elimination for Zn and Ni with removal efficiencies of 99.9% and 80%, respectively, and corrosion inhibition by 30% even after ten treatment batches. Another research compared the performance of iron plate anodes coated with PANI-MMT (polyaniline-montmorillonite) and compared their performance with one coated with PPy/PTS. The results showed that the PANI-MMT coating was better in mechanical strength, corrosion resistance, and prevention of gas and liquid leakage at a current density of 20 mA/cm^2, a treatment duration of 100 min, and a pH of 7, giving higher removal efficiencies for Zn and Ni. Furthermore, one of the strategies implemented to improve the performance of electrochemical treatment for wastewater from the textile industry by applying a thin layer of activated carbon extracted from banana peels (BPAC) on stainless steel electrodes by dip coating, spin coating, and spraying, which has provided an effective and sustainable method for treating and reusing industrial (Sadaf et al., 2024).

Studies have shown that nanopolymer composites effectively fill polymer pores and improve barrier properties. For example, the addition of nanoparticles like TiO_2, Si, SiO_2, ZnO, and AlO_3 to conducting polymers enhances anticorrosion behavior, adhesion, mechanical strength, and thermal stability. A study investigated the corrosion resistance of nanocomposite PANI/epoxy/Zn coated at an iron electrode; the results successfully offered high mechanical strength and characteristics of a barrier that provides sacrificial protection, filling of pores, and PANI's redox characteristics. Also, nano PANI-DBSA coatings were investigated on iron, which are dodecylbenzene sulfonic acid-doped polyaniline nanoparticles, which offer strong anticorrosion qualities as the DBSA starts to react with the iron cations when corrosion is started (Sadaf et al., 2024).

2.3.4 Combination of Electrodialysis With Membrane Technology, Designing, and Application

Research emphasizes the importance of developing sustainable wastewater purification methods that support resource recovery. Electrochemical oxidation, flocculation, coagulation, and electro-flotation, which cooperate to accomplish specific different treatments, are mechanisms that can be used in electrocoagulation to integrate with membrane technology (Dube et al., 2025; Chukeaw et al., 2025). A study confirmed that the membrane fouling issue can be overcome to raise the efficiency of the membrane in organic recovery by integration of the membrane with pretreatment methods like electrocoagulation, combining electrochemical oxidation, electro-flotation, and in-situ metallic coagulant generation such as Al^{+3} and Fe^{+3}. Combining membrane with electrocoagulation enhances water treatments in many ways due to the formation of metal hydroxide flocs in situ that destabilize and collect foulants using controlled current and sacrificial metal anodes and reducing the amount of added chemicals. Overall contaminant removal performance is improved with the integration system (Chukeaw et al., 2025)

According to recent studies, material compatibility between membranes and electrochemical environments is a critical determinant of system longevity, selectivity, and energy efficiency. Conventional polymeric membranes, such as polysulfone, polyamide, and poly(vinylidene fluoride) (PVDF), dominate water treatment applications due to their low cost and ease of fabrication. However, their structural and chemical stability can be compromised under electrochemical operating conditions, including exposure to electric fields, redox gradients, extreme pH environments, and reactive oxidizing species. To overcome these drawbacks, research has focused on the incorporation of electrochemically stable nanomaterials such as titanium dioxide (TiO_2), graphene oxide (GO), and carbon nanotubes (CNTs) into membrane matrices, leading to the development of nanocomposites and thin-film nanocomposites that can be combined with electrochemical systems. These advanced materials enhance mechanical strength, electrochemical stability, and charge flexibility while preserving high separation performance (Foorginezhad et al., 2025).

Several nanocomposite membrane configurations, such as surface-located nanocomposites, mixed-matrix, and thin-film, have been developed to be used in integration systems aiming to enhance mechanical strength, thermal stability, and overall performance in water treatment applications. The separation efficiency and performance of these membranes are greatly influenced by the structural and physicochemical properties of the support, inner, and active layers that comprise them. Besides the significant mechanical stability provided by the support layer, it should be highly porous and hydrophilic, reducing hydraulic resistance and improving permeability, while the inner layer, which is typically a polymeric matrix designed to reduce polarization caused by internal concentration, improves water flux and selectivity. In terms of the active layer, which is mostly made of polyamide or other selective polymers, their thickness, surface morphology, and chemical functionalization control solute rejection and fouling resistance for the separation of salt and heavy metals. In this regard, nanocomposite membranes can achieve superior performance in advanced water treatment processes by strategically optimizing these layered components (Foorginezhad et al., 2025).

Surface charge compatibility represents another critical design consideration in electrochemical–membrane systems. Donnan exclusion, a charge-based separation mechanism that selectively blocks co-ions and draws counter-ions while controlling ion distribution and transportation across the membrane, requires electrostatic interactions between the ionic species and the membrane's fixed charge groups ((Foorginezhad et al., 2025; Nabila Chaabani et al., 2025). Negatively charged membranes are favored in most electrochemical separation procedures due to their favorable interaction with Donnan exclusion under applied electric fields. According to data reported, rejection efficiencies for divalent ions, including Mg^{2+}, SO_4^{2-}, Pb^{2+}, and Cd^{2+}, frequently exceed 97–99.9%, whereas the rejection of monovalent salts is generally lower unless membrane pore size and surface charge density are optimized (Foorginezhad et al., 2025; El-kordy et al., 2025).

Electrode-membrane compatibility is another important factor that determines system efficiency, according to quantitative performance patterns. Electrochemically relevant membrane systems have consistently shown high selectivity, with reported salt rejection efficiencies of 99.88% for NaCl, 99.80% for Na_2SO_4, and 99.50% for $MgSO_4$. Furthermore, a variety of species show almost total rejection of heavy metal separation: Cu (II), Cs (I), and As (V) typically approach 100% removal; Pb (II), Cd (II), and Ni (II) achieve approximately 99% removal, while Zn (II) shows a slightly lower yet still substantial removal efficiency of approximately 92% (Foorginezhad et al., 2025). Regarding heavy metals, Table (4) shows heavy metal removal using the most effective nanocomposite membranes in wastewater from industries. Integration of these highly effective nanomembranes in heavy metal removal with electrochemical processes is expected to give significant enhancement in water treatment.

Table 4. The most promising nanocomposite membranes for the removal of heavy metals (Foorginezhad et al., 2025)

Membrane matrix	Nanoparticle	Water permeability ($L\ m^{-2}\ h^{-1}\ bar^{-1}$)	Pollutant	Removal
polyethersulfone /polyamide	Titania nanosheet	1.57	Cs (II)	100%
polyethersulfone /polyamide	MCM-41	4.05	Cu (II)	100%
polyethersulfone	Amine, thiol-SiO2	6.70	Pb (II)	99%
polyethersulfone	WO3@GO Tungsten trioxide immobilized at (GO) Sheets	445	Ni (II)	99%
Polyethylene/polyamide	TiO2	5	As (II)	100%

Water permeability data show a broad range of performance: optimized nanocomposite membranes achieve significantly higher permeabilities of up to 445 $L{\cdot}m^{-2}{\cdot}h^{-1}{\cdot}bar^{-1}$, while conventional thin-film nanocomposite (TFN) membranes exhibit much lower permeabilities compared with nanocomposite membranes, between 1 and 35 $L{\cdot}m^{-2}{\cdot}h^{-1}{\cdot}bar^{-1}$. The importance of electrostatic interactions in electrochemical membrane processes is demonstrated by the fact that over 90% of high-performance cases use negatively charged membranes, and over 40% of reported systems use polysulfone (PSF) substrates because of their favorable mechanical and chemical properties. **Figure 18** summarizes the published data with the highest level of heavy metal removal efficiency using nanocomposite membrane between 2023 and 2024 (Foorginezhad et al., 2025).

Figure 18. The most recent studies on nanocomposite membranes with the highest removal percentage for heavy metal removal

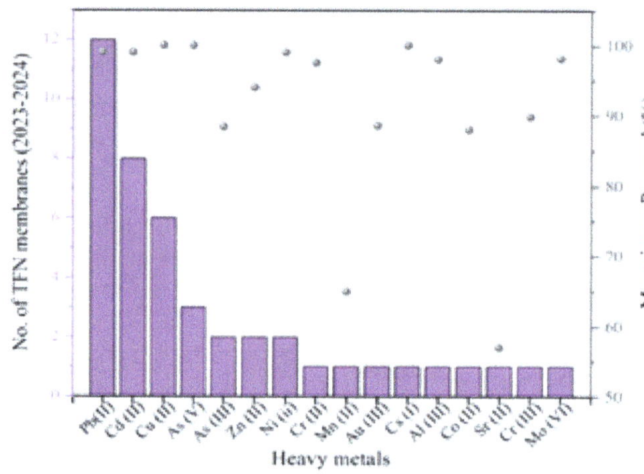

(Foorginezhad et al., 2025)

Integration of electrocoagulation technology with membrane pre-concentration (MPC) destabilizes and aggregates contaminants, leading to the formation of larger flocs that reduce pore blockage, decrease membrane fouling, and enhance filtration efficiency. A conducted review aimed at developing municipal wastewater (MWW) treatment and getting the benefit of collected biodegradable organic matter by converting it to energy using the integration of electrocoagulation and (MPC). MWW contains a significant fraction of biodegradable organic matter that might be converted to energy via anaerobic digestion; however, the presence of these materials is lower than ideal, restricting the direct conversion (Chukeaw et al., 2025). Anaerobic digestion is considered a viable method for generating energy into bioenergy, especially methane, from converted organics present in wastewater with concentrations above 1500–2000 mg/L. The membrane pre-concentration method is suitable for collecting and concentrating organics before energy recovery (Chukeaw et al., 2025). Typically, this wastewater contains 200–600 mg/L of chemical oxygen demand (COD), of which about 70–85% are organic materials that decompose readily, generating about ~0.35 kWh (kilowatt-hour) of electricity per cubic meter by applying anaerobic digestion technology. The study concludes that this combination is enhancing organic removal quantity, forming larger flocs, improving energy recovery, and lowering the possibilities of membrane fouling, enabling greater solid-liquid separation efficiency, resulting in a higher amount of biogas production. The efficiency of fouling suppression and floc formation during treatment is influenced by factors related to operations like pH, current density, and electrode material. Although that integration provides higher quality and flux of water, energy optimization, applying it on a large scale, and controlling electrode-related impacts are still significant challenges (Chukeaw et al., 2025).

Hydraulic fracturing generates large amounts of produced water that contains a lot of organic matter and other contaminants. This wastewater needs to be treated effectively to protect the environment and allow for reuse. A study examines the efficiency and performance of the treatment of produced water by membrane after electrocoagulation (EC) as a pretreatment step by using a continuous EC reactor with iron and aluminum sacrificial electrodes under varying configurations and current intensities. Applying the EC method to form metal hydroxide floc makes the performance better with fewer fouling possibilities and easy separation without the need for chemical coagulants (Cao et al., 2022). The study reported that using aluminum electrodes successfully decreased fouling during the pre-step at higher current around 5 A and achieved greater total organic carbon removal up to 42% while producing larger, more compressible flocs that improved TOC removal with more sludge management challenges, whereas iron electrodes produced denser, faster-settling sludge.

After membrane distillation (MD) of untreated produced water, a noticeable brownish fouling layer was observed on the membrane surface, as shown in Figure 19b. On the other hand, using membrane-electrocoagulation pretreated produced water at a current intensity of 5 A in 5 minutes generates white surface contaminants presented in Figure 19c, suggesting that the brown-colored pollutants of organic and colloidal foulants were effectively removed. These observations confirm that by removing colloidal and suspended organic matter through applying the membrane method, EC pretreatment significantly reduces membrane fouling, resulting in cleaner membrane surfaces and enhanced operational performance. Overall, the study demonstrates that EC is a successful pretreatment method in produced water purification, is effective in enhancing the total performance, and reduces membrane fouling considering electrode material, operating conditions, and sludge characteristics (Cao et al., 2022).

Figure 19. Digital images of membranes before and after MD using produced water: A) unutilized; B) produced water without pretreatment; C) produced water with pretreatment water with 5 A for a duration of 5 minutes

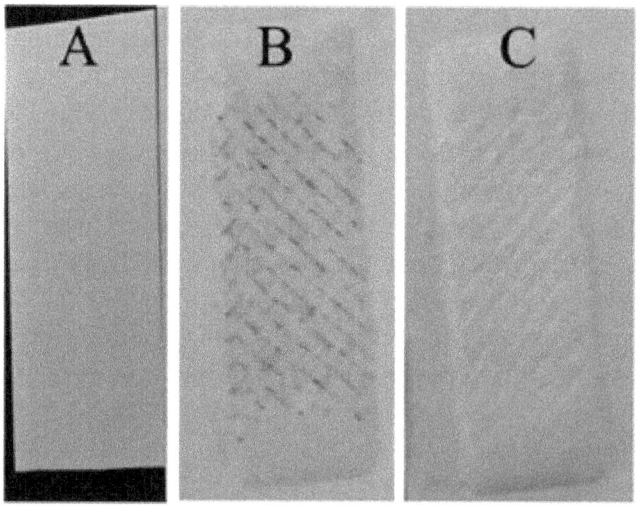

The combination of a nano-based membrane with an electrochemical system at low pressures generates reactive oxygen species for disinfection to prevent the formation of toxic byproducts, as a study reported that flow-through electrochemical systems have demonstrated exceptional results in elimination applications for heavy metal ions, bacteria, and pathogens from water. In general, chemical use and pretreatment processes can be decreased with electrochemical filtration technology, as it has significant benefits in terms of process optimization and operation cost reduction (Qi et al., 2023).

REFERENCES

Abdullah, Md., Obayedullah, M., Shariful Islam Shuvo, Md., Abul Khair, Md., Hossain, D. and Nahidul Islam, Md. (2025). A review on multifunctional applications of nanoparticles: Analyzing their multi-physical properties. Results in Surfaces and Interfaces, [online] 21, p.100635. doi:.DOI: 10.1016/j.rsurfi.2025.100635

Alam, M. M., Mustehasan., Akram, U., Ahmad, M., & Ahmed, N. Z. (2025). Nanotechnology in the Unani system of medicine: A comprehensive review and hypothesis. *Advances in Integrative Medicine*, *12*(4), 100564. DOI: 10.1016/j.aimed.2025.100564

Alam, M. M., Mustehasan., Akram, U., Ahmad, M., & Ahmed, N. Z. (2025). Nanotechnology in the Unani system of medicine: A comprehensive review and hypothesis. *Advances in Integrative Medicine*, *12*(4), 100564. DOI: 10.1016/j.aimed.2025.100564

Alazaiza, M. Y. D., Albahnasawi, A., Ali, G. A. M., Bashir, M. J. K., Copty, N. K., Amr, S. S. A., Abushammala, M. F. M., & Al Maskari, T. (2021). Recent Advances of Nanoremediation Technologies for Soil and Groundwater Remediation: A Review. *Water (Basel)*, *13*(16), 2186. DOI: 10.3390/w13162186

Alkhadra, M.A., Su, X., Suss, M.E., Tian, H., Guyes, E.N., Shocron, A.N., Conforti, K.M., de Souza, J.P., Kim, N., Tedesco, M., Khoiruddin, K., Wenten, I.G., Santiago, J.G., Hatton, T.A. and Bazant, M.Z. (2022). Electrochemical Methods for Water Purification, Ion Separations, and Energy Conversion. Chemical Reviews, [online] 122(16), pp.13547–13635. doi:.DOI: 10.1021/acs.chemrev.1c00396

Al-Marri, J. S., Abouedwan, A. B., Ahmad, M. I., & Bensalah, N. (2023). Electrocoagulation using aluminum electrodes as a sustainable and economic method for the removal of kinetic hydrate inhibitor (polyvinyl pyrrolidone) from produced wastewaters. *Frontiers in Water*, *5*, 1305347. Advance online publication. DOI: 10.3389/frwa.2023.1305347

Altammar, K. A. (2023). A review on nanoparticles: Characteristics, synthesis, applications, and challenges. [PubMed]. *Frontiers in Microbiology*, *14*(1155622), 1155622. Advance online publication. DOI: 10.3389/fmicb.2023.1155622

Ameer, S., Abdelrahman, E.A., Batool, F., Batool, A., Akhtar, T., Alhamzani, A.G. and Abou-Krisha, M.M. (2025). Bimetallic nanoparticles loaded on waste derived carbon quantum dots: A green approach for degradation of azo dye. International Biodeterioration & Biodegradation, [online] 205, p.106184. doi:.DOI: 10.1016/j.ibiod.2025.106184

Ariho, A., Aja, L., Muhammad, T., & Mohammad, L. (2024). Water pollution: causes, impacts and current efforts to address the issues of water pollution along river Meizimera-kihihi, Kanugu District,Uganda. F1000Research, [online] 13, pp.1298–1298. doi:DOI: 10.12688/f1000research.155162.1

Aroor, G., Khan, M.A., Shetty, A.R., Rai, R., Ganesha, H. and Navada, M.K. (2025). From chemistry to performance: How nano additives are transforming bio-lubricants for enhanced tribological applications. Journal of Molecular Liquids, [online] 425, p.127242. doi:.DOI: 10.1016/j.molliq.2025.127242

Asif, Z., Chen, Z., An, C. and Dong, J. (2022). Environmental Impacts and Challenges Associated with Oil Spills on Shorelines. Journal of Marine Science and Engineering, [online] 10(6), p.762. doi:.DOI: 10.3390/jmse10060762

Astruc, D. (2020). Introduction: Nanoparticles in Catalysis. [PubMed]. *Chemical Reviews, 120*(2), 461–463. DOI: 10.1021/acs.chemrev.8b00696

Aydin, D., Gübbük, İ. H., & Ersöz, M. (2024). Recent advances and applications of nanostructured membranes in water purification. *Turkish Journal of Chemistry, 48*(1), 1–20.

Azzaoui, K., Barboucha, B., Hammouti, B., & Touzani, R. (2022). Nanotechnology: History and Various Applications, a Mini Review. PRSM. [online] doi:DOI: 10.34874/prsm.ehei-jst-vol2iss1.36531

Bayda, S., Adeel, M., Tuccinardi, T., Cordani, M. and Rizzolio, F. (2019). The History of Nanoscience and Nanotechnology: From Chemical–Physical Applications to Nanomedicine. Molecules, [online] 25(1), p.112. doi:.DOI: 10.3390/molecules25010112

Burlec, A. F., Corciova, A., Boev, M., Batir-Marin, D., Mircea, C., Cioanca, O., Danila, G., Danila, M., Bucur, A. F., & Hancianu, M. (2023). Current Overview of Metal Nanoparticles' Synthesis, Characterization, and Biomedical Applications, with a Focus on Silver and Gold Nanoparticles. [PubMed]. *Pharmaceuticals (Basel, Switzerland), 16*(10), 1410–1410. DOI: 10.3390/ph16101410

Cai, X., Wu, L., Li, Y., Lei, S., Xu, J., Lyu, H., Li, J., Wang, H., Dong, X., Zhu, Y. and Wang, G. (2023). Remote sensing identification of urban water pollution source types using hyperspectral data. Journal of Hazardous Materials, [online] 459, p.132080. doi:.DOI: 10.1016/j.jhazmat.2023.132080

Cao, Y., Malmali, M., Qian, X., & Wickramasinghe, S. R. (2022). Continuous electrocoagulation-membrane distillation unit for treating hydraulic fracturing produced water. *Journal of Water Process Engineering*, *50*, 103219. DOI: 10.1016/j.jwpe.2022.103219

Chaturvedi, S. and Dave, P.N. (2019). Water Purification Using Nanotechnology an Emerging Opportunities. Chemical Methodologies, [online] 3(1), pp.115–144. doi:.DOI: 10.22034/chemm.2018.143461.1069

Chukeaw, T., Fang, K., Xiong, Q., Gu, Y., He, C., Wang, K. and Liu, G. (2025). Synergistic integration of electrocoagulation and membrane pre-concentration: A critical review on advanced strategies for organic recovery and fouling mitigation in municipal wastewater treatment. Separation and Purification Technology, [online] 382, p.135994. doi:.DOI: 10.1016/j.seppur.2025.135994

Das, A. K., & Chen, L. (2024). A Review on Electrochemical Advanced Oxidation Treatment of Dairy Wastewater. *Environments (Basel, Switzerland)*, *11*(6), 124–124. DOI: 10.3390/environments11060124

Del Prado-Audelo, M. L., García Kerdan, I., Escutia-Guadarrama, L., Reyna-González, J. M., Magaña, J. J., & Leyva-Gómez, G. (2021). Nanoremediation: Nanomaterials and Nanotechnologies for Environmental Cleanup. *Frontiers in Environmental Science*, *9*, 793765. Advance online publication. DOI: 10.3389/fenvs.2021.793765

Drexler, K. E. (1981). Molecular engineering: An approach to the development of general capabilities for molecular manipulation. Proceedings of the National Academy of Sciences, [online] 78(9), pp.5275–5278. doi:DOI: 10.1073/pnas.78.9.5275

Dube, A., Malode, S.J., Alshehri, M.A. and Shetti, N.P. (2025). Electrochemical water treatment: Review of different approaches. Journal of Environmental Management, [online] 373, p.123911. doi:.DOI: 10.1016/j.jenvman.2024.123911

Dubey, S. S., Singh, A. K., & Tripathi, U. N. (2022). Water Quality - A Review. International Journal of Chemical Sciences, [online] 20(3), pp.1–7. Available at: https://www.tsijournals.com/abstract/water-quality--a-review-15778.html [Accessed 11 Nov. 2025].

El Hamdouni, Y., El Hajjaji, S., Szabó, T., Trif, L., Felhősi, I., Abbi, K., Labjar, N., Harmouche, L., & Shaban, A.Youssra El Hamdouni. (2022). Biomass valorization of walnut shell into biochar as a resource for electrochemical simultaneous detection of heavy metal ions in water and soil samples: Preparation, characterization, and applications. *Arabian Journal of Chemistry*, *15*(11), 104252–104252. DOI: 10.1016/j.arabjc.2022.104252

El-Kalliny, A.S., Abdel-Wahed, M.S., El-Zahhar, A.A., Hamza, I.A. and Gad-Allah, T.A. (2023). Nanomaterials: a review of emerging contaminants with potential health or environmental impact. Discover Nano, [online] 18(1), p.68. doi:.DOI: 10.1186/s11671-023-03787-8

El-kordy, A., Elgamouz, A., Khalil, A.K.A., Tijani, N., Kawde, A.-N. and Laoui, T. (2025). Development and electrochemical evaluation of Bismuth/Clay-based membranes for heavy metal rejection from wastewater. Surfaces and Interfaces, [online] 72, p.106953. doi:.DOI: 10.1016/j.surfin.2025.106953

Fadillah, G., Hidayat, R., Saputra, A., Wasisto, H.S., Rani, R., Saeki, K., Toda, K. and Ohira, S.-I. (2025). Advanced electrodialysis techniques for analytical separation: A comprehensive review. Analytica Chimica Acta, [online] 1380, p.344637. doi:.DOI: 10.1016/j.aca.2025.344637

Feculak, M., Loureiro, S., White, J.C., Xing, B., Wu, K.C.-W. ., Sheteiwy, M.S., Gao, Y., Oleszczuk, P. and Jośko, I. (2025). Engineered nanoparticle transformations: Rethinking toxicity in water. Nano Today, [online] 65, p.102804. doi:.</unknown>DOI: 10.1016/j.nantod.2025.102804

Foorginezhad, S., Zerafat, M.M., Ismail, A.F. and Goh, P.S. (2025). Emerging membrane technologies for sustainable water treatment: a review on recent advances. Environmental Science: Advances, [online] 4(4), pp.530–570. doi:.DOI: 10.1039/D4VA00378K

Fouda-Mbanga, B.G., Nyoni, B., Mnyango, J.I., Phiri, C., Hlangothi, S.P., Miracle, A., Boluwatife, A., Celestina, O., Voukeng, S.N.K., Tsague, C.F., Nyamen, L.D., Mokolokolo, P.P., Mofokeng, J.P. and Ngobeni, P. (2025). The Nexuses of Micro(nano)plastics at Environmental Interfaces: Challenges in Wastewater Treatment and Sustainable Water Management. Journal of Environmental Chemical Engineering, [online] p.119952.

Iijima, S. (1991). Helical microtubules of graphitic carbon. *Nature*, *354*(6348), 56–58. DOI: 10.1038/354056a0

Iqbal, A., Cevik, E., Mustafa, A., Qahtan, T. F., Zeeshan, M., & Bozkurt, A. (2024). Emerging developments in polymeric nanocomposite membrane-based filtration for water purification: A concise overview of toxic metal removal. *Chemical Engineering Journal*, *481*, 148760.

Jiang, M., Althomali, R. H., Ansari, S. A., Abdu, E., & Gupta, J. (2023). Advances in preparation, biomedical, and pharmaceutical applications of chitosan-based gold, silver, and magnetic nanoparticles: A review. [PubMed]. *International Journal of Biological Macromolecules*, *251*, 126390–126390. DOI: 10.1016/j.ijbiomac.2023.126390

Joudeh, N. and Linke, D. (2022). Nanoparticle classification, physicochemical properties, characterization, and applications: a comprehensive review for biologists. Journal of Nanobiotechnology, [online] 20(1). doi:.DOI: 10.1186/s12951-022-01477-8

Klimonda, A., & Kowalska, I. (2021). Sequential process: Membrane filtration and ion exchange as an effective method for water solution purification containing cationic surfactants. *Desalination and Water Treatment*, *214*, 232–241. DOI: 10.5004/dwt.2021.26663

Kumar, R., Rose, P. K., Sharma, P. K., Lamba, J., Kumar, M., & Bhattacharya, P. (2025). Micro(nano)plastic and per- and polyfluoroalkyl substances in soil/sediment–water ecosystems: Sources, transport, interactions, and challenges. *Current Opinion in Chemical Engineering*, *48*, 101125. DOI: 10.1016/j.coche.2025.101125

Liang, C., Wei, D., Zhang, S., Ren, Q., Shi, J. and Liu, L. (2021). Removal of antibiotic resistance genes from swine wastewater by membrane filtration treatment. Ecotoxicology and Environmental Safety, [online] 210, p.111885. doi:.DOI: 10.1016/j.ecoenv.2020.111885

Magdy, G., Aboelkassim, E., Abd Elhaleem, S. M., & Belal, F. (2024). A comprehensive review on silver nanoparticles: Synthesis approaches, characterization techniques, and recent pharmaceutical, environmental, and antimicrobial applications. *Microchemical Journal*, *196*, 109615–109615. DOI: 10.1016/j.microc.2023.109615

Mao, Y., Zhao, Y. and Cotterill, S. (2023). Examining Current and Future Applications of Electrocoagulation in Wastewater Treatment. Water, [online] 15(8), p.1455. doi:.DOI: 10.3390/w15081455

Misman, N. A., & Sharif, M. F. Jalal, A. and Nur Hafizah Azizan (2023). Water pollution and the assessment of water quality parameters: a review. DESALINATION AND WATER TREATMENT, 294, pp.79–88. doi:DOI: 10.5004/dwt.2023.29433

Morsy, A., Hatem, N., Ewais, H.A., Mohamed, S., Morsy, A., Fathi, H., Tarek, A. and Nassef, E. (2025). High-performance self-cleaning cellulose acetate membranes reinforced with nanochitosan for heavy metal removal. Environmental Research, [online] 288, p.123197. doi:.DOI: 10.1016/j.envres.2025.123197

Nabila Chaabani, Ncib, S., Lasâad Dammak, Larchet, C., Wided Bouguerra and Elimame Elaloui (2025). High-Efficiency Removal of Copper and Nickel via Donnan Dialysis Using Fujifilm Cation-Exchange Membranes: Process Optimization Through Response Surface Methodology. Membranes, [online] 15(12), pp.346–346. doi:.DOI: 10.3390/membranes15120346

Norén, A., Modin, O., Rauch, S., Fedje, K.K., Björklund, S., Thoutam, A.K., Markocsan, N., Andersson-Sköld, Y. and Strömvall, A.-M. (2025). Enhanced photoelectrocatalysis for oxidation of organic pollutants and metal recovery from polluted water and sediments. Journal of Hazardous Materials, [online] 500, p.140375. doi:. DOI: 10.1016/j.jhazmat.2025.140375

Othman, N. H., Alias, N. H., Fuzil, N. S., Marpani, F., Shahruddin, M. Z., Chew, C. M., & Ismail, A. F. (2021). A review on the use of membrane technology systems in developing countries. *Membranes*, *12*(1), 30.

Patel, S. K., Shukla, S. C., Natarajan, B. R., Asaithambi, P., Dwivedi, H. K., Sharma, A., & Prajapati, A. K. (2025). State of the art review for industrial wastewater treatment by electrocoagulation process: Mechanism, cost and sludge analysis. *Desalination and Water Treatment*, *321*, 100915.

Puri, N., Gupta, A., & Mishra, A. (2021). Recent advances on nano-adsorbents and nanomembranes for the remediation of water. *Journal of Cleaner Production*, *322*, 129051. DOI: 10.1016/j.jclepro.2021.129051

Qi, Y., Li, D., Zhang, S., Li, F., & Hua, T. (2023). Electrochemical filtration for drinking water purification: A review on membrane materials, mechanisms and roles. [online]. Journal of Environmental Sciences (China). Advance online publication. doi: PubMedDOI: 10.1016/j.jes.2023.06.033

Reihanifar, M., Takallou, A., Taheri, M., Gholizadeh Lonbar, A., Ahmadi, M., & Sharifi, A. (2024). Nanotechnology advancements in groundwater remediation: A comprehensive analysis of current research and future prospects. *Groundwater for Sustainable Development*, *27*, 101330. DOI: 10.1016/j.gsd.2024.101330

Sadaf, S., Roy, H., Fariha, A., Rahman, T.U., Tasnim, N., Jahan, N., Sokan-Adeaga, A.A., Safwat, S.M. and Islam, M.S. (2024). Electrocoagulation-based wastewater treatment process and significance of anode materials for the overall improvement of the process: A critical review. Journal of Water Process Engineering, [online] 62, p.105409. doi:.DOI: 10.1016/j.jwpe.2024.105409

Saleem, H., Zaidi, S. J., Ismail, A. F., & Goh, P. S. (2022). Advances of nanomaterials for air pollution remediation and their impacts on the environment. [PubMed]. *Chemosphere*, *287*, 132083. DOI: 10.1016/j.chemosphere.2021.132083

Sun, H. (2019). Grand Challenges in Environmental Nanotechnology. *Frontiers in Nanotechnology*, *1*, 2. Advance online publication. DOI: 10.3389/fnano.2019.00002

Sun, H. (2019). Grand Challenges in Environmental Nanotechnology. *Frontiers in Nanotechnology*, *1*, 2. Advance online publication. DOI: 10.3389/fnano.2019.00002

Tripathy, D. B., & Gupta, A.Divya Bajpai Tripathy and Gupta. (2023). Nanomembranes-Affiliated Water Remediation: Chronology, Properties, Classification, Challenges and Future Prospects. [PubMed]. *Membranes*, *13*(8), 713–713. DOI: 10.3390/membranes13080713

Ullah, A., Shahzada, K., Khan, S. W., & Starov, V. (2020). Purification of produced water using oscillatory membrane filtration. *Desalination*, *491*, 114428. DOI: 10.1016/j.desal.2020.114428

Valjarević, A. (2025). Oil field impacts on Venezuela's rivers and water stress with environmental challenges. Journal of South American Earth Sciences, [online] 164, p.105675. doi:.DOI: 10.1016/j.jsames.2025.105675

Yusof, N. F., Raffi, A. A., Yahaya, N. Z. S., Abas, K. H., Othman, M. H. D., Jaafar, J., & Rahman, M. A. (2023). Surface Modification of UiO-66 on Hollow Fibre Membrane for Membrane Distillation. [PubMed]. *Membranes*, *13*(3), 253. DOI: 10.3390/membranes13030253

Zahid, M., Rashid, A., Akram, S., Rehan, Z.A. and Razzaq, W. (2018). A Comprehensive Review on Polymeric Nano-Composite Membranes for Water Treatment. Journal of Membrane Science & Technology, [online] 08(01). doi:.DOI: 10.4172/2155-9589.1000179

Zhang, J., Li, G., Yuan, X., Li, P.-P., Yu, Y., Yang, W., & Zhao, S. (2023). Reduction of Ultrafiltration Membrane Fouling by the Pretreatment Removal of Emerging Pollutants: A Review. [PubMed]. *Membranes*, *13*(1), 77–77. DOI: 10.3390/membranes13010077

Chapter 3
Nanostructure-Based Materials for Sorption and Photoelectrochemical Degradation of Water Pollutants:
Water Purification Techniques

Marwah Ahmed Alabdulrahem
Imam Abdulrahman bin Faisal University, Dammam, Saudi Arabia

Maha Aqeel Alaqeel
Imam Abdulrahman bin Faisal University, Dammam, Saudi Arabia

Tawfik Abdo Saleh
https://orcid.org/0000-0002-3037-5159
King Fahd University of Petroleum and Minerals, Saudi Arabia

ABSTRACT

Water pollution from persistent organic contaminants and toxic heavy metal ions continues to pose a critical threat to environmental and public health worldwide. Conventional treatment strategies often lack efficiency and sustainability when adsorption or photoelectrochemical (PEC) technologies are applied in isolation. Adsorption effectively concentrates pollutants but does not eliminate them, whereas PEC offers strong oxidative and reductive pathways for pollutant degradation, yet suffers from limited surface interaction with target species. This chapter discusses polymer-functionalized carbon-based nanocomposites as promising materials for

DOI: 10.4018/979-8-3373-5886-4.ch003

adsorption, and nanostructures of metal oxides for photodegradation of pollutants. The chapter will present some examples of multifunctional platforms capable of coupling PEC processes to achieve enhanced water purification. An example of the adsorption mechanisms will be discussed with highlights on the Langmuir and the Freundlich isotherm models.

1.0 INTRODUCTION

1.1 Water Contamination and types of Pollutants

Water pollution refers to the contamination of water bodies such as rivers, lakes, oceans, and groundwater by physical, chemical, or biological substances that degrade water quality and disrupt ecological and human health. It occurs when harmful materials enter aquatic systems beyond their natural assimilation capacity (Lin et al., 2022). Water pollution originates from a wide range of natural and human-induced sources. For example, Industrial effluents from textile, mining, metal plating, and pharmaceutical factories often release toxic chemicals, heavy metals, and synthetic dyes into nearby water bodies. Agricultural runoff is another major contributor, carrying pesticides, herbicides, and fertilizers rich in nitrogen and phosphorus that cause eutrophication and algal blooms. Domestic sewage and urban wastewater introduce detergents, pharmaceuticals, and microplastics, while mining and oil extraction activities add suspended solids and hydrocarbons. In addition, climate-related flooding and poor waste management in developing regions exacerbate pollutant transport into freshwater systems. Collectively, these sources lead to persistent contamination, threatening both drinking-water quality and ecosystem resilience (Lin et al., 2022; Ariho et al., 2024).

Pollutants found in contaminated water can be broadly grouped into inorganic, organic, and emerging categories. Inorganic pollutants include heavy metals such as lead (Pb), cadmium (Cd), mercury (Hg), and arsenic (As), which are non-biodegradable and bioaccumulate in aquatic organisms. Organic contaminants, including dyes, phenols, pesticides, and petroleum hydrocarbons, can cause oxygen depletion and toxicity to aquatic life. Nutrient pollutants, mainly nitrates and phosphates from fertilizers, trigger eutrophication and harmful algal blooms. In recent years, emerging pollutants such as pharmaceuticals, endocrine-disrupting compounds, microplastics, and antibiotic-resistant bacteria have drawn increasing concern because conventional treatment plants often fail to remove them effectively. These pollutants vary in their persistence, mobility, and toxicity, making water purification increasingly complex (Lin et al., 2022; Ariho et al., 2024; Alvizuri-Tintaya et al., 2022).

The consequences of water pollution are severe and multifaceted. For humans, contaminated water is linked to gastrointestinal infections, neurological damage, and cancer risks resulting from long-term exposure to toxic metals or organic pollutants. Heavy metals like lead and mercury can cross the blood–brain barrier, affecting cognitive development in children, while nitrates in drinking water cause methemoglobinemia. Environmentally, polluted water disrupts aquatic ecosystems, reduces biodiversity, and impairs oxygen transfer, leading to fish kills and habitat loss. Persistent pollutants such as microplastics can also accumulate through the food chain, ultimately reaching humans. These combined effects threaten sustainable development, food security, and overall ecosystem health (Babuji et al., 2023; Li et al., 2021). In this regard, urgent action should be adopted to prevent pollution and remediation to safeguard both health and nature using effective quantitative and qualitative analysis techniques that can help to control this problem.

1.2 Nanoparticles

1.2.1 Definition, Properties, and History of Nanoparticles:

In general, a nanoparticle (NP) is a particle having at least one external dimension in the nanoscale, typically from 1to 100 nm. In many chemical applications, the key point is that when materials are reduced to the nanoscale, they offer a higher surface area-to-volume ratio and exhibit significantly different physical and chemical properties compared to bulk (Jeevanandam et al., 2018; Čitaković, 2019). NPs catalytic activity, for instance, strongly depends on size and shape, where smaller NPs provide a large surface area means more atoms/molecules are at or near surfaces (greater number of active sites), and shape controls the exposure of active crystal facets, edges, and charge-transport pathways, thus enhancing catalytic efficiency and lower activation energy (Čitaković, 2019; Cheong et al., 2010). The electrical properties of NPs, including conductivity and resistivity, differ significantly from bulk materials and are influenced by particle diameter, shape, and lattice strain, as observed in carbon nanotubes (CNTs). Magnetic properties are governed by finite-size and surface effects, which modify the magnetic behavior relative to bulk counterparts. Some metal and metal oxide nanoparticles exhibit a magnetic moment under an external magnetic field similar to paramagnetic materials, but it is significantly stronger, up to 10^4 times larger, involving multiple atoms. The optical properties of NPs are affected by quantum confinement, where decreasing particle size increases the band gap energy, resulting in a blue shift in absorption and emission spectra, whereas larger particles exhibit a red shift; for example, gold nanoparticles display a red color at approximately 20 nm and progressively shift to gray or black as their size increases. Also, the quantum confinement effect leads to narrow fluorescence emission and broad absorption bands, making NPs excellent optical transducers. Their emission wavelength can be tuned by altering size, shape, or composition, allowing multiplex detection of different analytes. Because of their broad absorption spectrum, various QDs can be excited by a single energy source, enabling simultaneous multi-analyte sensing (Čitaković, 2019; Barhoum et al., 2022; Willner & Vikesland, 2018). Moreover, the ability of surface functionalization or modification (changing nanoparticle surfaces, attaching them to supports, or adding them to chemical systems) makes them useful for many research and industrial uses. For instance, adding Ag nanoparticles to polycyclic Aromatic Hydrocarbons (PAHs) solutions decreases the detection limits (Guo et al., 2023).

Humans have been using nanomaterials for thousands of years. Over 4,500 years ago, natural asbestos nanofibers were added to ceramic matrices to improve their strength. Similarly, the Ancient Egyptians developed an early synthetic process more than 4,000 years ago to produce lead sulfide (PbS) nanoparticles, about 5 nm in size, which they used as a hair dye (Walter et al., 2006; Heiligtag & Niederberger, 2013). The modern era of nanoscience began in 1959, when Richard Feynman delivered his visionary lecture *"There's Plenty of Room at the Bottom,"* proposing the manipulation of matter at the atomic and molecular scale (Jeevanandam et al., 2018; Joudeh & Linke, 2022). After that, practical chemical and industrial applications rise significantly with the development of advanced synthetic, characterization, and instrumentation methods in the late 20th century. The 1970s, for example, saw the discovery of surface-enhanced Raman scattering (SERS) on silver nanoparticles, one of the first analytical applications of nanomaterials. Adding these NPs increases the Raman signal, making even trace amounts of organic compounds (e.g., pyridine, polyaromatic hydrocarbon) detectable in water (Guo et al., 2023; Jeanmarie, 1977). The 1980s were a turning point in nanoscience, the decade when nanoparticles began to be used intentionally in chemistry, physics, and materials science. It shows the development of NPs synthesis and the usage as of metal NPs as a catalyst. For example, gold nanoparticles (Au NPs), previously thought to be chemically inert, are highly active catalysts of CO at low temperature when their size is below ~10 nm and supported on metal oxides (Ji et al., 2018). The next decade presents the development of semiconductor quantum dots (very small nanoparticles) and the recognition that their electronic and optical properties depend strongly on the size of the band gap, which relates directly to the size reduction of the semiconductor (Han et al., 2020). Also, the discovery of carbon nanotube (CNTs), cylindrical nanostructures made of rolled-up sheets of graphene, with exceptional strength, electrical conductivity, and thermal stability (Iijima, 1991). Since the early 2000s, nanoparticles have become essential tools in many areas of chemistry due to their unique surface and size-dependent properties. Nanoparticles have enhanced the analytical detection and sensitivity of electrochemical and optical sensors through surface functionalization (Sorkin & Zhang, 2011), increasing organic transformations and environmental reaction rates by providing high surface area and tunable active sites (Cheong et al., 2010). More recently, nanoparticles play an increasingly important role in energy production, the oil industry, and environmental pollution control. In the energy sector, to improve fuel cells, hydrogen generation, and solar energy conversion (Naoki Toshima, 2013). In the oil and gas industry, nanoparticles enhance oil recovery, drilling fluid stability, and corrosion resistance (Alsaba et al., 2020). For environmental applications, metal and carbon-based nanomaterials, because of their large surface area, abundant active site, and functionalization flexibility, enhance the intermolecular interaction with organic pollutants, dyes, and heavy metals, enabling efficient removal and degradation of pollutants from environmental systems (Asghar et al., 2024; Rezania et al., 2024). Derived from environmental science, nanoparticles are shown to improve modern agriculture in several key ways. Nano-fertilizers provide controlled and slow nutrient release to improve plant growth while minimizing waste, and nano-pesticides offer targeted control with reduced chemical use and environmental impact. Nano-sensors enable real-time monitoring of soil health, moisture, and nutrient levels, supporting precision farming practices, while nanomaterials enhance seed germination, stress resistance, and crop yield (Sri Renukadevi Balusamy et al., 2023). Furthermore, Nanoparticles enhance drug delivery by improving how drugs reach and act inside the body. Their small size allows them to cross biological barriers and enter target cells more easily. Their large surface area enables the attachment of drugs. Nanoparticles can also protect drugs from degradation, increase the solubility of poorly soluble medicines, and reduce side effects, improving treatment efficiency. The same concept can be applied in the food industry, where NPs improve nutrient delivery, preserve freshness from degradation, and in food packaging to provide antimicrobial protection and detect spoilage. In the cosmetic industry, nanoparticles such as liposomes, zinc oxide (ZnO), and titanium dioxide (TiO_2) NPs improve skin absorption, UV protection, and stability of active compounds. They help deliver moisturizing or anti-aging agents deeper into the skin, make sunscreens more transparent and effective, and increase the overall performance and safety of cosmetic products. Additionally, silver NPs are used in some cosmetic products as an antibacterial component (Yusuf et al., 2023).

1.2.2 Classification of Nanoparticles

Nanoparticles can be classified according to their origin, shape, dimension, phase, dispersion, and composition, which determine their physical, chemical behavior and potential applications. The sources of nanoparticles (NPs) can be grouped into three main categories: natural, incidental, and engineered. Natural sources include nanoparticles formed through geological and biological processes such as volcanic eruptions, forest fires, ocean spray, mineral weathering, and biological synthesis by microorganisms. These processes release nanoscale particles like clays, metal oxides, soot, and natural fibers into the environment. Incidental sources are unintended by-products of human activities, mainly from industrial processes, combustion, and vehicle emissions. Examples include fly ash, welding fumes, and soot particles, which often enter the atmosphere or water systems during manufacturing or energy production. Engineered (synthetic) sources involve intentional fabrication of nanoparticles for specific applications in medicine, energy, agriculture, and environmental remediation. These are produced using chemical, physical, or biological synthesis methods. The synthesis method plays a crucial role in determining the specific features and properties of nanomaterials, which are generally produced through top-down, bottom-up, or hybrid approaches. In the bottom-up approach, nanoparticles are formed by assembling atoms, ions, or molecules into larger structures through self-assembly and chemical interactions such as hydrogen bonding, ionic forces, and van der Waals interactions, which enable the synthesis of nanostructures with high uniformity, controlled composition, and minimal defects. However, these methods often require chemically compatible precursors and surfaces and may lack precise control over atomic manipulation (Jeevanandam et al., 2018; Čitaković, 2019; Barhoum et al., 2022).

The top-down approach, in contrast, involves breaking down bulk materials into nanosized particles using physical or mechanical methods. It is especially useful for fabricating thin films and nanostructures larger than 100 nm, commonly applied in electronics and microfabrication due to its ability to produce integrated and interconnected systems. Its major limitation lies in the resolution limit of current technology, which restricts size control at the atomic level. Finally, hybrid approaches combine both methods, employing top-down fragmentation followed by bottom-up assembly to generate complex nanostructures. For instance, volcanic activity naturally produces nanoscale fragments through top-down processes, which can later act as precursors for bottom-up reactions. In engineered systems, lithography is a typical hybrid example where etching serves as the top-down step and ion layering or deposition acts as the bottom-up step. Although hybrid methods provide versatility and precision, they often require expensive equipment and toxic reagents, which limit large-scale application (Jeevanandam et al., 2018; Čitaković, 2019; Barhoum et al., 2022).

In general, NPs exhibit a variety of geometries, including spherical, nano-rod, cubic, nanotube, nano-sheet or plate, and irregular forms, which provide essential insight into how morphology governs their performance across chemical, environmental, and biological applications. Spherical nanoparticles (e.g., Au, Ag, TiO_2, fullerene, and QDs) are the most common and provide uniform surface area, making them ideal for catalysis, drug delivery, imaging, and exhibiting strong quantum confinement effects. Nano-rod (nanowire) found in materials like Au, ZnO, and Fe_2O_3, with high aspect ratios, which is useful in optical devices, photothermal therapy, and energy storage. Nanotubes (e.g., carbon nanotubes) offer high-surface-area structures made of rolled graphene sheets, making them effective in adsorption, energy storage, and nanocomposites. Nano-sheet or plate nanoparticles (e.g., graphene) provide extensive surface exposure, enhancing their utility in energy storage, catalysis, and environmental remediation. Cubic nanoparticles are typical of metal oxides (CeO_2, Fe_3O_4) and perovskite materials (ABX_3 structure, where *A* and *B* are cations, and *X* is an anion (usually oxygen or a halide)) (Irshad et al., 2022). This shape affects surface reactivity and catalytic performance because of its well-defined geometry that can favor specific reaction pathways, improving selectivity in catalytic and adsorption reactions compared to irregular shapes. Irregular nanoparticles are often produced naturally or incidentally, such as soot or mineral dust, resulting from uncontrolled formation, exhibit heterogeneous morphology and variable reactivity (Jeevanandam et al., 2018; Čitaković, 2019; Barhoum et al., 2022).

Based on the number of dimensions that fall within the nanoscale range (1–100 nm), nanoparticles are generally divided into zero-, one-, two-, and three-dimensional (0D–3D) categories. Zero-dimensional (0D) nanoparticles, such as nanospheres, possess all dimensions within the 1–100 nm range, and offer isotropic properties leading to equal surface energy and reactivity all around the surface. One-dimensional (1D) nanostructures, including nanorods, nanowires, and nanotubes, that compose only one dimension (length), exceed the nanoscale, while the other two remain in the nano range. In two-dimensional (2D) nanomaterials, two dimensions are in the nanoscale; the third (thickness) is very thin, such as nano sheets. Also, 2D NPs exhibit direction-dependent (anisotropic) properties, behavior changes along different axes, that enhance their performance in catalysis, sensing, and energy applications. Three-dimensional (3D) nanostructures, such as nanocomposites, cubic, and nanoporous frameworks, consist of interconnected nanoscale building blocks with high surface area, and multiple active sites offering enhanced mechanical strength, catalytic activity, and porosity (Jeevanandam et al., 2018; Barhoum et al., 2022; Joudeh & Linke, 2022).

Nanoparticles can also be classified by their dispersion and the number of phases they contain. In this regard, single-phase nanoparticles are composed of one continuous material phase, such as metallic (Au, Ag) or metal oxide (TiO_2, ZnO) nanoparticles. In contrast, multi-phase systems contain two or more distinct phases, which may be solid–solid, solid–liquid, or solid–gas structures, including examples like core–shell nanoparticles, colloidal suspensions, or aerosols, respectively. Nanomaterials are classified into dispersed and aggregated types depending on their dispersibility and the solvent used. Their dispersibility is commonly assessed through dynamic light scattering (DLS), which provides the polydispersity index (PDI), a measure of particle size distribution. Based on PDI values, nanomaterials can be categorized as well-dispersed (0.1–0.3), moderately dispersed (0.3–0.5), or aggregated (0.5–1). They can also be further divided into isometric (uniform size) and inhomogeneous (varying sizes) particles. The zeta potential is another key parameter for evaluating dispersion stability; higher absolute zeta potential values indicate stronger electrostatic repulsion between nanoparticles, resulting in more stable dispersions, while lower values lead to coagulation or aggregation (Barhoum et al., 2022)

Figure 1. Classification of nanoparticles according to composition

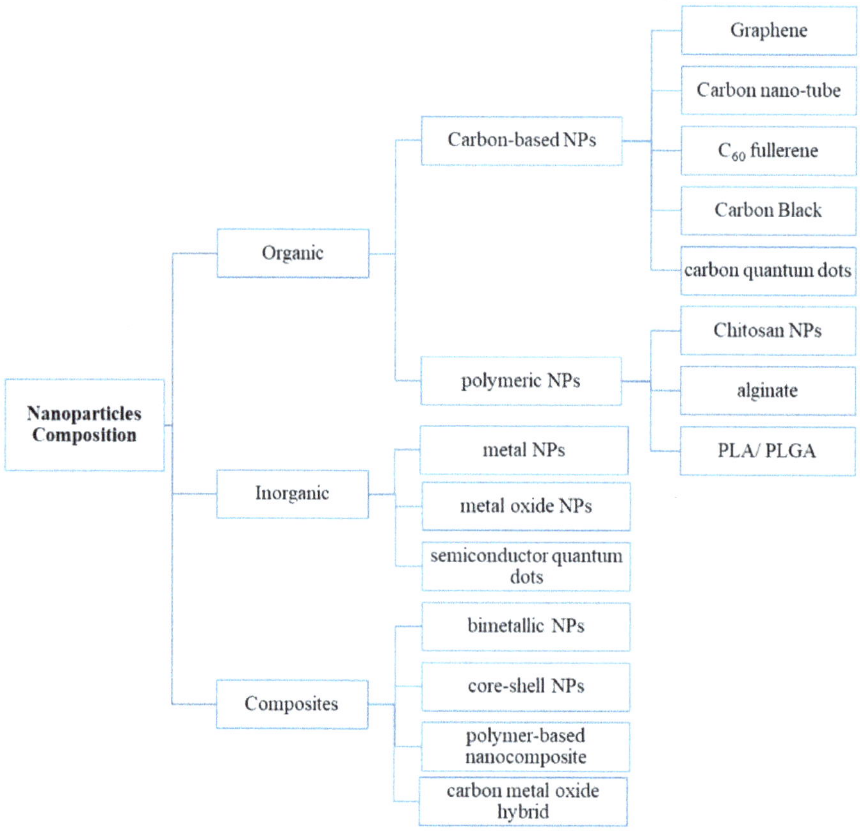

Nanoparticles can be broadly classified by their composition into organic, inorganic, and composite categories, each with distinct structural and functional characteristics (**Figure 1**). Organic nanoparticles are mainly composed of carbon atoms and are divided into *carbon-based* and *polymeric* systems. The first category, Carbon-based nanoparticles (NPs), is highly valued in nanotechnology due to carbon's ability to form strong and stable chains. They exhibit exceptional chemical stability, electrical and thermal conductivity, optical absorption, and luminescence. Common types include graphene, CNTs, fullerenes, carbon black, and quantum dots. Graphene is a single atomic layer of carbon arranged in a two-dimensional lattice, providing exceptional electrical conductivity, mechanical strength, and surface area, which makes it ideal for sensors, energy storage devices, and biomedical applications. Carbon nanotubes (CNTs) are cylindrical structures formed from rolled graphene sheets that possess outstanding tensile strength and adsorption capacity, making them effective for catalysis, pollutant removal, and drug delivery. Fullerenes, composed of carbon atoms arranged in hollow spheres (like C_{60}), demonstrate unique electronic and antioxidant behavior and are employed in photodynamic therapy, organic solar cells, and biomedical imaging. Carbon quantum dots (CQDs) are nanosized, luminescent carbon particles known for their biocompatibility, water solubility, and fluorescence, enabling their use in bioimaging, biosensing, and photocatalysis. Finally, carbon black, consisting of amorphous carbon nanoparticles, is valued for its high conductivity and surface area, finding widespread application in rubber reinforcement, conductive coatings, and energy materials. The second one is Polymeric nanoparticles (PNPs) are made from repeated monomer units, giving them great structural and functional diversity. Their composition and flexible synthesis make them highly suitable for various applications because many PNPs are derived from natural sources that are biocompatible, biodegradable, and exhibit low toxicity, making them ideal for drug and gene delivery systems. Common polymers used include natural polymers like chitosan and alginate, and synthetic polymers such as poly lactic acid (PLA), poly glycolic acid (PGA), and their copolymer poly(lactide co-glycolide) (PLGA), each chosen based on its specific physicochemical and biomedical properties (Jeevanandam et al., 2018; Čitaković, 2019; Barhoum et al., 2022; Joudeh & Linke, 2022; Kumari & Sarkar, 2021).

Inorganic nanoparticles are primarily composed of metallic, metal oxide, or semiconductor quantum dots (QDs) elements, known for their optical, magnetic, and catalytic properties. They are generally more stable under harsh conditions than organic NPs. Metal nanoparticles (e.g., gold (Au), silver (Ag), platinum (Pt)) exhibit high catalytic activity, magnetic properties, and strong surface plasmon resonance, which occur when conduction electrons on the surface of metal nanoparticles (typically gold or silver) oscillate collectively in response to incident light at a specific wavelength. This resonance leads to a strong absorption and scattering of light, giving metal nanoparticles their intense color and size-dependent optical properties. This makes metallic NPs useful in chemical sensing, catalysis, and antimicrobial coatings. Metal oxide nanoparticles (e.g., TiO_2, ZnO, Fe_3O_4) possess excellent photocatalytic, semiconducting, magnetic, and ceramic properties. Ceramic nanoparticles, mainly composed of alumina or silica, show excellent chemical resistance, thermal and electrical insulation, and high hardness, though they remain mechanically brittle.

Even at the nanoscale, they retain their hardness, heat and corrosion resistance, and insulating properties. They are widely applied in environmental remediation, solar energy conversion, and magnetic drug targeting. Semiconductor quantum dots are traditionally made from semiconductor materials such as CdSe, CdTe, PbS, or ZnS. These are inorganic nanocrystals used extensively in LEDs, solar cells, quantum computing, and bio-imaging because they give high brightness due to quantum confinement, size-dependent fluorescence, but are limited in biomedical use due to toxicity concerns. (e.g., cadmium) (Jeevanandam et al., 2018; Čitaković, 2019; Barhoum et al., 2022; Joudeh & Linke, 2022; Kumari & Sarkar, 2021).

On the other hand, composite nanoparticles bridge the advantages of multiple materials, providing tunable mechanical strength, catalytic efficiency, and surface functionality. Core–shell nanoparticles consist of an inner core material (often metal or metal oxide) surrounded by an outer shell that enhances stability, biocompatibility, and controlled reactivity. These are widely applied in drug delivery, catalysis, and imaging, as the shell can regulate surface interaction or protect sensitive cores. Bimetallic nanoparticles, formed by combining two different metals such as Au–Ag or Pt–Pd, exhibit enhanced catalytic activity, selectivity, and optical performance compared with their single-metal counterparts, making them valuable in electrocatalysis, sensing, and environmental remediation. Polymer-based nanocomposites, in which inorganic nanoparticles are embedded within polymer matrices, combine mechanical flexibility and chemical functionality, extending their use to biosensors, coatings, and controlled-release systems. Similarly, carbon–metal oxide hybrids integrate the conductivity and stability of carbon materials with the redox activity of metal oxides, offering superior performance in energy storage and photocatalytic degradation of pollutants (Jeevanandam et al., 2018; Čitaković, 2019; Barhoum et al., 2022; Joudeh & Linke, 2022; Kumari & Sarkar, 2021).

1.3 Toxicity and Mitigation

Nanoparticles (NPs) have become vital in modern technology due to their unique size-dependent properties, but their growing use has raised concerns about environmental and human health impacts. Their small size, high surface area, and chemical reactivity enable them to move easily through air, water, and soil, influencing ecosystems and living organisms at multiple levels. In the atmosphere, nanoparticles such as carbon black, TiO_2, and metal oxides can contribute to aerosol formation and airborne particulate matter. These particles can remain suspended for long periods, affecting air quality, cloud formation, and sunlight reflection, which in turn influences climate patterns. When inhaled, they penetrate deep into the lungs, enter the bloodstream, and may cause respiratory, cardiovascular, and inflammatory diseases. In soil, nanoparticles interact with minerals, organic matter, and microorganisms. While some (e.g., Fe_3O_4 or TiO_2) can help in soil remediation and pollutant removal, others, such as Ag or ZnO NPs, can disrupt microbial communities, hinder nutrient cycling, and reduce soil fertility. Prolonged accumulation can lead to soil acidification and toxicity to plants and beneficial organisms. In aquatic environments, nanoparticles from industrial discharge or runoff can accumulate in sediments or be taken up by aquatic organisms. Metal and metal oxide NPs often generate reactive oxygen species (ROS), causing oxidative stress, DNA damage, and reproductive toxicity in algae, fish, and crustaceans. They can also bioaccumulate through the food chain, posing long-term risks to ecosystems and humans. For human health, exposure occurs mainly through inhalation, ingestion, or dermal contact. Nanoparticles can cross biological membranes and accumulate in vital organs, leading to oxidative stress, inflammation, and genotoxic effects. Certain nanoparticles containing heavy metals (e.g., Cd, Pb, or Ni) are associated with neurological, hepatic, and renal toxicity (Čitaković, 2019; Barhoum et al., 2022; Joudeh & Linke, 2022).

To mitigate these effects, researchers emphasize the importance of green and safer-by-design synthesis methods, using biological or plant-based precursors that eliminate toxic reagents. Surface functionalization and coating with biocompatible materials (e.g., polymers, lipids, or silica) can reduce direct interaction with biological tissues, lowering cytotoxicity and enhancing stability. Additionally, controlled dosage, particle size optimization, and aggregation prevention are critical to reducing harmful effects. The implementation of toxicological assessment frameworks and eco-safety regulations is also necessary to monitor nanoparticle exposure and ensure environmental and human health protection (Čitaković, 2019; Barhoum et al., 2022; Joudeh & Linke, 2022).

2.0. ADSORPTION PROCESS

2.1 Adsorption for Pollutant Removal

Adsorption is a surface process in which molecules, ions, or gases accumulate at the interface of a solid adsorbent, forming a surface-bound layer. It is widely regarded as one of the simplest water-treatment and separation techniques because it operates through a straightforward surface interaction between contaminants and a solid adsorbent, without requiring complex chemical reactions or elaborate equipment. The process typically involves only mixing the adsorbent with the contaminated solution and allowing sufficient contact time for pollutants to attach to the surface, making it easy to implement at both laboratory and industrial scales. Most adsorption processes run under ambient temperature and pressure, further simplifying operation and reducing energy requirements. Moreover, adsorption shows the ability to remove a broad range of contaminants from aqueous media. Adsorption performance is greatly enhanced using nanomaterials as adsorbents, whose defining feature is their exceptionally high surface-area-to-volume ratio, which dramatically increases the number of accessible active sites for pollutant binding. This structural advantage enables faster adsorption kinetics and higher removal capacities for contaminants compared with conventional adsorbents. Consequently, incorporating nanomaterials into adsorption processes significantly elevates their efficiency and broadens their applicability to trace-level and diverse pollutants (Chandran et al., 2023; Mogashane et al., 2024; Yagub et al., 2014).

2.1.1 Adsorption Mechanism

Adsorption interactions are broadly classified into physisorption and chemisorption, each governed by distinct bonding forces and adsorption behaviors. Physisorption is driven by weak intermolecular forces without chemical bond formation, making it generally reversible and fast. It is favored at low to moderate temperatures and depends primarily on surface area, porosity, and molecular polarity. Several important mechanisms fall under physisorption: electrostatic attraction, which governs the binding of oppositely charged pollutants and adsorbent surfaces and is strongly influenced by pH and surface charge (Yagub et al., 2014); π–π stacking, a non-covalent interaction between π-electron systems of aromatic rings on the adsorbent (e.g., graphene, CNTs) and planar aromatic pollutants; hydrogen bonding, which facilitates the adsorption of polar organic molecules on functionalized carbons; and hydrophobic interactions, which drive the removal of nonpolar contaminants such as hydrocarbons and aromatic organics due to the affinity of hydrophobic surfaces for nonpolar species (Mogashane et al., 2024). Additionally, pore-filling involves adsorption driven by the diffusion of molecules into micro-, meso-, or nanopores, where geometric confinement enhances van der Waals interactions, enabling efficient removal of gases such as CH_4 and CO_2 (Wood et al., 2012).

In contrast, chemisorption involves the formation of strong, often irreversible, chemical bonds—covalent, coordination, or ionic—between adsorbate molecules and active sites on the adsorbent surface and is typically favored at higher temperatures. Chemisorption can occur through surface complexation, ion exchange, and surface precipitation. In surface complexation, metal ions share or accept electrons from functional groups on the adsorbent (e.g., –OH, –COOH, $-NH_2$), forming stable metal–ligand complexes (Daneshyar et al., 2017; Kulkarni et al., 2014). Ion exchange is driven by maintaining charge balance and affinity differences, leading to the substitution of surface-bound ions by ions in solution without altering the solid's overall framework (Kulkarni et al., 2014). Finally, surface precipitation occurs when dissolved species, such as CO_2, undergo chemical reactions at the interface to form new solid phases (e.g., hydroxides, carbonates, oxides) that deposit on the surface. This mechanism becomes dominant at higher concentrations or specific pH conditions and results in strong, often irreversible pollutant immobilization (Kulkarni et al., 2014; Aruni Gankanda et al., 2016).

2.1.2 Adsorption Application

2.1.2.1 Adsorption of PAH

Carbon-based nano materials such as tubes (CNTs) and graphene are pure carbon allotropes in which three of the four valence electrons participate in strong σ bonds within a two-dimensional sp^2-hybridized network, while the remaining electron is delocalized to form π bonds above and below the carbon plane (Apul & Karanfil, 2015). This delocalized π-electron system enables CNTs and graphene to interact with organic molecules, such as polyaromatic hydrocarbon (PAH), primarily through non-covalent mechanisms, such as π–π stacking, van der Waals interactions, and electrostatic forces. These physical interactions allow effective adsorption of organic contaminants without significantly altering the intrinsic chemical or physical properties of the carbon nanomaterials. Therefore, they show high affinity towards PAH. In the reviewed studies, a broader mixture of PAHs, including naphthalene, acenaphthene, was adsorbed onto granular activated carbon (GAC), where pseudo-first-order kinetics suggested physisorption governed mainly by π–π stacking, hydrophobic forces, and pore filling in the porous carbon matrix. Magnetic carbon nanomaterials also showed high affinity for phenanthrene, combining π–π stacking with the practical benefit of magnetic recovery during treatment. Surface functionalization of carbon nanomaterials introduces hydrophilicity and additional binding sites. This structural modification enables multiple adsorption mechanisms and broadens affinity toward both hydrophobic and hydrophilic organic contaminants without suppressing π–π interactions. Notably, the review highlights that lower–lower-molecular-weight PAHs (e.g., naphthalene and acenaphthene) exhibit weaker π–π stacking interactions due to their smaller, less planar aromatic systems, leading to comparatively lower adsorption affinities than higher-ring PAHs. Consistently, phenanthrene (PHEN) and pyrene (PYR), larger, more hydrophobic PAHs were efficiently captured by three-dimensional graphene wool (GW), achieving up to 99% removal, with adsorption driven principally by strong π–π interactions between the planar PAHs and graphene sheets; PSO kinetics indicated partial chemisorption while Sips isotherms reflected mixed monolayer–multilayer behavior. These interaction mechanisms strengthen with increasing PAH molecular weight and planarity. Functionalized GO similarly captured fluoranthene and anthracene methanol, **Figure 2**, where oxygenated groups (–OH, –COOH, epoxy) introduced additional hydrogen bonding and dipolar interactions alongside the dominant π–π mechanism. Collectively, these findings demonstrate that graphene-based nanomaterials remove PAHs primarily through π–π stacking and hydrophobic interactions while functional groups enhance selectivity and capacity. However, challenges such as nanomaterial aggregation, competition from natural organic matter, and reduced performance upon regeneration remain important considerations for practical wastewater treatment applications (Mogashane et al., 2024; Bellier et al., 2022).

Figure 2. (a) graphene, (b) graphene oxide, and (d) reduced graphene oxide structure

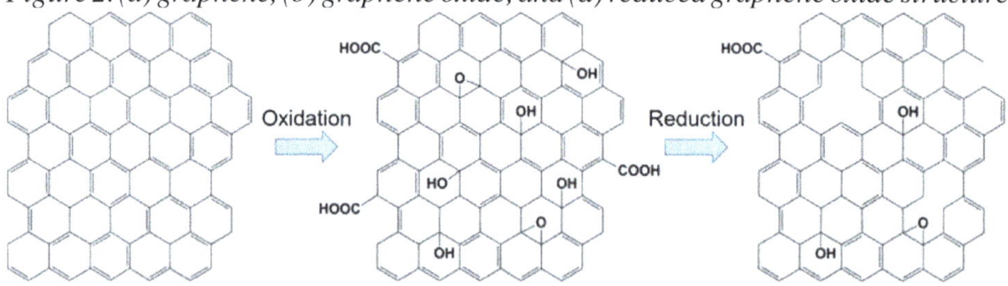

(Bellier et al., 2022)

2.1.2.2 Adsorption of Dyes

The structural characteristics of adsorbents, such as the distribution of functional groups, play a critical role in guiding the rational design and functionalization of nanomaterials, enabling strong affinity toward target pollutants. Acidic dyes, for example, are negatively charged in aqueous solution due to functional groups such as sulfonate ($-SO_3^-$), carboxyl ($-COO^-$), and nitro groups, and they commonly belong to azo, anthraquinone, nitro, nitroso, and triphenylmethane families. Their anionic nature gives them a strong affinity toward protonated or positively charged adsorbent surfaces, enabling adsorption through electrostatic attractions, often supported by hydrogen bonding (Mogashane et al., 2024). A representative example is methyl orange (MO), an anionic azo dye containing sulfonate groups, which was effectively removed using a chitosan–silica nanocomposite (CHS). The nanocomposite provides protonated ($-NH_3^+$) groups at acidic pH, enabling strong electrostatic binding with MO, while hydrogen bonding and surface porosity further enhance uptake; equilibrium was achieved rapidly (20 min) with effective removal in the pH 2.5–5 range (Mohammed et al., 2020). By contrast, cationic dyes (basic dyes) are positively charged molecules containing protonated amine chromophores (e.g., $-NR_3^+$). Their positive charge drives strong interactions with negatively charged adsorbent through electrostatic attraction, π–π stacking (when dyes contain aromatic rings), and hydrogen bonding (Mogashane et al., 2024). For example, crystal violet (CV), a cationic triarylmethane dye with extended aromatic rings, exhibited exceptionally high adsorption on a montmorillonite–graphene oxide nanocomposite (MtGO). The negative surface charge of MtGO (−25.9 mV), its layered porosity, and graphene π-electron system enabled synergistic electrostatic and π–π interactions, resulting in a very high monolayer capacity of 746 mg g^{-1} and rapid pseudo-second-order kinetics (Puri & Gajjala Sumana, 2018). Similarly, MB, a planar cationic thiazine dye, was removed efficiently (93–98%) by eco-friendly rGO. The adsorption mechanism, **Figure 3**, was identified as a mixed physisorption–chemisorption process, dominated by π–π stacking between the aromatic MB molecules and the restored sp^2 domains of rGO, together with electrostatic attraction and hydrogen bonding involving residual oxygen-containing functional groups (Arias Arias et al., 2020). Together, these examples demonstrate that acid adsorbates favor positively charged, protonated nano adsorbents, while cationic compounds show strong affinity for negatively charged carbon-based and GO nanostructures, especially when aromatic interactions (π–π stacking) are available. This distinction in molecular charge and structure governs the adsorption mechanism and determines the optimal nanomaterial design for effective adsorption.

Figure 3. Adsorption mechanism of MB dye on rGO

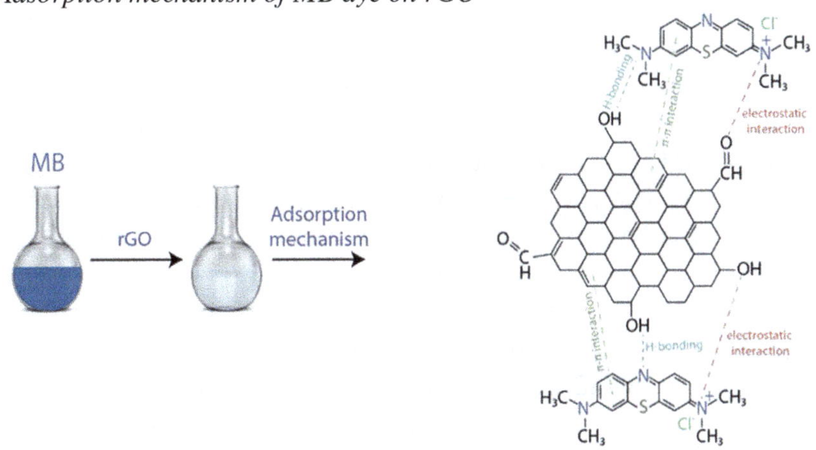

(Arias Arias et al., 2020)

2.1.2.3 Adsorption of Heavy Metals

Heavy metals exist mainly as positively charged ions in water and interact with surfaces through charge-driven and coordination-based mechanisms rather than hydrophobic effects. Their adsorption behavior is controlled by ionic radius, hydration energy, oxidation state, and ability to form stable complexes with electron-donating groups. Therefore, the most suitable adsorbents are high-surface-area nanomaterials enriched with functional groups such as –COOH, –OH, $–NH_2$, or –SH, which enable strong electrostatic attraction, ion exchange, and surface complexation. Functionalized carbon nanomaterials and metal oxide nanoparticles are the most widely investigated systems for heavy metal adsorption because they provide high affinity and selectivity toward heavy metal ions. Graphene, a two-dimensional sp^2-hybridized carbon sheet, offers an exceptionally high theoretical surface area and excellent electronic properties. Nevertheless, its tendency to restack and its limited surface functionality restrict direct metal binding, making adsorption mainly governed by weak physical interactions. In contrast, GO contains abundant oxygenated functional groups distributed on its basal planes and edges, which serve as effective coordination sites for metal ions. Consequently, GO demonstrates superior adsorption capacities for heavy metals such as Pb^{2+}, Cd^{2+}, and Cu^{2+}, primarily via surface complexation and electrostatic attraction. The complexation occurs through M–O coordination with carboxyl and hydroxyl groups, with adsorption fitting pseudo-second-order kinetics and Langmuir isotherms, confirming chemisorption as the dominant mechanism (Chandran et al., 2023). Carbon nanotubes (CNTs) possess higher surface area structure with accessible inner and outer surfaces. Pristine CNTs exhibit limited affinity toward metal ions because of their hydrophobicity and lack of functional groups; however, oxidation or chemical functionalization introduces oxygen-containing groups (–COOH, –OH), significantly enhancing adsorption through electrostatic attraction, ion exchange, and surface complexation. For example, **Figure 4**, oxidized multi-walled CNTs showed markedly enhanced adsorption of Pb^{2+}, Cd^{2+}, Cu^{2+}, and Ni^{2+}, where FTIR and XPS analyses confirmed coordination between metal ions and surface carboxyl groups. Also, ion exchange occurs through the replacement of surface-bound protons from carboxyl and hydroxyl groups by divalent metal cations (e.g., Pb^{2+}, Cu^{2+}, Cd^{2+}), resulting in a pH after adsorption from the H^+ release, which contributes significantly to chemisorption alongside surface complexation. These metals show strong affinity for oxidized CNTs due to soft-acid/soft-base interactions with S-containing and O-containing groups. For example, Thiol-functionalized graphene forms strong coordination complexes with Hg^{2+}, giving exceptionally high uptake even at low concentrations (Chandran et al., 2023).

Figure 4. Heavy metals adsorption on modified CNT

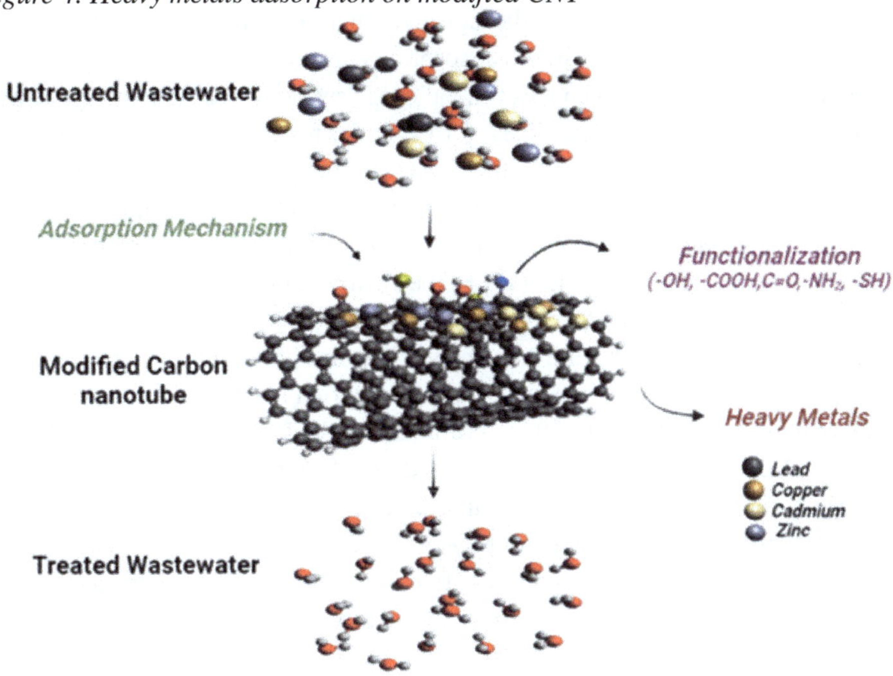

(Chandran et al., 2023)

To further improve performance, surface functionalization, metal oxide, and nanocomposite design have been widely adopted. Functionalizing GO with polymers or biopolymers introduces additional chelating groups and enhances structural stability. A representative example is graphene oxide functionalized with hyperbranched polyamide-amine (HPAMAM) and microcrystalline cellulose, which showed exceptionally high adsorption capacity for Pb^{2+} (up to ~680 mg g^{-1}). In this system, adsorption was governed mainly by chelation between Pb^{2+} and amine groups, supported by electrostatic attraction and ion exchange, as evidenced by FTIR band shifts and pseudo-second-order kinetics (Liu et al., 2022). Similarly, metal oxide–based nano adsorbents, such as iron oxide nanoparticles, have been successfully used for heavy-metal removal. In one study, Pb(II) adsorption onto iron oxide nanoparticles, as a core-shell, immobilized in alginate–fungal beads, occurred via surface complexation, ion exchange, and electrostatic attraction. Complexation where Pb^{2+} ions coordinate with functional groups such as carboxyl ($-COO^-$), hydroxyl (–OH), and amide (–CONH–) groups present on the fungal cell wall and alginate matrix. Ion exchange also contributes, as Pb^{2+} replaces Ca^{2+} or H^+ ions within the alginate structure. In addition, electrostatic attraction between positively charged Pb^{2+} ions and negatively charged functional sites enhances initial uptake, while iron oxide nanoparticles provide additional binding sites (Fe–O groups) and promote internal diffusion of Pb(II) into the porous structure. These adsorbents present Langmuir behavior and a maximum adsorption capacity of approximately 176 mg g^{-1}, highlighting the effectiveness of metal oxide surfaces rich in hydroxyl groups (Xu et al., 2012). ZnO nanoparticles, for example, readily form surface hydroxyl (≡Zn–OH) groups upon contact with water due to the polar nature of Zn–O bonds. Heavy metal ions such as Pb^{2+}, Cd^{2+}, and Cr^{3+} initially interact with the ZnO surface through electrostatic attraction and surface complexation. As adsorption progresses, particularly at near-neutral to alkaline pH, the local accumulation of metal ions at the ZnO interface promotes the formation of insoluble metal hydroxide or metal oxide surface, supported by the appearance of new crystalline phases in XRD patterns. This surface-induced precipitation leads to strong and often irreversible immobilization of the metals (Olawade et al., 2024). Similarly, nano-TiO_2 was investigated as an effective adsorbent for mercury Hg^{2+} removal from water. Hg^{2+} adsorption occurred primarily through surface complexation between mercury ions and hydroxyl groups (≡Ti–OH) present on the TiO_2 nanoparticle surface. Kinetic analysis typically follows pseudo-second-order behavior, confirming chemisorption-controlled uptake, while negative Gibbs free energy ($\Delta G° < 0$) demonstrated the spontaneous nature of the process (Ghasemi et al., 2011).

2.1.2.4 Adsorption of Dissolved Gases

Dissolved gases such as Methane (CH_4), Carbon dioxide (CO_2), and hydrogen sulfide (H_2S) pose significant environmental and industrial challenges in aqueous systems. Their removal depends strongly on molecular size, polarity, hydration behavior, and chemical reactivity, which govern interactions with adsorbent surfaces. Nanomaterials, with high surface area and tunable surface chemistry, enable effective capture through different mechanisms. CH_4 is the simplest hydrocarbon molecule, consisting of a tetrahedral structure with a central carbon atom bonded to four hydrogen atoms through strong σ (C–H) bonds. Its molecular geometry is highly symmetric, resulting in zero permanent dipole moment and extremely low polarity. Consequently, CH_4 exhibits very weak interactions with water molecules, forming only a minimal hydration shell in aqueous environments. This nonpolar character renders methane chemically inert under ambient conditions, with negligible tendency to undergo chemical reactions or form surface complexes during adsorption. Due to its low polarizability and lack of functional groups, methane adsorption is governed almost exclusively by physical adsorption (physisorption) through van der Waals interactions rather than chemisorption. In aqueous or moist systems, CH_4 preferentially partitions from water into hydrophobic environments, such as nonpolar nanopores or carbonaceous surfaces, where confinement enhances dispersion forces. The small kinetic diameter of methane (~0.38 nm) allows it to readily diffuse into micropores and nanopores, making pore size and hydrophobicity the dominant factors controlling its adsorption. However, the presence of water strongly competes for adsorption sites, particularly in hydrophilic pores (polar functionalized surface), often reducing methane uptake by pore blocking (Wood et al., 2012; Fan et al., 2018).

CO_2 is a linear molecule (O=C=O) with no permanent dipole moment; however, it possesses a significant quadrupole moment arising from the uneven electron distribution between the electron-rich oxygen atoms and the electron-deficient carbon atom. In aqueous environments, this property commonly promotes CO_2 hydration, leading to the formation of carbonic acid (H_2CO_3), a well-known and ubiquitous equilibrium in water. This hydrated form strongly influences CO_2 interactions with polar and hydroxylated adsorbent surfaces, such as metal oxides or functionalized nanomaterials. As shown in **Figure 5** the presence of water enhances CO_2 uptake on ZnO nanoparticles compared with dry conditions due to water-induced activation of surface hydroxyl sites. Dissolved CO_2 first hydrates to H_2CO_3, which then interacts with surface OH groups to form surface bicarbonate (HCO_3^-) and carbonate (CO_3^{2-}) species through chemisorption. At higher water activity, this behavior may further evolve into crystallization-like processes, resulting in the formation of ordered solid phases such as metal carbonate structures stabilized by hydrogen-bonded water networks near the adsorbent surface, confirmed by new crystalline features in XRD, as illustrated in **Figure 5 d,e** (Wood et al., 2012; Aruni Gankanda et al., 2016).

H_2S is a weak acid and a polar, highly reactive gas that readily dissolves in water and partially dissociates into $H_2S/HS^-/S^{2-}$ species, depending on pH. This dissociation significantly enhances its affinity toward metal-based nano adsorbents, particularly those containing transition metals such as Cu, Zn, Ni, and Ti, which exhibit strong chemical interactions with sulfur species through soft acid–soft base (HSAB) interactions.

Figure 5. CO_2 adsorption, High-resolution TEM images of (a) ZnO NPs (b) after exposure to CO_2 under dry conditions (c) after exposure under humid conditions (d) dissolved CO2 adsorption on metal oxide NPs

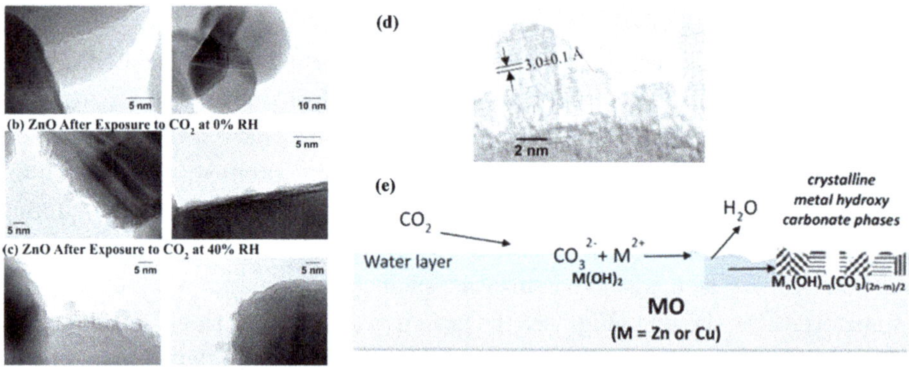

(Aruni Gankanda et al., 2016)

In aqueous environments, H_2S adsorption on Cu-, Zn-, Ni-, and TiO_2-based nanomaterials proceeds predominantly through chemisorption via surface complexation, where sulfide species coordinate with surface metal or metal–oxygen sites to form stable metal–sulfur bonds. A representative example is the TiO_2/CNT nanohybrid, in which TiO_2 nanoparticles provide reactive metal–oxygen sites while CNTs enhance surface area and dispersion; in this system, adsorption follows Langmuir isotherm behavior and pseudo-second-order kinetics, confirming chemisorption-controlled uptake and achieving ≈93% removal efficiency. At higher surface coverage or favorable pH conditions, surface-complexed sulfide species may further evolve into precipitation-like immobilization, resulting in the formation of insoluble metal sulfide phases (e.g., CuS, ZnS, NiS) directly on the nanoparticle surface. This transition from surface complexation to precipitation leads to strong and largely irreversible H_2S removal, distinguishing its adsorption behavior from non-reactive gases such as CH_4 and weakly reactive gases such as CO_2 (Daneshyar et al., 2017; Khorshidi et al., 2020).

Overall, based on the observations summarized in **Table 1**, carbon-based nanomaterials emerge as the most versatile and widely used adsorbents across diverse pollutant classes, including PAHs, dyes, heavy metals, and dissolved gases. Their dominance arises from a unique combination of high specific surface area, tunable surface chemistry, and extended π-electron systems, which enable multiple adsorption mechanisms such as π–π stacking, electrostatic attraction, surface complexation, and pore filling. Materials such as graphene, graphene oxide, carbon nanotubes, and activated carbon can be readily functionalized to target both organic and inorganic contaminants, allowing simultaneous interaction with hydrophobic molecules, charged dyes, and metal ions. This adaptability explains why carbon-based nanomaterials are consistently reported as promising, common, and high-performance adsorbents for water and wastewater remediation compared with single-mechanism inorganic adsorbents.

2.1.3 Adsorption Factors

2.1.3.1 Functionalization on Carbon-based Nanomaterial

Bare CNTs possess largely inert, hydrophobic sp^2-carbon surfaces, limiting adsorption mainly to weak physical interactions and resulting in low metal uptake. Oxidized CNTs (o-MWCNTs) contain introduced oxygenated functional groups (–OH, –COOH) that generate surface charge and act as coordination sites, enabling stronger adsorption through electrostatic attraction, ion exchange, and surface complexation. Further decoration with metal oxides, such as MnO_2, introduces additional hydroxyl-rich and redox-active sites, which significantly enhance inner-sphere complexation and ion-exchange capacity. Consequently, adsorption capacity follows the trend bare CNTs < o-MWCNTs < MnO_2–o-MWCNTs (Chandran et al., 2023). A study systematically evaluated the adsorption of MB using GO and rGO. The initial dye concentration was fixed at 350 mg L^{-1} for all experiments, allowing direct comparison of the two adsorbents under identical conditions. Contact time had a strong effect on performance, with GO demonstrating extremely rapid adsorption, achieving more than 95% removal within the first 5–7 minutes, and reaching equilibrium in approximately 60 minutes, while rGO required nearly 4.5 hours to reach equilibrium, reflecting slower kinetics due to its lower content of oxygen functional groups. Functionalization of GO or its reduced form has been shown to substantially enhance the adsorption of MB dye. In one study, citrate-modified GO demonstrated a maximum adsorption capacity of 222.22 mg g^{-1} under mild batch conditions (6 mg adsorbent in 50 mL solution, pH 7, 5 min) through carboxylate ($-COO^-$) sites, which interact electrostatically with cationic MB^+ (Abd-Elhamid et al., 2022).

Table 1. Summary of different adsorbates and showing how the adsorbate structure influences the mechanism of adsorption

Adsorption Category	Adsorbate (Examples)	Structural Features of Adsorbate	Adsorption Mechanisms	Representative Nano adsorbents
PAHs (Organic pollutants)	Naphthalene, Acenaphthene, Phenanthrene, Pyrene	Polycyclic aromatic rings; nonpolar, hydrophobic; increasing planarity with molecular weight	ð–ð stacking, van der Waals forces, hydrophobic interactions, pore filling (physisorption)	CNTs, graphene, graphene wool, activated carbon, magnetic carbon nanomaterials
Dyes – Anionic (Acid dyes)	Methyl orange (MO)	Aromatic structure; negatively charged (–SO_3^-, –COO^-); azo linkage	Electrostatic attraction, hydrogen bonding, pore diffusion (mixed physisorption–chemisorption)	Chitosan–silica nanocomposites, protonated GO, metal oxides
Dyes – Cationic (Basic dyes)	Methylene blue (MB), Crystal violet (CV)	Planar aromatic systems; positively charged amine groups	Electrostatic attraction, ð–ð stacking, hydrogen bonding; (mixed physisorption–chemisorption)	GO, rGO, montmorillonite–GO composites, graphene-based nanomaterials
Heavy metals	Pb^{2+}, Cd^{2+}, Cu^{2+}, Ni^{2+}, Hg^{2+}	Hydrated metal cations; high charge density; coordination capability	Surface complexation, ion exchange, electrostatic attraction; precipitation at high pH	GO, oxidized CNTs, polymer-functionalized GO, metal oxide nanohybrids
Dissolved gases – Nonpolar	CH_4	Small, nonpolar; no dipole moment	Van der Waals interactions, pore confinement (physisorption)	Hydrophobic carbon nanopores, graphene, CNTs
Dissolved gases – Weakly reactive	CO_2	Linear molecule; quadrupole moment; hydrates to H_2CO_3 in water	Hydration-assisted chemisorption, bicarbonate/carbonate formation, crystallization	metal oxides NPs, hydroxylated nanomaterials
Dissolved gases – Reactive	H_2S	Polar weak acid; dissociates to HS^-/ S^{2-}	Surface complexation, HSAB interactions, and metal sulfide precipitation	TiO_2/CNT nanohybrids, Cu-, Zn-, Ni-based NPs, metal sulfide nanocomposites

Similarly, diethylenetriamine-functionalized rGO exhibited a significantly improved capacity of 291.4 mg g^{-1}(8 mg L^{-1} adsorbent, pH 9, 4 h), compared with only 77.1 mg g^{-1} for the unmodified rGO under the same conditions (Wang et al., 2024). These findings indicate primarily that chemical modification of the graphitic surface through the introduction of carboxylate, amine groups, or other functional moieties enhances and facilitates the adsorption by increasing surface charge density, creating additional active binding sites, and improving dispersibility in aqueous systems, maintaining a high accessible surface area. Further, modification with amine-rich groups such as polyethyleneimine (PEI), which are highly flexible and hydrophilic, offers greater swelling in water, preventing sheet aggregation and faster diffusion (Wang et al., 2024; Luo et al., 2022).

2.1.3.2 Polarizability Effect in Heavy Metal Adsorption

A comparison was conducted between different heavy metal adsorption with graphene–based nano adsorbent and its derivatives. It was found that Pb^{2+} generally exhibits higher adsorption capacities and stronger affinity than most other heavy metal ions, such as Cu, Sb, Hg, Fe, Zn, Co, and Ni. This preferential behavior is attributed to several intrinsic properties of Pb^{2+} discussed in the reviews. Pb^{2+} has a larger ionic radius and higher polarizability, which facilitates the formation of stable inner-sphere complexes with oxygen-containing functional groups (–COOH, –OH, epoxy) abundant on GO surfaces. In addition, Pb^{2+} possesses a lower hydration energy compared with smaller, more strongly hydrated ions (e.g., Ni^{2+} or Zn^{2+}), allowing it to more readily shed its hydration shell and interact directly with surface functional sites. These factors enhance surface complexation and ion-exchange interactions, while electrostatic attraction further contributes under favorable pH conditions. Consequently, Pb^{2+} adsorption on GO often follows pseudo-second-order kinetics and Langmuir-type behavior, indicative of chemisorption-dominated uptake. Nevertheless, as emphasized in the reviews, the magnitude of Pb^{2+} adsorption remains strongly influenced by GO functionalization degree, solution pH, competing ions, and experimental conditions, underscoring the need for cautious comparison across studies (Chandran et al., 2023).

2.1.3.3 Effect of pH on Adsorption

Solution pH plays a decisive role in adsorption performance across metal oxide and functionalized carbon-based nanomaterial systems by controlling surface charge, functional group ionization, and adsorbate speciation. For metal oxide (MO) nanoparticles, adsorption is minimal under acidic conditions due to surface protonation ($\equiv$MO–OH_2^+), which causes electrostatic repulsion and competition with excess H^+ ions for active sites when cationic adsorbates are present (Olawade et al., 2024). However, such protonated surfaces can favor the adsorption of anionic species, such as acidic dyes, through electrostatic attraction (Mohamed et al., 2020). As pH increases toward near-neutral conditions ($\approx$6–8), deprotonation of surface hydroxyl groups generates negatively charged ($\equiv$MO–O^-) sites, sharply enhancing adsorption of cationic adsorbates through electrostatic attraction and surface complexation. At excessively high pH, adsorption may plateau or decline due to the formation of metal hydroxide species in solution, indicating that precipitation-assisted immobilization may contribute to metal removal. In contrast, modified GO exhibits high adsorption capacity over a broader pH range (≈5–13), owing to its abundant oxygen-containing functional groups (–COOH, –OH, –O–) that progressively deprotonate with increasing pH. As a result, modified GO becomes increasingly negatively charged, promoting strong electrostatic attraction and complexation with cationic adsorbates without requiring precipitation mechanisms (Olawade et al., 2024; Jahan et al., 2022).

2.1.3.4 Effect of ionic strength on Adsorption

The effects of foreign ions and ionic strength on the adsorption of MB were systematically investigated by introducing electrolytes (NaCl and $CaCl_2$) at concentrations ranging from 0.001 to 0.100 mol L^{-1}. As the ionic strength increased, the equilibrium adsorption capacity of MB on GO decreased markedly. For instance, the adsorption capacity declined from approximately 351 mg g^{-1} in salt-free solution to about 260 mg g^{-1} in the presence of 0.10 mol L^{-1} NaCl and decreased even further in Ca^{2+}-containing solutions. This behavior is attributed to electrical double-layer compression, charge screening, and competitive occupation of active sites, with divalent Ca^{2+} exerting a stronger inhibitory effect than monovalent Na^+ (Nuengmatcha et al., 2014). In contrast, for systems involving the adsorption of anionic species, such as anionic dyes, onto metal oxide nanoparticles, GO, or polymer amine-modified GO under acidic conditions, the adsorbent surface becomes positively charged, favoring adsorption through electrostatic attraction. In such cases, a moderate increase in ionic strength (e.g., Na^+) can enhance adsorption by reducing electrostatic repulsion between anionic adsorbate molecules and compressing the electrical double layer, thereby promoting closer packing of the adsorbate on the surface rather than hindering uptake (Yagub et al., 2014).

2.1.3.5 Temperature Effect on Adsorption

The effect of temperature on adsorption has been extensively investigated across a wide range of nanomaterial-based systems, consistently demonstrating its strong influence on adsorption capacity and mechanism. For functionalized aromatic adsorbate, the adsorption capacity generally increases with rising temperature, indicating an endothermic adsorption process. For example, MB uptake increased from approximately 240–260 mg g^{-1} at 298 K to about 300–330 mg g^{-1} at 318 K (Jahan et al., 2022). Elevated temperatures enhance adsorbate diffusion, reduce solution viscosity, and activate surface functional groups, thereby strengthening electrostatic attraction and surface complexation interactions. Thermodynamic analysis revealed positive enthalpy ($\Delta H^\circ > 0$), confirming chemisorption-dominated uptake (Yagub et al., 2014; Jahan et al., 2022; Wang et al., 2024). In the case of heavy-metal ions, a representative example of Pb^{2+} removal by GO-based adsorbents showed a pronounced temperature dependence, with adsorption capacity increasing from approximately 210–230 mg g^{-1} at 298 K to 280–300 mg g^{-1} at 318 K. This enhancement was associated with increased metal-ion mobility and stronger coordination with surface –COOH and –OH groups, again reflected by positive $\Delta H°$ values and spontaneous adsorption behavior ($\Delta G° < 0$) (Chandran et al., 2023; Wang et al., 2024). However, in the adsorption of phenanthrene (a representative PAH) onto graphene-based nanomaterials, adsorption capacity was found to decrease with increasing temperature, confirming an exothermic physisorption process. Specifically, the adsorption capacity decreased from approximately 42–45 mg g^{-1} at 298 K to about 30–33 mg g^{-1} at 318 K, indicating that elevated temperature weakens π–π stacking and van der Waals interactions between the aromatic PAH molecules and the sp^2-carbon surface. Thermodynamic analysis reported negative enthalpy changes ($\Delta H° < 0$), further confirming that PAH adsorption is energetically favored at lower temperatures (Mogashane et al., 2024). Overall, increasing adsorption capacity with temperature is a strong indicator of endo reaction and chemisorption, whereas decreasing the uptake is characteristic of physisorption-dominated systems. Collectively, these studies confirm that adsorption on nanomaterials is frequently endothermic and thermodynamically favorable at elevated temperatures, particularly when chemisorption mechanisms such as surface complexation and ion exchange dominate over weak physical adsorption (Mogashane et al., 2024). Overall, temperature-dependent adsorption behavior provides a clear distinction between chemisorption- and physisorption-dominated systems. Adsorption of functionalized aromatic and heavy-metal ions on nanomaterials is typically endothermic and enhanced at higher temperatures, reflecting the dominance of electrostatic attraction and surface complexation mechanisms. In contrast, adsorption of nonpolar organic contaminants such as PAHs is exothermic and favored at lower temperatures, as it is governed by weak π–π interactions and van der Waals forces that are destabilized by thermal agitation.

2.1.3.6 Effect of Initial Concentration

The initial concentration of adsorbates plays a critical role in controlling adsorption performance by influencing the driving force for mass transfer and the degree of surface site saturation. For dye adsorption, studies on MB uptake by GO show that adsorption capacity increases with increasing initial MB concentration because a large number of active surface sites are available, and mass transfer from the bulk solution to the adsorbent surface is the dominant driving force. For example, MB adsorption capacity increased from approximately 120–150 mg g^{-1} at low initial concentrations (≈50 mg L^{-1}) to over 300 mg g^{-1} at higher concentrations (≈300–400 mg L^{-1}), before approaching saturation as active sites became fully occupied. A similar trend is observed for heavy-metal adsorption, where Pb^{2+} uptake on GO-based adsorbents increased from around 100–130 mg g^{-1} at low initial Pb^{2+} concentrations to 280–300 mg g^{-1} at higher concentrations, reflecting increased surface complexation and ion-exchange interactions until monolayer saturation was reached. In contrast, for physisorption-dominated systems, such as the adsorption of phenanthrene (PAH) on graphene-based nanomaterials, adsorption capacity increases with initial concentration at low levels but plateaus rapidly due to limited pore volume and weak interaction forces, with saturation occurring at comparatively lower uptake values. In chemisorption-dominated systems, such as the adsorption of MB and Pb^{2+} ions on graphene oxide and functionalized carbon materials, deviations from linearity occur at relatively higher adsorption capacities. This reflects the presence of strong surface interactions (electrostatic attractions, surface complexation, ion exchange) and a higher density of chemically active sites, often leading to Langmuir-type monolayer saturation. Overall, increasing initial concentration enhances adsorption capacity by strengthening the mass-transfer driving force, while deviations from linearity at high concentrations indicate site saturation and isotherm-controlled behavior, distinguishing between chemisorption- and physisorption-dominated systems. In chemisorption-dominated systems, such as the adsorption of methylene blue (MB) and Pb^{2+} ions on graphene oxide and functionalized carbon materials, deviations from linearity occur at relatively higher adsorption capacities. This reflects the presence of strong surface interactions (electrostatic attractions, surface complexation, ion exchange) and a higher density of chemically active sites, often leading to Langmuir-type monolayer saturation. In contrast, for physisorption-dominated systems, such as phenanthrene (PAH) adsorption on graphene-based nanomaterials, deviations from linearity occur at much lower uptake values. Because adsorption relies mainly on weak van der Waals forces and π–π interactions, the available adsorption sites are fewer and energetically less heterogeneous, resulting in early saturation and Freundlich-type behavior (Chandran et al., 2023; Mogashane et al., 2024; Jahan et al., 2022; Wang et al., 2024; Arora, 2019).

2.1.4 Desorption and Regeneration

Desorption is the reverse process of adsorption, in which previously adsorbed molecules or ions are released from the adsorbent surface back into the surrounding phase. Desorption efficiency is a key parameter for evaluating the reusability and sustainability of adsorbents, and it strongly depends on the dominant adsorption mechanism. For chemisorption-dominated heavy-metal systems, desorption is commonly achieved through acidic regeneration, which reverses electrostatic attraction and disrupts surface complexes. Pb^{2+} and Cd^{2+} adsorbed on graphene oxide and functionalized carbon materials exhibited desorption efficiencies of 80–95% using dilute HCl (0.01–0.1 M), with adsorption capacity retained above 85% after 4–5 cycles, confirming that surface complexation and ion exchange are largely reversible when precipitation is avoided (Arora, 2019). Also, in heavy-metal adsorption systems dominated by surface complexation, Pb^{2+} ions bind to oxygen-containing functional groups (–COO^-, –OH) on GO through inner-sphere metal–ligand complexes. Desorption can be effectively achieved by introducing a strong chelating agent such as EDTA, which forms highly stable soluble Pb–EDTA complexes (log K $\approx$ 18). The chelating agent competes with surface ligands and pulls the metal ion away from the adsorbent surface, resulting in reverse complexation, desorption efficiencies of 85–95% for Pb^{2+} using 0.01–0.05 M EDTA, with adsorption capacity largely preserved after multiple cycles (Arora, 2019). In contrast, systems involving surface precipitation show limited regeneration, as metal hydroxide or sulfide formation results in desorption efficiencies below 40%, even under strong acidic conditions, indicating quasi-irreversible adsorption (Kulkarni et al., 2014) For functionalized adsorbates such as dyes, desorption is generally more efficient due to the prevalence of weaker interactions. Methylene blue adsorbed on graphene-based materials showed desorption efficiencies exceeding 90% using solvents (ethanol) or mild pH adjustment, with minimal loss in adsorption capacity over multiple cycles. These results confirm that dye adsorption governed by electrostatic attraction and π–π interactions is highly reversible (Yagub et al., 2014). Similarly, physisorption-dominated systems, such as PAH adsorption on graphene and CNTs, exhibit effective thermal desorption; increasing temperature from 298 K to 318 K resulted in 25–30% desorption of phenanthrene, consistent with weak van der Waals and π–π interactions and negative adsorption enthalpy values (Mogashane et al., 2024). Overall, these quantitative results demonstrate that desorption efficiency decreases in the order: physisorption > electrostatic attraction > surface complexation > precipitation, emphasizing that adsorption mechanisms directly determine regeneration feasibility and long-term adsorbent performance.

2.1.5 Challenges and Limitations

Despite the simplicity and high efficiency of adsorption, several challenges limit the practical application of nanomaterial-based adsorbents. In real aqueous systems, the presence of competing ions, natural organic matter, and mixed pollutants can significantly reduce adsorption selectivity and capacity. Strong adsorption mechanisms such as surface complexation and precipitation often hinder desorption and regeneration, leading to decreased adsorbent reusability and increased operational cost. Adsorption performance is highly sensitive to solution pH, ionic strength, and temperature, which may fluctuate in real wastewater and alter adsorption mechanisms. In addition, nano adsorbents may suffer from aggregation, surface instability, or leaching, reducing accessible surface area and raising environmental and health concerns. Finally, challenges related to large-scale synthesis, recovery of nano adsorbents after use, and safe disposal of spent materials remain critical obstacles to the industrial implementation of adsorption-based nanotechnologies (Chandran et al., 2023; Mogashane et al., 2024; Yagub et al., 2014; Kulkarni et al., 2014; Arora, 2019).

2.2 Photoelectrochemistry

Photoelectrochemistry (PEC) for water treatment combines light irradiation (photocatalyst) and a small electrical potential (electrochemistry) to enhance pollutants removal from water. Unlike adsorption-based methods that only transfer contaminants from water to a solid phase, PEC enables the chemical degradation of persistent organic pollutants, including dyes, pharmaceuticals, and other toxic compounds, into smaller and less harmful molecules and toxic heavy metal ions into less harmful or stable forms. In a simple PEC system, **Figure 6**, a photo electrode is illuminated by light, which generates charged particles electrons (e^-) and holes (h^+). The applied electrical potential plays a crucial role in separating these charges and suppressing their recombination, allowing photogenerated holes to drive water oxidation to produce oxygen and protons at photoanode. On the other hand, the photogenerated electrons (e^-) are transferred through the external circuit to the cathode (counter electrode), where they reduce protons, leading to hydrogen gas evolution. n wastewater treatment applications, PEC operates primarily through the generation of highly reactive oxygen species, including hydroxyl radicals (•OH) and superoxide radicals ($\bullet O_2^-$), formed via reactions between charge carriers, water, and dissolved oxygen. These reactive species are capable of attacking and decomposing organic pollutants and in some cases, can completely mineralize them into carbon dioxide, while photogenerated electrons can induce the reduction of toxic metal ions at the counter electrode (Kudo & Miseki, 2009; Zhou et al., 2025). Compared with conventional photocatalysis, PEC offers several important advantages, primarily due to the application of an external electrical bias that promotes efficient charge separation. In photocatalytic systems, a large fraction of photogenerated electrons and holes recombine rapidly, which limits pollutant degradation efficiency. In photocatalysts, construction of heterostructures has been developed to suppress electron–hole recombination, including the doping of metals, metal oxides, non-metals, or carbon-based materials. These approaches improve charge separation by transferring photogenerated electrons to lower-energy states, thereby enhancing photocatalytic activity. However, such electron transfer often results in energy loss and a reduced redox potential, which limits the overall oxidation and reduction capability of conventional photocatalytic systems. In contrast, PEC systems actively drive electrons away from the photoelectrode, extending the lifetime of charge carriers and increasing the production of reactive species. Quantitatively, the PEC system achieved organic pollutant degradation efficiencies approximately 2.3 times higher than photocatalysis alone, reaching 98% degradation of methylene blue within 2 hours. This significant improvement clearly demonstrates the role of electrical bias in enhancing PEC performance (Zhou et al., 2025). In addition to the higher degradation efficiency, PEC offers better control over reaction pathways through applied current, reducing charge recombination and improving performance for refractory pollutants, making it a promising and versatile technology for advanced water and wastewater treatment (Zhou et al., 2025)

Figure 6. Simple water splitting by PEC (Kudo & Miseki, 2009)

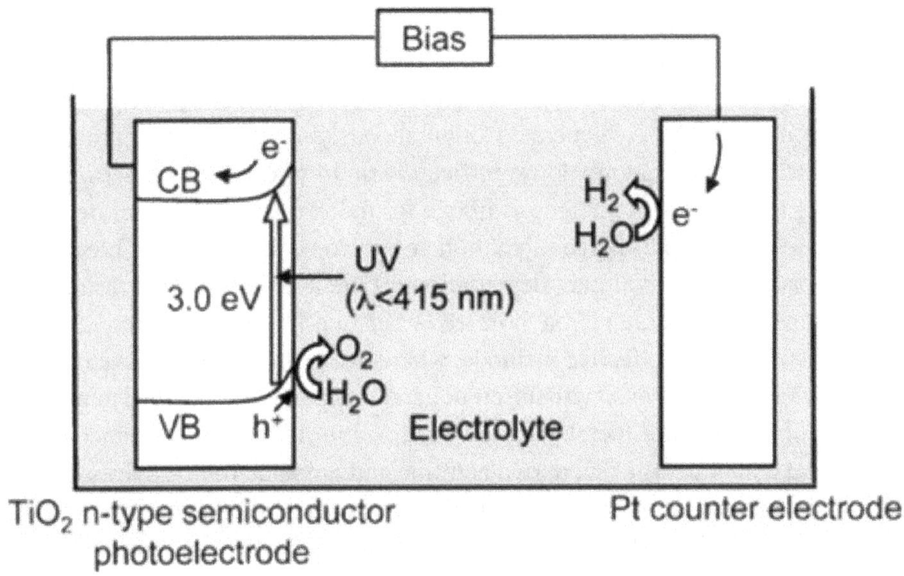

2.2.1 Type of Electrodes and Charge Separation

A typical photoelectrochemical (PEC) cell for water treatment consists of a photoelectrode, a counter electrode, and, in laboratory studies, a reference electrode. The photoelectrode is commonly fabricated from semiconductor materials or composite semiconductors. A semiconductor is a solid material whose electrical conductivity lies between that of an insulator and a metal, and whose electronic structure allows it to absorb light and generate charge carriers, and is irradiated with light to generate electrons and holes. Upon light irradiation, the semiconductor photoelectrode produces photogenerated electrons and holes, which drive photoelectrochemical reactions. A suitable semiconductor for PEC applications must possess an appropriate band gap that allows efficient light absorption while maintaining sufficient redox potential. Typically, semiconductor photoelectrodes used in PEC systems exhibit band gaps ranging from those of wide-band-gap materials such as TiO_2 (~3.2 eV) to narrower-band-gap semiconductors such as CdSe (~1.8 eV), enabling ultraviolet or visible-light-driven photoelectrochemical reactions (Kudo & Miseki, 2009; Zhou et al., 2025). The counter electrode, often made of conductive and chemically stable materials such as carbon or graphite, receives the photogenerated electrons and facilitates reduction reactions. In research-scale PEC systems, a reference electrode (e.g., Ag/AgCl) is introduced to accurately control and monitor the applied potential, particularly when precise measurements of current or electrode potential are required (Zhu et al., 2025).

The separation of photogenerated charge carriers is a critical factor governing the efficiency of photocatalytic and photoelectrochemical systems. Upon light irradiation, electrons and holes are generated simultaneously in semiconductor materials; however, their rapid recombination significantly limits photocatalytic performance by reducing the number of charges available for redox reactions. Therefore, effective strategies to promote charge separation and prolong charge-carrier lifetime are essential for enhancing pollutant degradation efficiency. Several approaches have been developed to suppress electron–hole recombination. Heterojunction construction (n-n) is one of the most effective methods, where two semiconductors, the majority charge carriers are electrons, with different band structures, are coupled to promote directional charge transfer, thereby improving separation efficiency. p–n junctions are fundamental components of modern electronic and optoelectronic devices, as they enable efficient separation of photoexcited electrons and holes through an internal built-in electric field. A p-type semiconductor, in which the majority charge carriers are holes positively charge, is a material usually created by doping a semiconductor with elements that have fewer valence electrons than the host material. They complement n-type materials and enable strong internal electric fields. Doping metal oxide semiconductors with different p-type materials, including inorganic semiconductors, organic semiconductors, and perovskites, is an effective strategy for achieving high-performance photoelectric devices that introduce additional energy levels or electron sinks that facilitate charge migration away from the recombination sites. However, these methods may lead to partial loss of redox potential due to electron transfer to lower-energy states. Further, introducing an insulating layer between p-type and n-type semiconductors to form a p–i–n junction further enhances photodetector performance by improving charge control. The insulation layer acts as an electron-blocking spacer that prolongs carrier transport and adjusts the spectral response. Such p–i–n heterostructures have demonstrated broadband photo response and high responsivity due to effective band alignment and controlled carrier transport. The application of an external electric field in photoelectrochemical systems represents a highly effective strategy for charge separation, as it actively drives electrons and holes toward different electrodes, significantly reducing recombination without sacrificing redox capability. Additionally, a Schottky junction is formed at the interface between a metal (or conductor) and a semiconductor, where a potential barrier is created due to the difference in their work functions. This barrier generates a depletion region that suppresses dark current and enables fast charge separation and transport under illumination. As a result, Schottky-junction-based devices exhibit advantages such as fast response speed, low dark current, high UV–visible rejection ratio, and potential self-powered operation, making them highly suitable for high-performance photodetectors. Moreover, decorating semiconductor surfaces with metal nanoparticles or graphene can form localized Schottky junctions that further enhance photoelectric performance by improving charge separation and light–matter interaction. Advanced charge-transfer mechanisms, such as S-scheme systems, selectively retain high-energy electrons and holes while allowing low-energy carriers to recombine, thereby maintaining strong oxidation and reduction ability. Overall, improving charge-carrier separation is fundamental to advancing photocatalytic and photoelectrochemical technologies, and combining structural design, material modification, and electrical assistance offers the most effective pathway toward high-efficiency pollutant removal (Ouyang et al., 2019).

2.2.2 Application

2.2.2.1 UV light photoelectrochemistry

PEC can be conducted under UV light, which requires high band gap semiconductors as illustrated above. Here is an example of PEC water purification device designed for the simultaneous removal of organic pollutants and heavy metal ions (**Figure 7**). The system consists of a Cu counter electrode, TiO_2 photoelectrode deposited on FTO glass, a transparent conductive substrate that allows light transmission while providing an efficient pathway for electron transport, and an Ag/AgCl reference electrode, forming a typical three-electrode PEC configuration. When wastewater flows into the reactor, it first contacts the TiO_2/FTO photoelectrode. Under light irradiation, TiO_2 absorbs photons and generates electron–hole pairs. The photogenerated holes remain at the TiO_2 surface, where they oxidize organic pollutants in the wastewater, leading to their degradation. At the same time, the photogenerated electrons are transported through the external circuit toward the Cu counter electrode, where they participate in reduction reactions that convert toxic heavy metal ions into less harmful or recoverable forms. The Ag/AgCl electrode serves only as a reference to control and monitor the applied potential and does not directly participate in pollutant removal. As a result of these coupled oxidation and reduction processes, contaminated water entering the system is treated and exits as clean water, demonstrating the dual-function capability of the PEC system (Zhou et al., 2025).

Figure 7. PEC schematic device for wastewater treatment

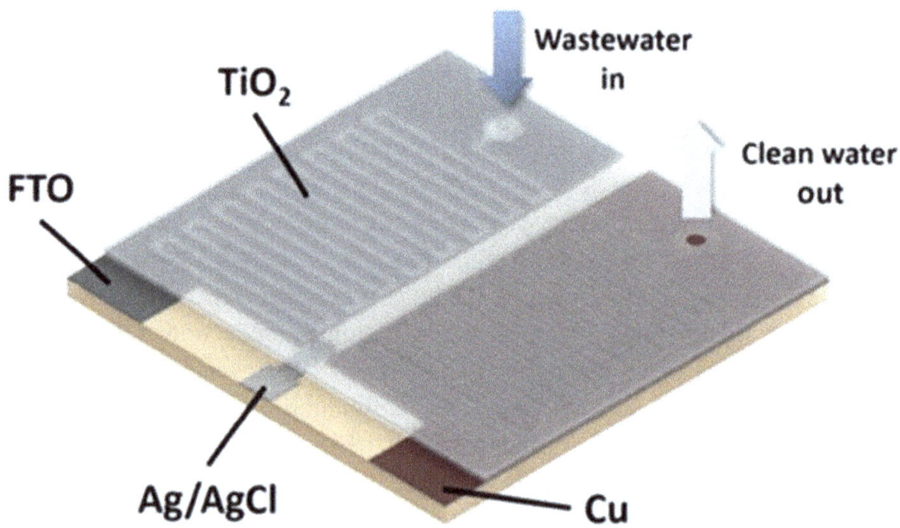

(Zhou et al., 2025)

Another study reported the development of a UV-light-responsive Ag/TiO_2/PVA nanocomposite for the efficient removal of heavy metal ions from industrial wastewater. In this system, TiO_2 acts as the primary wide-band-gap semiconductor, while metallic Ag nanoparticles are introduced to enhance charge separation through the formation of Schottky junctions and plasmonic effects, which effectively trap photogenerated electrons under UV irradiation. Polyvinyl alcohol (PVA) serves as a stabilizing and immobilizing matrix that suppresses electron–hole recombination and prevents nanoparticle agglomeration. Prior to photocatalytic operation, the adsorption behavior of heavy metal ions (Cr^{6+}, Cu^{2+}, Zn^{2+}, and Ni^{2+}) on the Ag/TiO_2/PVA nanocomposite was evaluated under dark equilibrium conditions, and the adsorption data were best described by the Langmuir isotherm, indicating predominant monolayer adsorption on finite and relatively homogeneous surface sites. This adsorption step concentrated metal ions at the catalyst surface, thereby facilitating subsequent photoinduced reactions. Under UV-C irradiation (253 nm), the nanocomposite exhibited strong photocatalytic performance, achieving more than 94%

removal of Cr^{6+} and approximately 70–90% removal of Cu^{2+}, Zn^{2+}, and Ni^{2+} within 180–236 min, even when treating real electroplating wastewater. Optical analysis revealed that the incorporation of Ag and PVA reduced the band gap of TiO_2 from approximately 3.5 eV to 2.9 eV, thereby lowering the energy required for excitation and improving light absorption and charge-carrier utilization, **Figure 8**. From this example, pollutants removal is governed by a synergistic combination of adsorption and photoelectrochemical processes. The presence of polyvinyl alcohol (PVA) provides abundant surface hydroxyl groups that enhance the initial adsorption of heavy metal ions onto the nanocomposite surface, effectively concentrating pollutants near the photoactive TiO_2 sites. This adsorption step increases pollutant–electrode contact and facilitates subsequent photoelectrochemical reactions (Dehghani & Delnavaz, 2024).

Figure 8. Ag/TiO$_2$/PVA nanocomposite for efficient generation of e^- and h^+

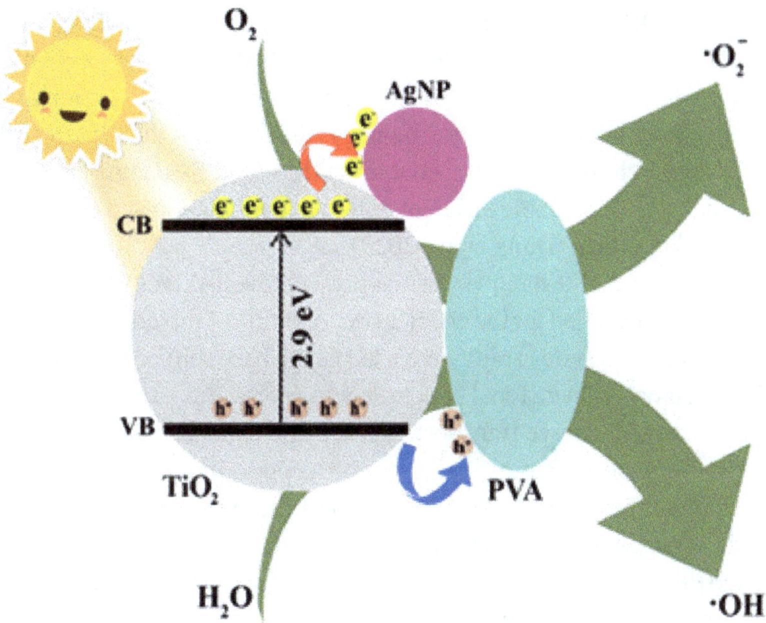

(Dehghani & Delnavaz, 2024)

2.2.2.2 Visible Light Photoelectrochemistry

The study reports the development of an efficient photoelectrochemical (PEC) system based on an S-scheme N-doped carbon dots (NCDs) anchored g-C_3N_4/Fe_2O_3 shell/core composite for simultaneous antibiotic degradation and hydrogen production. In this system, α-Fe_2O_3 acts as the core semiconductor with a deep valence band suitable for oxidation reactions, while g-C_3N_4 forms a protective shell that enhances stability and visible-light absorption. Nitrogen-doped carbon dots are anchored onto the composite surface to further improve light harvesting, charge transfer, and electron storage capability, leading to the formation of an S-scheme heterojunction that effectively suppresses electron–hole recombination while preserving strong redox ability. The g-C_3N_4/Fe_2O_3/NCD photoelectrode is mainly activated under visible light, owing to the suitable band gaps of g-C_3N_4 (~2.7 eV) and Fe_2O_3 (~2.0–2.2 eV), while N-doped carbon dots further enhance visible-light absorption and charge transfer efficiency. The PEC photoanode exhibited a high photocurrent density of 3.07 mA cm^{-2} at 1.6 V, which was 4–15 times higher than that of the individual components. When applied to water treatment, the system achieved up to 95% degradation of trimethoprim in aqueous solution and 90% removal in lake water, demonstrating strong applicability under realistic conditions. Reactive species analysis confirmed that both hydroxyl radicals (•OH) and superoxide radicals ($•O_2^-$) played dominant roles in pollutant degradation. In parallel, photogenerated electrons were utilized at the cathode for hydrogen evolution, with a production rate of 550 μmol cm^{-2} h^{-1}, highlighting the dual functionality of the PEC system. Overall, enhanced performance is attributed to the synergistic effects of the shell/core architecture, N-doped carbon dots, and S-scheme charge transfer pathway, which together enable efficient pollutant degradation and energy recovery in a single PEC platform (Dang et al., 2023).

The study on CdSe/TiO_2 nanowire arrays presents a self-driven photoelectrochemical (PEC) system designed for the efficient degradation of mixed organic pollutants. In this work, TiO_2 nanowire arrays were used as the base semiconductor photoelectrode due to their one-dimensional structure, which provides a large surface area and efficient pathways for charge transport, while CdSe nanoparticles were introduced to form a heterojunction that extends light absorption into the visible region and enhances charge separation. A key innovation of the system is the integration of a triboelectric nanogenerator, which supplies the required electrical bias and enables self-powered PEC operation without external electricity input. The system was applied to the simultaneous degradation of methylene blue, methyl orange, and tetracycline hydrochloride, representing a realistic mixed-pollutant scenario. Quantitatively, the CdSe/TiO_2 PEC device achieved a total degradation efficiency of 99.4% within 4 h, significantly outperforming pure TiO_2 photocatalysis and non-modified PEC systems, while also demonstrating good operational stability. The enhanced performance was attributed to the synergistic effects of heterojunction-induced charge separation, rapid electron transport along TiO_2 nanowires, and photoelectrochemical assistance from the self-generated electrical field, which together promoted the formation of reactive species responsible for organic pollutant oxidation (Li et al., 2013b).

2.2.3 Photocatalytic Factors

2.2.3.1 Contact Time

The photocatalytic removal of heavy metal ions was found to be strongly influenced by reaction time. The removal efficiency of metal ions increases progressively with reaction time, indicating that prolonged light exposure enhances photocatalytic activity by generating more photogenerated charge carriers. For example, a rapid increase in removal efficiency of Zn, Cu, Ni, and Cr was observed during the initial 150–200 min, followed by a gradual approach toward equilibrium, with Cr(VI) being removed at the highest rate among the investigated metal ions, **Figure 9a** (Dehghani & Delnavaz, 2024).

2.2.3.2 Effect of pH

The effect of pH on metal removal was investigated. A study demonstrates the strong dependence of metal removal on solution pH. The highest removal efficiencies are achieved under acidic conditions (pH $\approx$ 3), where Cr(VI), Cu^{2+}, Zn^{2+}, and Ni^{2+} show removal efficiencies exceeding 85–95%. As the pH increases from 3 to 7, the removal efficiency decreases for all metal ions. This behavior is attributed to reduced electron availability, increased competition with hydroxide ions, and possible metal hydroxide formation at higher pH values, which hinder photocatalytic reduction processes, **Figure 9b** (Dehghani & Delnavaz, 2024).

2.2.3.3 Photocatalytic Dosage

photocatalyst dosage significantly affects photocatalytic performance, where increasing the catalyst amount leads to a significant enhancement in metal removal due to the increased number of active sites and higher generation of reactive charge carriers. However, further excessive catalyst loading may cause light scattering or shielding effects that limit light penetration, **Figure 9c** (Dehghani & Delnavaz, 2024).

2.2.3.4 Light Source

The effect of light source on the photocatalytic removal of heavy metal ions was systematically evaluated by comparing UV-C, UV-A, and sunlight irradiation under identical reaction conditions on Ag/TiO_2/PVA photocatalyst experiment. It was observed that significantly higher removal efficiencies were achieved under UV-C irradiation for all investigated metal ions, confirming that high-energy ultraviolet light plays a crucial role in activating the Ag/TiO_2/PVA photocatalyst. Under UV-C illumination, removal efficiencies of approximately 99% for Cr(VI), 94.5% for Cu^{2+}, 89.5% for Zn^{2+}, and 86.7% for Ni^{2+} were obtained within 3 h. In contrast, when sunlight was used as the irradiation source, the removal efficiencies were reduced to moderate levels, ranging from about 46.4% to 63.7%, while the lowest removal efficiencies were observed under UV-A irradiation, with values generally below 50% for all metal ions. This trend was attributed to the higher photon energy of UV-C light, which more effectively excites the wide bandgap TiO_2 semiconductor and generates a greater number of photogenerated charge carriers. Overall, the results demonstrate that the photocatalytic performance of the Ag/TiO_2/PVA system is highly dependent on the irradiation source, with UV-C providing the most favorable conditions for efficient heavy-metal removal (Dehghani & Delnavaz, 2024).

Figure 9. (a) time (b) pH (c) catalyst dosage parameters, effect of heavy metal removal

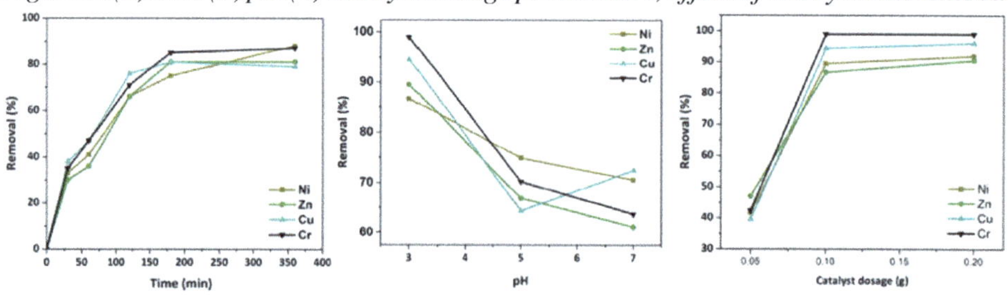

(Dehghani & Delnavaz, 2024)

2.2.3.5 Photoelectrode Material Properties

The intrinsic properties of the photoelectrode, particularly the band gap energy and band-edge positions, determine light absorption capability and redox driving force in PEC systems. Wide-band-gap semiconductors such as TiO_2 (Eg ≈ 3.2 eV) exhibit high chemical stability but require UV irradiation, which limits solar utilization. For example, Zhou et al. demonstrated that a TiO_2-based PEC system achieved >90% degradation of organic pollutants under UV light, whereas performance under visible light was negligible, highlighting the importance of semiconductor band structure in PEC efficiency (Zhou et al., 2025)

2.2.3.6 Nanostructure and Morphology

Nanostructure enhances PEC performance by increasing surface area and providing efficient charge-transport pathways. One-dimensional architectures such as TiO_2 nanowires or nanotubes facilitate directional electron transport and reduce recombination losses. In a self-powered CdSe/TiO_2 nanowire array PEC system, the nanowire morphology contributed to a 99.4% degradation efficiency of mixed organic pollutants within 4 h, significantly outperforming planar TiO_2 electrodes (Li et al., 2013b).

2.2.3.7 Surface Chemistry and Heterojunction Design

Surface modification and heterojunction engineering improve PEC efficiency by promoting charge separation and suppressing electron–hole recombination. In the CdS/TiO_2 heterojunction, Supersonic Cluster Beam Deposition (SCBD) further enhances this effect by precisely controlling how CdS nanoparticles are deposited onto TiO_2 nanotubes. Unlike conventional wet-chemical prepared CdS/TiO_2 systems, which often lead to particle agglomeration, uneven coverage, and weak interfacial contact, this controlled deposition maximizes the interfacial contact area between CdS and TiO_2, thereby increasing the number of effective heterojunction sites available for charge separation. At the same time, the homogeneous and nanoscale dispersion of CdS clusters ensures short electron transfer distances, allowing photogenerated electrons to rapidly migrate from CdS into the conduction band of TiO_2, which suppresses electron–hole recombination. In addition, precise control over CdS loading prevents excessive coverage that could block TiO_2 nanotube channels or introduce recombination centers. As a result of these combined effects, SCBD-CdS/TiO_2 photoelectrodes exhibit significantly higher photocurrent density and improved photoelectrochemical activity compared with conventionally prepared CdS/TiO_2, where non-uniform deposition and poor interface quality limit charge separation efficiency **Figure 10** (Li et al., 2013b). Similarly, Ag/TiO_2/PVA nanocomposites benefited from Schottky junction formation and polymer stabilization, leading to enhanced interfacial charge transfer (Dehghani & Delnavaz, 2024).

Figure 10. Schematic comparison between SCBD and conventional CdS/TiO_2 NTs

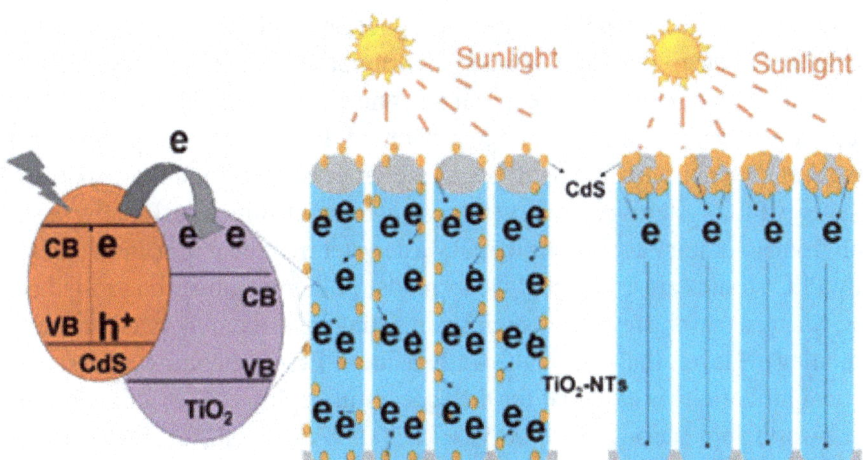

(Li et al., 2013b)

2.2.4 Challenges

Despite the significant potential of photoelectrochemistry (PEC) for water treatment and environmental remediation, several challenges still limit its large-scale application. One of the main challenges is the recombination of photogenerated electrons and holes, which reduces the efficiency of charge utilization and limits the formation of reactive species, even in systems where an external bias is applied. Although heterojunctions, cocatalysts, and nano-structuring strategies have been developed to mitigate this issue, complete suppression of recombination remains difficult. Another major challenge is the limited light utilization efficiency, particularly for wide-band-gap semiconductors such as TiO_2, which require ultraviolet light for activation. Since UV light represents only a small fraction of the solar spectrum, reliance on UV irradiation restricts the energy efficiency and practical applicability of PEC systems. While visible-light-responsive materials have been explored, many of them suffer from reduced stability, weaker redox potential, or complex synthesis procedures. Material stability and durability also present significant challenges for PEC technology. Photoelectrodes may undergo photo corrosion, surface fouling, or structural degradation during long-term operation, especially in real wastewater containing complex mixtures of organic and inorganic species. The loss of catalytic activity over repeated cycles limits the lifetime of PEC systems and increases operational costs. In addition, electrode fabrication and system complexity hinder large-scale implementation. Many high-performance PEC photoelectrodes rely on sophisticated nanostructures, noble metals, or multi-component composites, which can be expensive, difficult to reproduce, and challenging to scale up. The need for precise control of electrode thickness, morphology, and electrical contact further complicates device fabrication. Energy consumption is another concern, as PEC systems often require external electrical bias and controlled light sources. Although the applied potential improves charge separation, it increases energy input, which may offset the environmental benefits of the process. Developing self-powered or low-energy PEC systems remains an important research direction. Finally, the translation from laboratory conditions to real wastewater treatment remains challenging. Many PEC studies are conducted using model pollutants under controlled conditions, whereas real wastewater contains competing ions, natural organic matter, and fluctuating pH levels that can significantly reduce PEC efficiency. Addressing these practical limitations is essential for the successful deployment of PEC technology in real-world applications (Kudo & Miseki, 2009; Zhou et al., 2025; Ouyang et al., 2019; Dehghani & Delnavaz, 2024l Dang et al., 2023; Li et al., 2013b).

REFERENCES

Abd-Elhamid, A. I., El Morsy, S. M., Soliman, H. M. A., & El-Sayed, M. E. A. (2022). Superior adsorption performance of citrate modified graphene oxide as nanomaterial for removal of organic and inorganic pollutants from aqueous solution. *Scientific Reports*, *12*(1), 1863. DOI: 10.1038/s41598-022-13111-6 PMID: 35654871

Alsaba, M. T., Al Dushaishi, M. F., & Abbas, A. K. (2020). A comprehensive review of nanoparticles applications in the oil and gas industry. *Journal of Petroleum Exploration and Production Technology*, *10*(4), 1389–1399. Advance online publication. DOI: 10.1007/s13202-019-00825-z

. Alvizuri-Tintaya, P.A., Villena-Martínez, E.M., Avendaño-Acosta, N., Lo-Iacono-Ferreira, V.G., Torregrosa-López, J.I. and Lora-García, J. (2022). Contamination of Water Supply Sources by Heavy Metals: The Price of Development in Bolivia, a Latin American Reality. *Water*, [online] 14(21), p.3470. .DOI: 10.3390/w14213470

Apul, O. G., & Karanfil, T. (2015). Adsorption of synthetic organic contaminants by carbon nanotubes: A critical review. *Water Research*, *68*, 34–55. DOI: 10.1016/j.watres.2014.09.032 PMID: 25462715

Arias Arias, F., Guevara, M., Tene, T., Angamarca, P., Molina, R., Valarezo, A., Salguero, O., Vacacela Gomez, C., Arias, M., & Caputi, L. S. (2020). The Adsorption of Methylene Blue on Eco-Friendly Reduced Graphene Oxide. *Nanomaterials (Basel, Switzerland)*, *10*(4), 681. DOI: 10.3390/nano10040681 PMID: 32260330

Ariho, A., Aja, L., Muhammad, T., & Mohammad, L. (2024). Water pollution: causes, impacts and current efforts to address the issues of water pollution along river... *F1000Research*, [online] 13, p.1298. DOI: 10.12688/f1000research.155162.2

Arora, R. (2019). Adsorption of Heavy Metals–A Review. *Materials Today: Proceedings*, *18*, 4745–4750. DOI: 10.1016/j.matpr.2019.07.462

Asghar, N., Hussain, A., Nguyen, D. A., Ali, S., Hussain, I., Junejo, A., & Ali, A. (2024). Advancement in nanomaterials for environmental pollutants remediation: A systematic review on bibliometrics analysis, material types, synthesis pathways, and related mechanisms. *Journal of Nanobiotechnology*, *22*(1), 26. Advance online publication. DOI: 10.1186/s12951-023-02151-3 PMID: 38200605

. Babuji, P., Thirumalaisamy, S., Duraisamy, K. and Periyasamy, G. (2023). Human Health Risks due to Exposure to Water Pollution: A Review. *Water*, [online] 15(14), p.2532. .DOI: 10.3390/w15142532

Balusamy, S. R., Joshi, A. S., Perumalsamy, H., Mijakovic, I., & Singh, P.Sri Renukadevi Balusamy. (2023). Advancing sustainable agriculture: A critical review of smart and eco-friendly nanomaterial applications. *Journal of Nanobiotechnology, 21*(1), 372. Advance online publication. DOI: 10.1186/s12951-023-02135-3 PMID: 37821961

Barhoum, A., García-Betancourt, M. L., Jeevanandam, J., Hussien, E. A., Mekkawy, S. A., Mostafa, M., Omran, M. M. S., Abdalla, M., & Bechelany, M. (2022). Review on Natural, Incidental, Bioinspired, and Engineered Nanomaterials: History, Definitions, Classifications, Synthesis, Properties, Market, Toxicities, Risks, and Regulations. *Nanomaterials (Basel, Switzerland), 12*(2), 177. DOI: 10.3390/nano12020177 PMID: 35055196

Bellier, N., Baipaywad, P., Ryu, N., Lee, J. Y., & Park, H. (2022). Recent biomedical advancements in graphene oxide- and reduced graphene oxide-based nanocomposite nanocarriers. *Biomaterials Research, 26*(1), 65. Advance online publication. DOI: 10.1186/s40824-022-00313-2 PMID: 36435846

Chandran, D. G., Loganathan, M., & Biswas, R. (2023). A review on adsorption of heavy metals from wastewater using carbon nanotube and graphene-based nanomaterials. *Environmental Science and Pollution Research International, 30*(51), 110010–110046. DOI: 10.1007/s11356-023-30192-6 PMID: 37804379

Cheong, S., Watt, J. D., & Tilley, R. D. (2010). Shape control of platinum and palladium nanoparticles for catalysis. *Nanoscale, 2*(10), 2045. DOI: 10.1039/c0nr00276c PMID: 20694209

. Čitaković, N. (2019). Physical properties of nanomaterials. *Vojnotehnicki glasnik*, 67(1), pp.159–171. .DOI: 10.5937/vojtehg67-18251

Daneshyar, A., Ghaedi, M., Sabzehmeidani, M. M., & Daneshyar, A. (2017). H2S adsorption onto Cu-Zn–Ni nanoparticles loaded activated carbon and Ni-Co nanoparticles loaded γ-Al2O3: Optimization and adsorption isotherms. *Journal of Colloid and Interface Science, 490*, 553–561. DOI: 10.1016/j.jcis.2016.11.068 PMID: 27923140

Dang, V., Thamilselvan, A., Khedulkar, A., Lin, J. T., Adorna, J. A., Wang, Y., Pandit, B., Huynh, T., & Doong, R. (2023). S-scheme N-doped carbon dots anchored g-C3N4/Fe2O3 shell/core composite for photoelectrocatalytic trimethoprim degradation and water splitting. 320, pp.121928–121928. DOI: 10.1016/j.apcatb.2022.121928

. Dehghani, M.T. and Delnavaz, M. (2024). UV-light-responsive Ag/TiO2/PVA nanocomposite for photocatalytic degradation of Cr, Ni, Zn, and Cu heavy metal ions. *Scientific Reports*, [online] 14(1). .DOI: 10.1038/s41598-024-56059-5

Fan, K., Li, Y., Elsworth, D., Dong, M., Yin, C., Li, Y., & Chen, Z. (2018). Three stages of methane adsorption capacity affected by moisture content. *Fuel*, *231*, 352–360. DOI: 10.1016/j.fuel.2018.05.120

Gankanda, A., Cwiertny, D. M., & Grassian, V. H. (2016). Role of Atmospheric CO_2 and H_2O Adsorption on ZnO and CuO Nanoparticle Aging: Formation of New Surface Phases and the Impact on Nanoparticle Dissolution. *Journal of physical chemistry. C. The Journal of Physical Chemistry. C, Nanomaterials and Interfaces*, *120*(34), 19195–19203. DOI: 10.1021/acs.jpcc.6b05931

Ghasemi, Z., Seif, A., Ahmadi, T. S., Zargar, B., Rashidi, F., & Rouzbahani, G. M. (2011). Thermodynamic and kinetic studies for the adsorption of Hg(II) by nano-TiO2 from aqueous solution. *Advanced Powder Technology*, *23*(2), 148–156. DOI: 10.1016/j.apt.2011.01.004

Guo, M., Li, M., Fu, H., Zhang, Y., Chen, T., Tang, H., Zhang, T., & Li, H. (2023). Quantitative analysis of polycyclic aromatic hydrocarbons (PAHs) in water by surface-enhanced Raman spectroscopy (SERS) combined with Random Forest. *Spectrochimica Acta. Part A: Molecular and Biomolecular Spectroscopy*, *287*, 122057. DOI: 10.1016/j.saa.2022.122057 PMID: 36332395

Han, C., Kim, H., & Yang, H. (2020). Quantum Dots and Applications. *Materials (Basel)*, *13*(4), 897. DOI: 10.3390/ma13040897 PMID: 32085387

Heiligtag, F. J., & Niederberger, M. (2013). The fascinating world of nanoparticle research. *Materials Today*, *16*(7-8), 262–271. DOI: 10.1016/j.mattod.2013.07.004

Iijima, S. (1991). Helical microtubules of graphitic carbon. *Nature*, *354*(6348), 56–58. DOI: 10.1038/354056a0

Irshad, M., Ain, Q., Zaman, M., Aslam, M. Z., Kousar, N., Asim, M., Rafique, M., Siraj, K., Tabish, A. N., Usman, M., Hassan Farooq, M., Assiri, M. A., & Imran, M. (2022). Photocatalysis and perovskite oxide-based materials: A remedy for a clean and sustainable future. *RSC Advances*, *12*(12), 7009–7039. DOI: 10.1039/D1RA08185C PMID: 35424711

Jahan, N., Roy, H., & Reaz, A. H. Shasawti Arshi, Rahman, E., Firoz, S.H. and Islam, M.S. (2022). A comparative study on sorption behavior of graphene oxide and reduced graphene oxide towards methylene blue. *Case Studies in Chemical and Environmental Engineering*, 6, pp.100239–100239. DOI: 10.1016/j.cscee.2022.100239

Jeanmarie, D. L. (1977). Surface Raman spectroelectrochemistry Part I. Heterocyclic, aromatic, and aliphatic amines adsorbed on the anodized silver electrode. *Journal of Electroanalytical Chemistry*, *84*(1), 1–20. DOI: 10.1016/0368-1874(77)80399-7

. Jeevanandam, J., Barhoum, A., Chan, Y.S., Dufresne, A. and Danquah, M.K. (2018). Review on Nanoparticles and Nanostructured materials: history, sources, Toxicity and Regulations. *Beilstein Journal of Nanotechnology*, [online] 9(1), pp.1050–1074. .DOI: 10.3762/bjnano.9.98

Ji, J., Lu, Z., Lei, Y., & Turner, C. H. (2018). Mechanistic insights into the direct propylene epoxidation using Au nanoparticles dispersed on TiO2/SiO2. *Chemical Engineering Science*, *191*, 169–182. DOI: 10.1016/j.ces.2018.06.064

. Joudeh, N. and Linke, D. (2022). Nanoparticle classification, physicochemical properties, characterization, and applications: a comprehensive review for biologists. *Journal of Nanobiotechnology*, [online] 20(1). .DOI: 10.1186/s12951-022-01477-8

Khorshidi, S., Sabbaghi, S., & Saboori, R. (2020). Removal of hydrogen sulfide from water-based drilling fluid using titania/carbon nanotubes nano-hybrid: Optimization, kinetics, and isotherms. *Desalination and Water Treatment*, *200*, 154–166. DOI: 10.5004/dwt.2020.26089

Kudo, A., & Miseki, Y. (2009). Heterogeneous photocatalyst materials for water splitting. *Chemical Society Reviews*, *38*(1), 253–278. DOI: 10.1039/B800489G PMID: 19088977

Kulkarni, S., Kaware, J., & Mumbai, N. (2014). Regeneration and Recovery in Adsorption- a Review. *International Journal of Innovative Science. Engineering & Technology*, *1*(8), 61–64.

. Kumari, S. & Sarkar, L. (2021). A Review on Nanoparticles: Structure, Classification, Synthesis & Applications. *Journal of Scientific Research,*65(8), pp.42-46.

Li, G., Wu, L., Li, F., Xu, P., Zhang, D., & Li, H. (2013b). Photoelectrocatalytic degradation of organic pollutants via a CdS quantum dots enhanced TiO2 nanotube array electrode under visible light irradiation. *Nanoscale*, *5*(5), 2118. DOI: 10.1039/c3nr34253k PMID: 23381869

. Li, P., Karunanidhi, D., Subramani, T. and Srinivasamoorthy, K. (2021). Sources and Consequences of Groundwater Contamination. *Archives of Environmental Contamination and Toxicology*, [online] 80(1), pp.1–10. .DOI: 10.1007/s00244-020-00805-z

Lin, L., Yang, H., & Xu, X. (2022). Effects of Water Pollution on Human Health and Disease Heterogeneity: a Review. *Frontiers in Environmental Science*, [online] 10. Available at: https://www.frontiersin.org/journals/environmental-science/articles/10.3389/fenvs.2022.880246/full

Liu, Z., Wang, Q., Huang, X., & Qian, X. (2022). Surface Functionalization of Graphene Oxide with Hyperbranched Polyamide-Amine and Microcrystalline Cellulose for Efficient Adsorption of Heavy Metal Ions. *ACS Omega*, *7*(13), 10944–10945. https://pubs.acs.org/journal/acsodf. DOI: 10.1021/acsomega.1c06647 PMID: 35415369

Luo, X., Liu, Y., Wang, X., Li, Y., & Zhang, L. (2022). Removal of methylene blue by crosslinked egg white protein/graphene oxide bionanocomposite aerogels. *Nanomaterials (Basel, Switzerland)*, *12*(3), 487. PMID: 35159833

Mogashane, T. M., Maree, J. P., & Mokoena, L. (2024). Adsorption of polycyclic aromatic hydrocarbons from wastewater using iron oxide nanomaterials recovered from acid mine water: A review. *Minerals (Basel)*, *14*(8), 826.

Mohammed, M. I., Ismael, M. K., & Gönen, M. (2020). Synthesis of Chitosan-Silica Nanocomposite for Removal of Methyl Orange from Water: Composite Characterization and Adsorption Performance. *IOP Conference Series. Materials Science and Engineering*, *745*(1), 012084. DOI: 10.1088/1757-899X/745/1/012084

Nuengmatcha, P., Mahachai, R., & Chanthai, S. (2014). Thermodynamic and kinetic study of the intrinsic adsorption capacity of graphene oxide for malachite green removal from aqueous solution. *Oriental Journal of Chemistry*, *30*(4), 1463–1474. DOI: 10.13005/ojc/300403

Olawade, D. B., Wada, O. Z., Egbewole, B. I., Fapohunda, O., Ige, A. O., Usman, S. O., & Ajisafe, O. (2024). Metal and metal oxide nanomaterials for heavy metal remediation: Novel approaches for selective, regenerative, and scalable water treatment. *Frontiers in Nanotechnology*, *6*, 1466721. Advance online publication. DOI: 10.3389/fnano.2024.1466721

Ouyang, W., Teng, F., He, J., & Fang, X. (2019). Enhancing the Photoelectric Performance of Photodetectors Based on Metal Oxide Semiconductors by Charge-Carrier Engineering. *Advanced Functional Materials*, *29*(9), 1807672–1807672. DOI: 10.1002/adfm.201807672

Puri, C., & Sumana, G. (2018). Highly effective adsorption of crystal violet dye from contaminated water using graphene oxide intercalated montmorillonite nanocomposite. *Applied Clay Science*, *166*, 102–112. DOI: 10.1016/j.clay.2018.09.012

Rezania, S., Darajeh, N., Rupani, P. F., Mojiri, A., Kamyab, H., & Taghavijeloudar, N. (2024). Recent Advances in the Adsorption of Different Pollutants from Wastewater Using Carbon-Based and Metal-Oxide Nanoparticles. *Applied Sciences (Basel, Switzerland)*, *14*(24), 11492–11492. DOI: 10.3390/app142411492

Sorkin, V., & Zhang, Y. W. (2011). Graphene-based pressure nano-sensors. *Journal of Molecular Modeling*, *17*(11), 2825–2830. DOI: 10.1007/s00894-011-0972-0 PMID: 21287214

Toshima, N. (2013). Metal nanoparticles for energy conversion. *Pure and Applied Chemistry*, *85*(2), 437–451. DOI: 10.1351/PAC-CON-12-08-17

. Walter, P., Welcomme, E., Hallégot, P., Zaluzec, N.J., Deeb, C., Castaing, J., Veyssière, P., Bréniaux, R., Lévêque, J.-L. and Tsoucaris, G. (2006). Early use of PbS nanotechnology for an ancient hair dyeing formula. *Nano Letters*, [online] 6(10), pp.2215–2219. .DOI: 10.1021/nl061493u

Wang, C., Zhou, J., Chu, L., Zhang, M., Xu, C., Liu, J., & Li, S. (2024). Diethylenetriamine-functionalized reduced graphene oxide having more amino groups for methylene blue removal. *RSC Advances*, *14*(5), 3280–3288. DOI: 10.1039/D3RA08767K PMID: 38249680

Wang, L., Chen, Y., Han, X., Zhang, J., & Zhao, L. (2024). Diethylenetriamine-functionalized reduced graphene oxide having more amino groups for methylene blue removal. *Journal of Water Process Engineering*, *59*, 103652. PMID: 38249680

Willner, M. R., & Vikesland, P. J. (2018). Nanomaterial enabled sensors for environmental contaminants. *Journal of Nanobiotechnology*, *16*(1), 95. Advance online publication. DOI: 10.1186/s12951-018-0419-1 PMID: 30466465

. Wood, B.C., Bhide, S.Y., Dutta, D., Kandagal, V.S., Amar Deep Pathak, Punnathanam, S.N., Ayappa, K.G. and Narasimhan, S. (2012). Methane and carbon dioxide adsorption on edge-functionalized graphene: A comparative DFT study. 137(5), pp.054702–054702. .DOI: 10.1063/1.4736568

Xu, P., Zeng, G. M., Huang, D. L., Lai, C., Zhao, M. H., Wei, Z., Li, N. J., Huang, C., & Xie, G. X. (2012). Adsorption of Pb(II) by iron oxide nanoparticles immobilized Phanerochaete chrysosporium: Equilibrium, kinetic, thermodynamic and mechanisms analysis. *Chemical Engineering Journal*, *203*, 423–431. DOI: 10.1016/j.cej.2012.07.048

Yagub, M. T., Sen, T. K., Afroze, S., & Ang, H. M. (2014). Dye and its removal from aqueous solution by adsorption: A review. *Advances in Colloid and Interface Science*, *209*, 172–184. DOI: 10.1016/j.cis.2014.04.002 PMID: 24780401

Yusuf, A., Almotairy, A. R. Z., Henidi, H., Alshehri, O. Y., & Aldughaim, M. S. (2023). Nanoparticles as drug delivery systems: A review of the implication of nanoparticles' physicochemical properties on responses in biological systems. *Polymers*, *15*(7), 1596. DOI: 10.3390/polym15071596 PMID: 37050210

Zhou, P., Xu, Y., & Cui, T. (2025). *A New Photoelectrocatalytical Water Purification System for Simultaneous Removal of Organic Pollutants and Heavy Metal Ions in Water*. Research., DOI: 10.34133/research.0654

Chapter 4
General Overview of Wastewater Types, Treatment Processes, and Sludge Valorization

El Mokhtar El Hafidi
https://orcid.org/0000-0003-0840-8572
Laboratory of Engineering Science for Energy (labSIPE) ENSA, Chouaib Doukkali University, El Jadida, Morocco

El Ghaouti Chahid
Polydisciplinary faculty of Sidi Bennour, Chouaib Doukkali University, El Jadida, Morocco

Said Laasri
https://orcid.org/0000-0002-2352-6828
Polydisciplinary faculty of Sidi Bennour, Chouaib Doukkali University, El Jadida, Morocco

ABSTRACT

The growing demand for clean water and increasing pollution from domestic, industrial, and agricultural sources necessitate effective wastewater treatment strategies. This study provides a comprehensive overview of conventional treatment methods and introduces impedance and dielectric spectroscopy as complementary tools for water quality analysis. It details wastewater classification and treatment stages—pretreatment, primary, secondary, and tertiary—and highlights the valorization of industrial by-products such as fly ash and bottom ash for pollutant removal. Emphasis is placed on water reuse in agriculture, industry, and urban sectors, as well as on sludge management and its integration into eco-friendly mortars. The work also

DOI: 10.4018/979-8-3373-5886-4.ch004

explores desalination technologies, including thermal and membrane processes, and their synergy with renewable energies. Impedance and dielectric spectroscopy offer non-destructive, frequency-dependent analyses, enhancing understanding of treatment efficiency and pollutant behavior. This dual approach promotes sustainability and resource recovery.

INTRODUCTION

Water, essential not only for industrial processes but also for the health of ecosystems and populations, often becomes contaminated with various chemical pollutants during its use in industrial and domestic activities. This contamination endangers aquatic environments and public health, making wastewater treatment indispensable to mitigate these impacts (Organization, 2022). However, this treatment also generates a significant amount of residual sludge, whose proper management is crucial (Chen et al., 2013).

This chapter is structured to explore the various dimensions of wastewater and sludge treatment. It begins with an analysis of domestic, industrial, and agricultural wastewater, highlighting their diverse characteristics and the specific challenges of their treatment. Domestic wastewater, originating from kitchens and toilets, differs from industrial wastewater, which varies significantly depending on the processes and products used (Pandey et al., 2021). Agricultural wastewater, influenced by the use of fertilizers and pesticides, along with stormwater capable of carrying various urban contaminants, adds to the complexity of treatment. Notably, wastewater from the detergent industry presents unique challenges due to the chemical compounds it contains (Bellarbi et al., 2017).

Wastewater treatment involves several stages: pretreatment, primary, secondary, and tertiary treatment, all aimed at removing pollutants and producing water that meets environmental standards. Additionally, the valorization of industrial by-products, such as fly ash from thermal power plants, offers opportunities to enhance wastewater treatment and reduce environmental impacts (Mortadi et al., 2022). This section of the chapter examines these stages in detail and explores innovative methods for efficiently and ecologically managing residual sludge.

The transition to renewable energy in desalination, which is essential for providing potable water in arid regions, represents a major opportunity to make this process more sustainable and cost-effective. Desalination technologies, such as multi-effect distillation, reverse osmosis, and multi-stage flash distillation, are traditionally energy-intensive and reliant on fossil fuels. Integrating renewable energy sources, such as solar, wind, or geothermal energy, helps reduce this dependence and optimize operational costs. This section analyzes the challenges and opportunities associated with this energy transition, exploring how these renewable energy sources can be effectively integrated into desalination processes (Ghaffour et al., 2015; Rovel, 2010).

Finally, impedance/dielectric spectroscopy proves to be a valuable technique for analyzing the electrical properties of materials, whether solid or liquid. Used in various fields, from electrochemistry to microelectronics, this technique enables material characterization based on dielectric properties such as dielectric constant, relaxation time, and electrical conductivity (Mortadi et al., 2019, 2020). Furthermore, wastewater assessment relies on conventional analytical methods based on key parameters such as suspended solids, biochemical and chemical oxygen demand, temperature, pH, and electrical conductivity (Chahid et al., 2013). These traditional methods are essential for water quality control and the design of treatment systems, and this section of the chapter details their critical aspects.

A- DOMESTIC WASTEWATER, INDUSTRY, AND RESIDUAL SLUDGE

I. General Overview of Domestic and Industrial Wastewater

I.1 Origins of wastewater

Wastewater is classified into four main categories based on its sources: domestic, industrial, agricultural, and stormwater.

I.2 Wastewater Treatment Stages

I.2.1 Wastewater Pretreatment

Screening:

Screening is a fundamental step in wastewater treatment, aimed at retaining large debris before the main treatment process. This is achieved by filtering materials through a screen, with the spacing determining the size of retained objects. Typically, screens have openings of about 10 millimeters, allowing them to capture items such as sticks and debris (See Figure 1). This stage is crucial as it removes a significant portion of pollutants. The collected waste is then directed to appropriate facilities for disposal.

Figure 1. Example of screening

Degreasing

Degreasing is the process of separating and removing fats and oils present in wastewater. It is typically carried out during the pretreatment stage of water treatment facilities to prevent the accumulation of grease and to avoid clogging issues in subsequent equipment (see Figure 2).

Figure 2. Example of degreasing in the pretreatment of a wastewater treatment plant

I.2.2 Primary Treatment

Sedimentation

Sedimentation, which occurs after pretreatment, is a crucial step in the wastewater treatment process. Its goal is to remove a portion of the remaining impurities in the water, particularly suspended solids and dissolved pollutants. The particles present vary in nature and density, making the treatment complex and requiring specific techniques. Primary treatment relies on the principle of gravitational sedimentation, where water flows slowly through settling tanks (see Figure 3). This slow movement allows settleable particles to accumulate at the bottom, forming primary sludge, while clarified water collects at the surface (Arslan-Alaton et al., 2008).

Figure 3. Principle of a settling basin

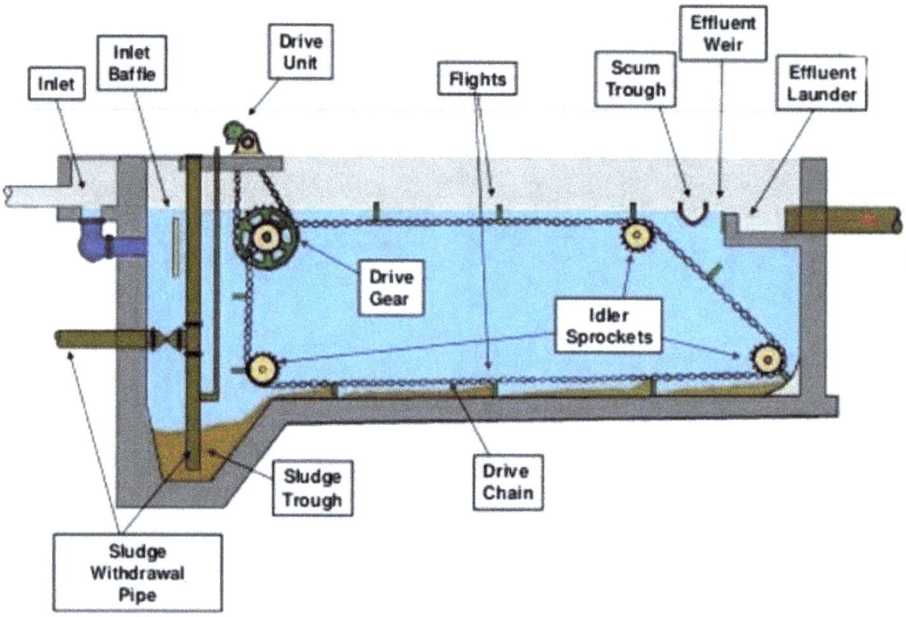

Flotation

Flotation is a process designed to separate solid phases from liquids by leveraging Archimedes' principle. In natural flotation, low-density particles form aggregates that rise to the surface. In contrast, assisted flotation uses injected air to generate bubbles that attach to solid particles, causing them to float. This method is particularly effective for removing grease: the injected air at the bottom of the system pushes the grease to the surface, where it is skimmed off and stored before disposal.

I.2.3. Secondary Treatment

Physicochemical treatment:

Physicochemical wastewater treatments, such as filtration and adsorption, effectively remove contaminants by trapping solid particles and biological organisms. Filtration purifies water by passing it through a porous medium like sand or activated carbon. Adsorption captures pollutants on the surface of a solid material, such as activated carbon, eliminating organic compounds and heavy metals. These methods are often combined with chemical additives, as in coagulation/flocculation, where particles aggregate and then settle. Using these techniques, wastewater is clarified and freed from suspended impurities. The terms coagulant and flocculant are sometimes confused; the former refers to chemicals that promote particle aggregation, while the latter describes the aggregation process itself [27]. Their definitions are presented below.

Coagulant:

A coagulant is a chemical substance, such as potassium alum or aluminum potassium sulfate, that induces the aggregation of fine particles in wastewater. This process facilitates the formation of a gelatinous precipitate, which can then be removed (see Figure 4). Coagulation works by neutralizing the electrical charges of suspended particles, allowing them to come together and form larger aggregates known as flocs. These flocs can then be separated from the water through sedimentation or filtration. Coagulants are essential in water and wastewater treatment processes, as they enhance the efficiency of subsequent purification steps, such as flocculation and sedimentation, by reducing turbidity and removing organic matter, heavy metals, and pathogens.

Figure 4. Simplified schematic of the coagulation process used in drinking water treatment

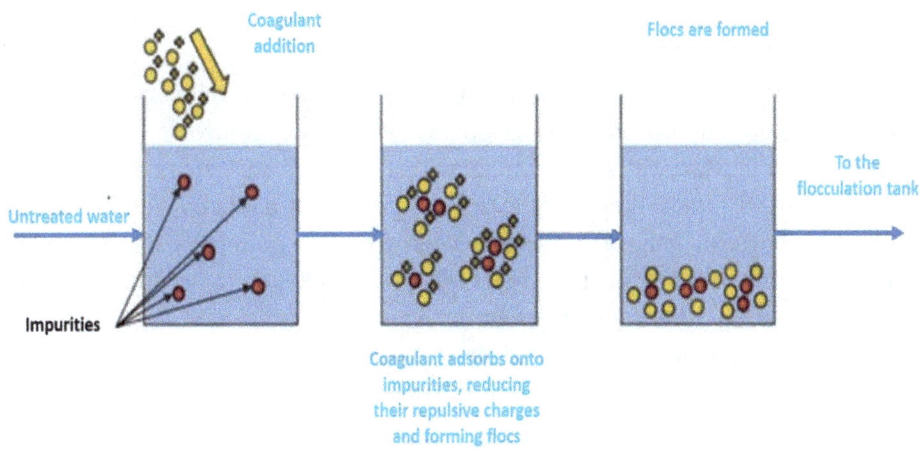

Flocculant:

This is a polymer that captures agglomerated colloidal matter, forming large flocs that settle by gravity. It is added after coagulation to further increase the size and cohesion of the flocs.

The coagulation-flocculation process can be used to remove various organic and inorganic substances, including fats, oils, phosphorus, suspended solids (SS), heavy metals, etc. . (EPA, 2000). Additionally, it also helps reduce bacterial populations.

However, a specific approach tailored to the type of effluent being treated must be developed to optimize the process. The required amounts of coagulant depend primarily on factors such as pH, alkalinity, conductivity, and the coagulant injection point (Casey & Casey, 1997). Once the process is complete, the precipitate, composed of solid micro-flocs, can be filtered or centrifuged for separate collection from the liquid effluent.

The resulting sludge can be directly landfilled or treated through various methods, including dewatering, incineration, chemical oxidation, and/or biological treatment. The choice of an appropriate coagulant or flocculant mainly depends on two factors: its effectiveness and its cost compared to other alternatives (Wang et al., 2004). Several types of coagulants and flocculants can be distinguished:

- ✓ Chemical coagulants and flocculants
- ✓ Metal salt-based coagulants
- ✓ Calcium oxide (CaO - lime)
- ✓ Alum ($Al_2(SO_4)_3 \cdot 14H_2O$)
- ✓ Ferric chloride ($FeCl_3$)

Biological Treatment:

Biological treatment replicates, within specific reactors, the self-purification process that naturally occurs in water bodies. The principle involves bringing the organic matter in wastewater into contact with a bacterial mass that consumes pollutants and breaks them down. There are two treatment methods: aerobic, which uses aerobic bacteria in the presence of oxygen to oxidize pollutants, and anaerobic, which relies on anaerobic bacteria in the absence of oxygen to decompose biodegradable compounds through fermentation. This treatment can be carried out with suspended microorganisms (activated sludge) or on supports such as plastic media (trickling filter) or biological discs.

I.3. Valorization of Industrial By-Products in Wastewater Treatment

I.3.1. Combustion Residues From Coal Incineration

I.3.1.1. Fly Ash

Physicochemical and mineralogical characteristics of fly ash

Fly ash exhibits varying physical, mechanical, and chemical properties depending on the different combustion processes used.

- **Physical Characteristics**

Fly ash is a powder with variable colors, ranging from light gray to dark gray or light gray to dark brown, depending on its carbon and iron oxide content. Its surface appears smooth due to its spherical structure, as illustrated in Figure 5. The density of fly ash typically ranges between 2.1 and 3 (Ahmaruzzaman, 2010; Chatterjee, 2011).

Figure 5. Morphology of fly ash

The main physical characteristics of silico-aluminous fly ash are presented in Table 1 below (Z Rakotoarisoa, L Tiruta-Barna, n.d.).

Table 1. Physical characteristics of silico-aluminous fly ash

Particle Size	Minimum Dimension	µm	0.5 – 5
	Maximum Dimension	µm	100 - 200
Density	Bulk Density		0.5 - 0.8
	Dry Density (After Compaction)		1 - 1.3
	Particle Density	g/cm^3	1.8 - 2.1
Specific Surface Area		Cm^2/g	2000 - 4500
Permeability (Proctor Density)		m/s	Around 10^{-7}

The granulometry and fineness of fly ash vary depending on combustion and desulfurization processes. Silico-aluminous fly ash generally has larger particle sizes than calcium-rich fly ash, but both types contain fine particles in similar proportions. The granulometric distribution of silico-aluminous fly ash is broader. Additionally, the specific surface area of calcium-rich fly ash is significantly higher due to the use of very fine lime as a desulfurizing agent (Fayet et al., 1999).

■ Chemical Characteristics

The chemical composition of fly ash is influenced by the type of coal used and the combustion process, including the presence of unburned carbon. Silica- and alumina-based fly ash predominantly contains these components, with low amounts of alkali and alkaline earth oxides, particularly calcium oxide in its combined form. The data in Table 2 summarize the results of chemical composition analyses of newly produced fly ash, which is primarily rich in silica and alumina (Adamiec et al., 2011).

Table 2. Major element composition of silico-aluminous fly ash

Element (%)	French Average	Dutch Average	Typical Composition
Al (as Al_2O_3)	27.4	26.1	22 - 32
Ca (as CaO)	2.8	3.4	1 - 8
Fe (as Fe_2O_3)	8.0	6.0	4 - 15
K (as K_2O)	2.5	1.3	2 - 5
Mg (as MgO)	1.9	1.3	1 - 3
Na (as Na_2O)	--	0.7	--
P (as P_2O_5)	--	1.6	--
S (as SO_3)	0.48	--	0.2 - 2
Si (as SiO_2)	53.9	49.4	43 - 54
Ti (as TiO_2)	--	1.3	--

- **Mineralogical Characteristics**

The mineralogical phases in fly ash generally originate from unburned minerals in coal, except in cases where limestone is injected to meet emission standards. These minerals may be inherited from coal, formed during combustion, or transformed within the boiler. Adamiec examined the relationships between these phases and coal minerals, as well as their transformation processes. The results indicate that quartz in fly ash is an inherited mineral, while the thermal treatment of kaolinite leads to the formation of mullite and cristobalite. Other coal minerals also undergo transformations during thermal processing, highlighting the complex processes involved in the formation of mineralogical phases in fly ash (Adamiec, 1998).

I.3.1.2. Bottom Ash

Bottom ash, also known as furnace ash, consists of solid residues from coal combustion in thermal power plants. Unlike fine fly ash collected by electrostatic precipitators, bottom ash has a granulometry similar to sand and is collected at the boiler outlet before being water-cooled. The term “mâchefer” originates from the phrase “mâche le fer”, historically referring to the agglomerates of furnace ash formed during the combustion of coal with low-melting-point ash, which caused corrosion of the grate bars in coal-fired furnaces (El Alami et al., 2011).

Physical characteristics and chemical properties

- **Physical characteristics**

The physical characteristics of bottom ash are influenced by the type and quality of coal used, as well as the operating conditions of the power plant. Typically, bottom ash particles are dark, angular, and exhibit porous textures, as illustrated in Figure 6.

Figure 6. Morphology of bottom ash

- **Chemical characteristics**

Bottom ash has a higher density than natural sand or gravel. Its main components include silicon dioxide (SiO_2), aluminum oxide (Al_2O_3), iron oxide (Fe_2O_3), calcium oxide (CaO), and carbon (C). Other elements are also present but generally in low concentrations (Herman et al., 2016; Obernberger & Supancic, 2009).

I.3.2. Use of Coal Ash in Wastewater Treatment

Heavy metal removal from wastewater

Fly ash, rich in alumina, silica, ferric oxide, calcium oxide, magnesium oxide, and carbon, presents promising applications in wastewater treatment. Its physical properties, such as porosity and specific surface area, enhance its efficiency. Numerous studies have documented its capability to remove heavy metals from wastewater (El Hafidi et al., 2023). Table 3 summarizes the results obtained regarding the removal of major heavy metals using fly ash.

Table 3. Summary of metal adsorption on fly ash

Metal	Adsorbent	Adsorption Capacity (mg/g)	Reference
As	fly ash	11	(Diamadopoulos et al., 1993)
Cr	fly ash	23.86	(Bhattacharya et al., 2008)
Cu	fly ash	0.63-0.81	(Janoš et al., 2003)
Zn	fly ash	6.5-13.3	(Ricou et al., 1999)

Removal of organic compounds

Various methods are used to remove pollutants from wastewater. Among them, coal ash demonstrates promising adsorption potential for reducing organic matter. Table 4 summarizes the adsorption capacity of various organic compounds on fly ash.

Table 4. Summary of organic pollutant adsorption on fly ash

Organic Compound	Adsorbent	Adsorption Capacity (mg/g)	Reference
2,4-dichlorophenol	fly ash	22	(Kao et al., 2000)
Cresol	fly ash	85.4–96.4	(Dutta et al., 2003)

Removal of dyes from wastewater

Various methods are available for treating dyes present in wastewater. Among them, adsorption using fly ash and bottom ash is an interesting alternative. Table 5 summarizes the adsorption capacity of different dyes on fly ash and bottom ash.

Table 5. Adsorption capacity of fly ash for dyes

Dye	Adsorption Capacity (mol/g)	Adsorbent	Reference
Methylene Blue	$1.89\ 10^{-5}$	Fly ash	(Janoš et al., 2003)
Crystal Violet	$9.76\ 10^{-5}$	Fly ash	(Mohan et al., 2002)

II. General Overview of Residual Sludge

1. Composition

1.1. Water

Water constitutes the majority of sludge, typically accounting for 95% to 97% after initial treatment (Colin & Gazbar, 1995). Its removal is therefore essential. From a physico-chemical perspective, water in sludge behaves differently depending on its interaction with the surrounding material. This interaction leads to the distinction of different water categories, such as "free" water and "bound" water, which have been the subject of extensive scientific research (Goodwin & Forster, 1985; Smith & Vesilind, 1995).

1.2. Solid Matter

The solid phase of sludge has a diverse composition, with organic matter content ranging from 30% to 80% of the dry matter (Ademe, 2001). It includes lipids, fibers (cellulose, hemicellulose, lignin), proteins, and sugars (starch, simple sugars, pectins) (Dieudé-Fauvel & Dentel, 2011), with a significant portion composed of polysaccharides, proteins, amino acids, lignin, metabolic by-products, and microbial bodies.

Sludge also contains mineral substances such as nitrogen, phosphorus, potash, magnesia, lime, as well as metallic ions in the form of trace elements, including cadmium, chromium, copper, mercury, nickel, lead, selenium, and zinc.

2. Floc structure

The structure of sludge has been studied using various techniques, such as microscopy, micro-tomography, and laser granulometry. Flocs, which are aggregates of the main sludge components, including microorganisms and polymers, serve as the fundamental building blocks of this architecture, as illustrated in Figure 7. Flocs exhibit a multi-level organization, consisting of primary particles, microflocs forming a local structure, and a tertiary structure where particles are bound by exopolymers, following the concept of fractal dimension, as shown in Figure 8. These flocs act as water reservoirs, containing the different forms of matter present in sludge (Léonard, 2003).

Figure 7. Schematic representation of a sludge floc

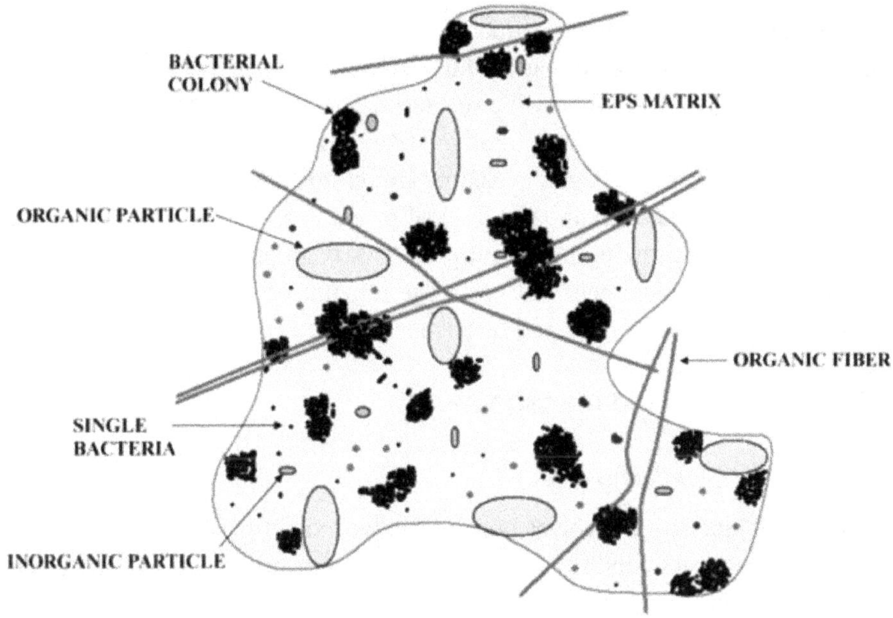

Figure 8. Different structural levels of sludge flocs

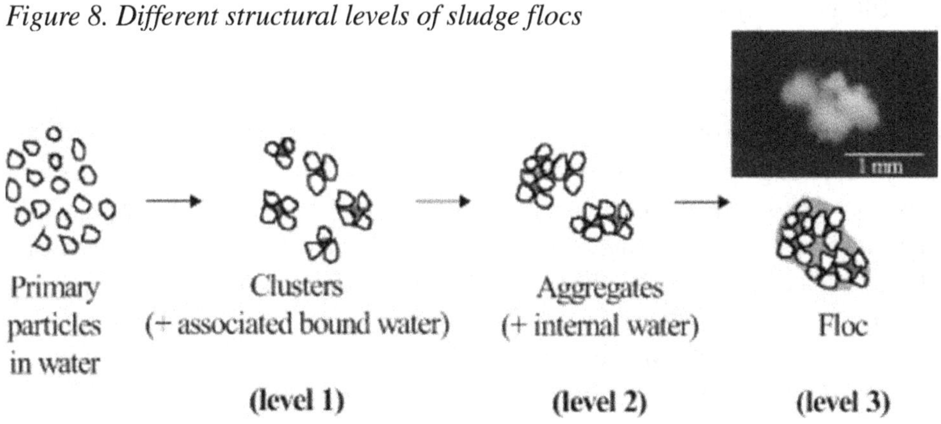

3. Physical Properties of Residual Sludge

Before dewatering, sludge is primarily composed of water, accounting for more than 95%, along with organic matter and bacteria that form flocs. These flocs are interactive suspended particles in which bacteria secrete extracellular polymers (EPS) that adsorb onto organic matter, acting as gels (Keiding et al., 2001; Sutherland, 2001).

Flocs generally carry a negative charge, similar to colloidal suspensions, which influences their mechanical and rheological properties. Controlling these properties is crucial for optimizing treatment processes (Campbell & Crescuolo, 1982). However, the rheological behavior of sludge remains complex to understand due to composition fluctuations caused by fermentation and the origin of biomaterials.

4. Sludge Control and Management

4.1 Controlled Landfilling

Controlled landfilling involves burying sludge, sometimes mixed with municipal waste, under strict conditions such as waste compaction, site sealing, leachate management, and site monitoring (Raynaud, 2010). Before disposal, sludge must be stabilized and dewatered to a maximum moisture content of 70%. However, this method is increasingly being abandoned due to financial costs and environmental impacts, including odor nuisances, insect proliferation, and the contamination of surface and groundwater by toxic elements (Allen, 2001; Kjeldsen et al., 2002). Modern landfills now only accept ultimate waste considered non-recyclable or hazardous.

4.2 Incineration

Incineration involves burning waste at high temperatures, producing heat that can be used to generate steam or electricity (Marttinen et al., 2003). In France, approximately 14% to 16% of urban sludge is incinerated, although this method has significant financial constraints and environmental concerns, such as the need to mix sludge with other waste to enhance combustion. In Morocco, sludge incineration has not yet been implemented.

Additionally, the disposal of incineration residues requires proper management. While incineration significantly reduces waste volume, it is criticized for emitting toxic gases and its environmental impact, leading to strict regulations to mitigate these harmful effects (Mininni et al., 2004; Prévot, 2000).

4.3 Agricultural Use of Sludge

The agricultural valorization of residual sludge is considered the most suitable recycling method to restore biogeochemical cycles (C, N, P), protect the environment, and offer substantial economic benefits. It aims to preserve natural resources and prevent organic matter wastage from incineration or landfill disposal (Nammari et al., 2004). Residual sludge can serve as a substitute for or reduce the excessive use of expensive fertilizers.

4.4 Valorization of Wastewater Filtration Sludge in Mortar Production.

In the current context of sustainable development and efficient resource management, the valorization of industrial by-products and waste has become a major concern. Among these underutilized resources are wastewater filtration sludge, often perceived as bulky and costly waste to dispose of. However, recent research has revealed its untapped potential in various fields, particularly in construction.

This study explores the use of sludge from wastewater filtration on different substrates such as sand, fly ash, and bottom ash in mortar production. This innovative approach seeks to transform a by-product often considered harmful into a valuable resource, providing both an ecological and economic solution for the construction industry. By examining the physico-chemical and mechanical properties of mortar mixtures containing these sludges, the study aims to assess their technical feasibility and environmental impact, paving the way for new sustainable practices in the construction sector (Graich et al., 2021; Mghaiouini et al., 2020).

B- USE OF RENEWABLE ENERGY IN DESALINATION

I. Main Desalination Technologies

There are 27 desalination processes, categorized into three major families: **distillation and freezing**, **membrane-based processes**, and **ion exchange and selective solvent extraction**. Among these, **distillation** and **reverse osmosis** are the most effective and widely adopted for seawater desalination on the global market. Other techniques have seen limited development due to high energy demands and substantial investment costs.

Desalination methods for seawater and brackish water can be classified into the following categories, as illustrated in Figure 9:

Despite advances in desalination technologies, energy consumption remains a critical challenge. Integrating **renewable energy sources**, such as solar, wind, and geothermal energy, into desalination processes is a promising avenue for reducing costs and environmental impacts while improving access to freshwater in arid regions.

Figure 9. Main desalination techniques

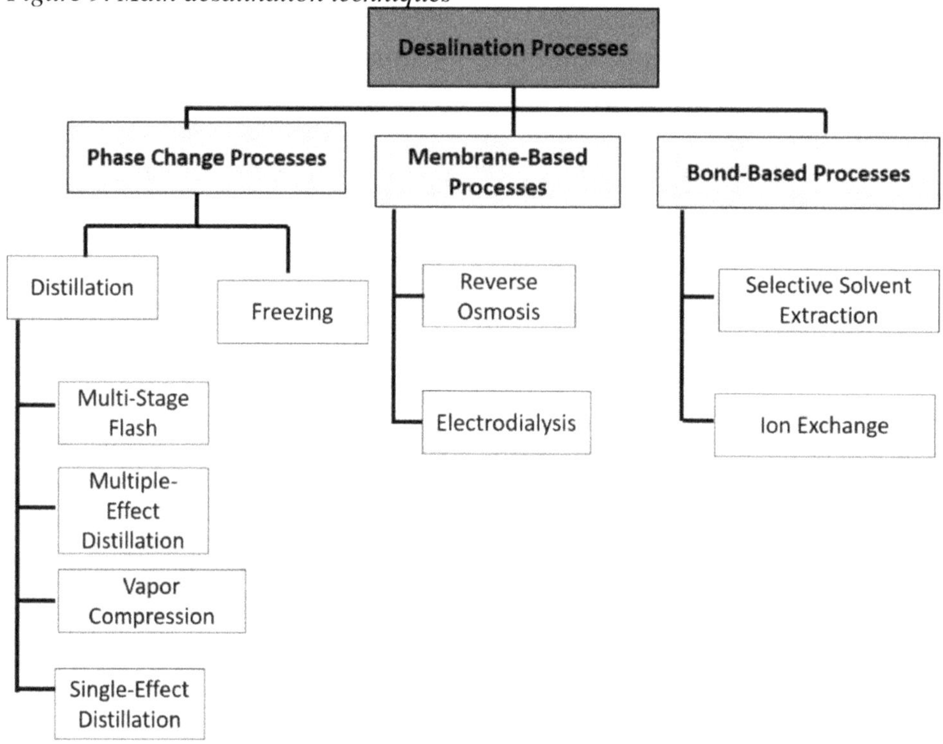

II. Distillation Processes

II.1. Single-Effect Distillation

The single-effect thermal desalination system proposed for distillation is outlined in Figure 10. It is a compact unit featuring a sleek design and constructed from inox (stainless steel) with a thickness of 3mm. This innovative system utilizes solar energy to desalinate water.

The evaporator tank is supplied with seawater through the hose. Subsequently, the seawater is heated to its boiling point by circulating hot steam through two shielded resistors positioned within the evaporator tank.

The resulting ascending vapor progresses towards the condenser tank. Within the forced-ventilation condenser chamber, a fan compels air through the condenser, causing the vapor to condense into liquid. This conversion process turns the water vapor into pure liquid water. The distilled water is then collected and stored in a dedicated storage tank. After each distillation cycle, any residual salt in the evaporator tank is eliminated or removed.

While simple in principle, single-effect distillation is highly energy-intensive, as it requires significant thermal input to maintain the phase change process. This limitation has led to the development of more advanced multi-effect and vapor compression distillation techniques, which enhance energy efficiency by reusing heat (El Hafidi et al., 2024).

Figure 10. The single-effect thermal desalination system designed for distillation

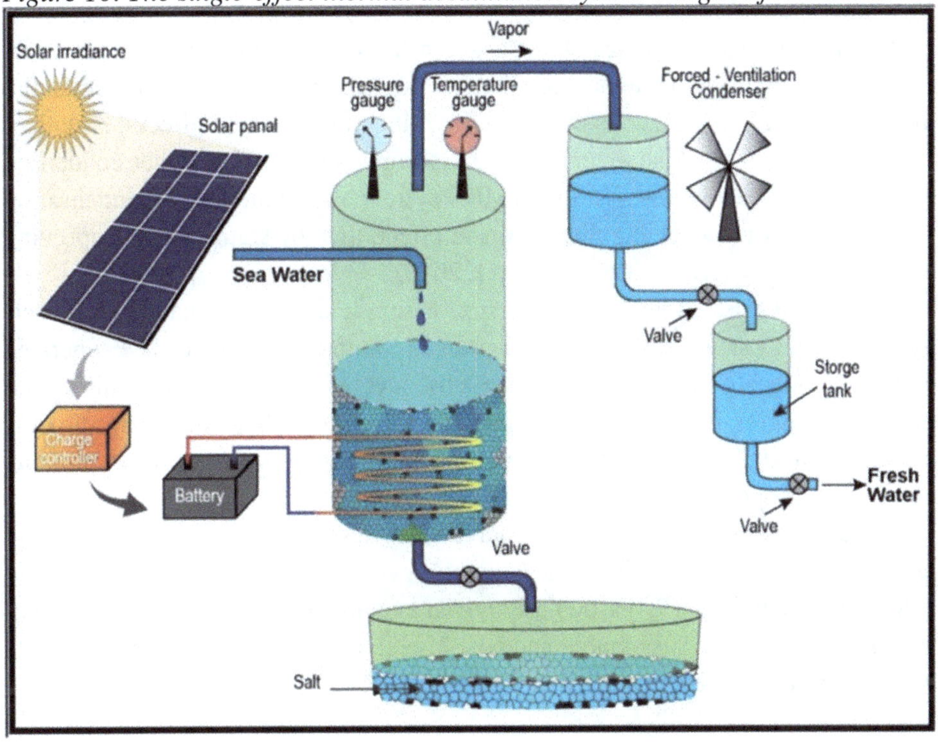

To minimize scaling, only a simple anti-scaling treatment is generally used. Manufacturers typically set the maximum operating temperature of this process between 60 and 65°C. Maintaining these moderate temperatures not only limits scaling but also significantly reduces corrosion, thereby extending the lifespan of the system components.

This approach is particularly beneficial in thermal desalination processes, where scale formation can impair heat transfer efficiency and lead to increased maintenance costs. By controlling temperature and using anti-scaling additives or surface coatings, the risk of mineral deposition is minimized, ensuring stable operation over time.

Furthermore, this method enables the production of highly pure condensed water, with salt concentrations as low as 1 mg/L. This level of purity is achieved through specialized equipment, such as droplet filters installed above the vaporization zone. These filters effectively capture entrained liquid droplets, preventing carryover of dissolved salts into the distillate. As a result, the produced water meets high-quality standards suitable for various applications, including industrial processes, laboratory use, and even potable water supply in certain cases (Rovel, 2010).

II.2. Multi-Effect Distillation (MED)

Multi-effect distillation (MED) is a water desalination process that consists of multiple interconnected evaporation chambers, or "effects," operating in a sequential manner. Each effect functions based on the principles of single-effect distillation, where seawater or brackish water is heated to produce vapor, which is then condensed to obtain fresh water. However, in the MED system, the latent heat of condensation from one effect is recycled to heat the next effect, thereby significantly improving energy efficiency (see Figure 11) (Rovel, 2010).

In a typical MED system, the first effect is supplied with external thermal energy, often from a steam source, which causes the water to evaporate. The generated steam is then used as a heat source for the subsequent effects, where the boiling point decreases progressively due to the reduction in pressure. This cascading process continues through multiple effects, maximizing water production while minimizing energy consumption.

MED technology is widely used in large-scale desalination plants due to its high energy efficiency, especially when integrated with waste heat recovery or renewable energy sources. It is particularly advantageous in regions facing freshwater scarcity, as it offers a sustainable and cost-effective method for producing potable water from saline sources.

Figure 11. Multi-effect distillation

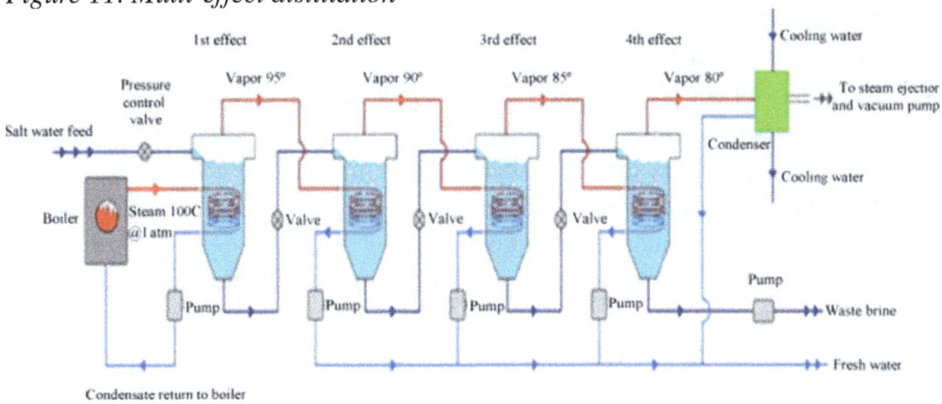

A multi-effect distillation (MED) system consists of multiple evaporation chambers, or effects, arranged in a sequential manner, each operating on the principle of single-effect distillation (see Figure 10). Initially, a heating fluid, typically steam, raises the temperature of seawater in the first chamber, where the highest temperature is maintained. The resulting vapor is then transferred to the next chamber, which operates at a slightly lower pressure. Since the boiling point of water decreases with pressure, the seawater in the second chamber also evaporates, forming the second effect. This process is repeated multiple times until the temperature reaches the lower limit determined by the cold seawater (see Figure 11).

To estimate the thermal consumption of this process, let n represent the number of effects and D the total fresh water production from these n chambers, with each chamber producing an equal amount of D/n. Assuming negligible heat losses and using the same notation as in single-effect distillation, the energy consumption of the multi-effect distillation process is significantly reduced compared to the single-effect method. This reduction is due to the efficient reuse of latent heat across multiple stages, making MED a more energy-efficient desalination technique.

$$Q = k.Dc_p\Delta t + \frac{Dr}{n}(I.1)$$

$$Y = \frac{Q}{D} = kc_p\Delta t + \frac{r}{n} \qquad (I.2)$$

Taking into account the defined parameters:

- o **n**: Number of effects
- o **D**: Total fresh water production from the **n** chambers
- o **Q**: Total thermal energy consumption
- o **k**: Proportionality coefficient relating to heat capacity
- o **cp**: Specific heat capacity of water
- o **Δt**: The temperature difference between the first and last effect
- o **Dr**: Total fresh water production
- o **r**: Latent heat of water vaporization
- o **Y**: Specific thermal energy consumption per unit of fresh water produced.

II.3. Mechanical Vapor Compression (MVC)

To enhance the efficiency of multi-effect distillation (MED), the generated vapor can be compressed and repurposed as a heating fluid through either mechanical or thermodynamic compression. This approach significantly reduces energy consumption by recycling thermal energy within the system.

When analyzing a distillation chamber, it becomes evident that the enthalpy of the emitted vapor is similar to that required for heating the process. This observation led to the concept of compressing the emitted vapor and using it as a heat transfer medium, as illustrated in Figure 12 (*Warsinger, D. M., Mistry, K. H., Nayar, K. G., Chung, H. W., & Lienhard V, J. H. (2015). Entropy Generation of Desalination Powered by Variable Temperature Waste Heat. Entropy, 17(11), 7530-7566*, n.d.). The mechanical vapor compression (MVC) process relies exclusively on electrical energy to power the compressor. However, it is possible to integrate a heat exchanger to recover low-grade waste heat available on-site. By utilizing this recovered energy, the system can preheat seawater at a reduced cost before it enters the evaporator, further improving overall energy efficiency.

The compressor used in MVC typically has a nominal power of approximately 300 kW for a unit capable of producing 600 m^3 of fresh water per day. This technology is particularly well-suited for small-scale desalination units, generally ranging between 100 and 1,000 m^3 per day (Rovel, 2010).

Figure 12. Single effect mechanical vapor compression process (MVC) with spray evaporation

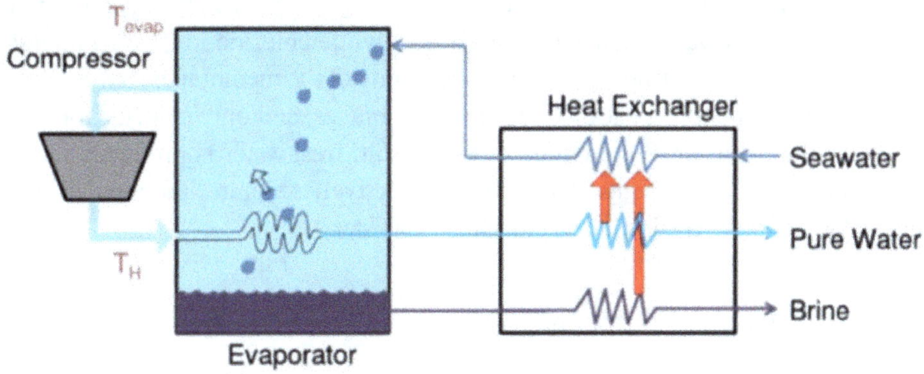

I.3.4. Process of distillation by successive expansions (Multi-Stage Flash distillation – MSF)

The Multi-Stage Flash (MSF) distillation process keeps water under pressure, where it instantly vaporizes at 120°C in a low-pressure chamber called a "flash" chamber. The water evaporates, condenses on tubes at the top of the chamber, and is collected below. This phenomenon is repeated at progressively lower pressures across successive stages—up to 40 stages in an industrial MSF unit. Seawater is heated to 120°C as it passes through the condensers of the different stages, being preheated by recovering the condensation heat of the water vapor, then brought to the required temperature using high-temperature steam produced by a boiler or power plant. Saltwater droplets that may be carried with the vapor are removed using mesh screens. The main energy required is thermal, supplied to the boiler, although electricity is also needed for the pumps. This MSF process lacks flexibility and is mainly suitable for large-scale desalinated water production.

III. Membrane Processes

III.1. Desalination by Electrodialysis

Electrodialysis desalination is an electro-membrane method that enables the selective transfer of ions through semi-permeable membranes. The main goal of this process is to remove salts from a solution, resulting in a highly concentrated brine on one side and a desalinated solution on the other (see Figure 13)(*Rezaee, R. (2024). The Potential of Salinity Gradient Energy Using Reverse Electrodialysis to*

Generate Electricity for Seawater Desalination Plants, an Example from Western Australia. Clean Energy and Sustainability, 2(2), 10006., n.d.).

When electric currents are applied to the external compartments, sodium ions (Na^+) are drawn toward the negative electrode, while chloride ions (Cl^-) migrate toward the positive electrode. Due to the specificity of the membranes, only sodium ions can pass through the cation-exchange membranes, and only chloride ions can cross the anion-exchange membranes. As a result, freshwater is produced in two out of the four compartments. This process effectively separates the ions, making it possible to obtain fresh water from saline seawater.

Figure 13. Parallel-fed electrodialysis unit

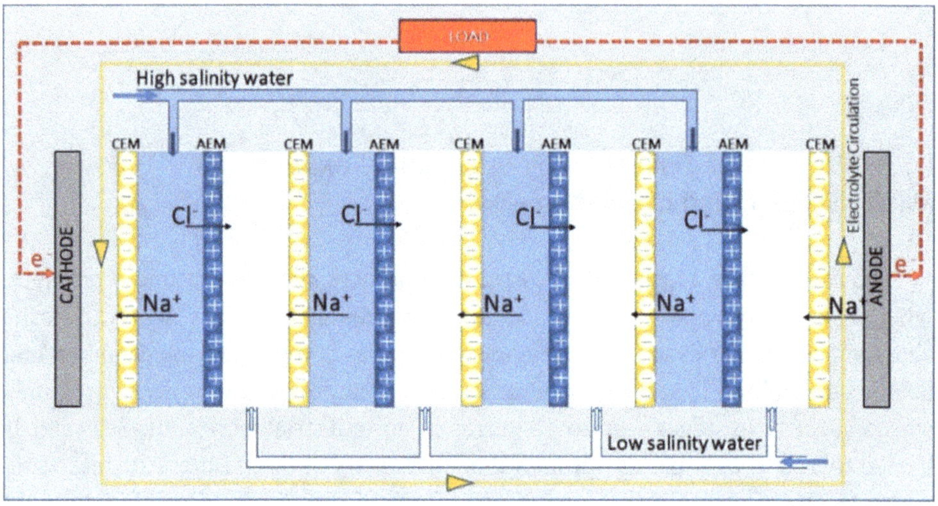

III.2. Reverse Osmosis (RO)

Reverse osmosis (RO) is a desalination process that relies on the natural principle of osmosis. Osmosis is a physical phenomenon observed in nature, where a semi-permeable membrane separates two solutions with different salt concentrations but at the same pressure. Under these conditions, diffusion occurs through the membrane, moving from the less saline solution to the more concentrated one. This diffusion is driven by the natural osmotic pressure, which can reach up to 24.1 bar for water with a salinity of 35 g/L (see Figure 14a) (Rovel, 2010).

Osmosis can be defined as the transfer of water (as a solvent) through a membrane due to a molecular concentration gradient. The osmotic pressure π is a measure of this force and can be expressed by the equation:

$$\pi = R \cdot T \cdot C \quad (I.3)$$

With: π = osmotic pressure; C = ionic concentration [mol/m^3]; R = ideal gas constant [8.314 J/mol·K]; and T = temperature [K].

The differential transmembrane osmotic pressure is given by:

$$\Delta \pi_{tm} = \frac{(\pi_{pure} - \pi_c)}{2} - \pi_p \quad (I.4)$$

With π_{pure} = pressure of pure water; π_c = pressure of the concentrated solution; and π_p = pressure of the permeate.

Indeed, when an external pressure greater than the osmotic pressure—typically between 50 and 75 bar—is applied to the concentrated solution during the reverse osmosis process, a notable physical phenomenon occurs. This phenomenon, known as "reverse osmosis," forces fresh water to flow in the opposite direction of the natural osmosis process, thus passing through the semi-permeable membrane. Meanwhile, the remaining saline solution becomes progressively more concentrated (see Figure 14 b, c).

This method enables the efficient separation of solutes, leading to the production of fresh water from saline sources such as seawater. The applied pressure must be sufficiently high to overcome the osmotic pressure and force water through the membrane. Additionally, the use of membranes specifically designed for reverse osmosis ensures effective filtration while minimizing solute leakage, thereby guaranteeing the quality and purity of the produced water (Djamel-Eddine, 2015).

Figure 14. Principle of osmosis and reverse osmosis

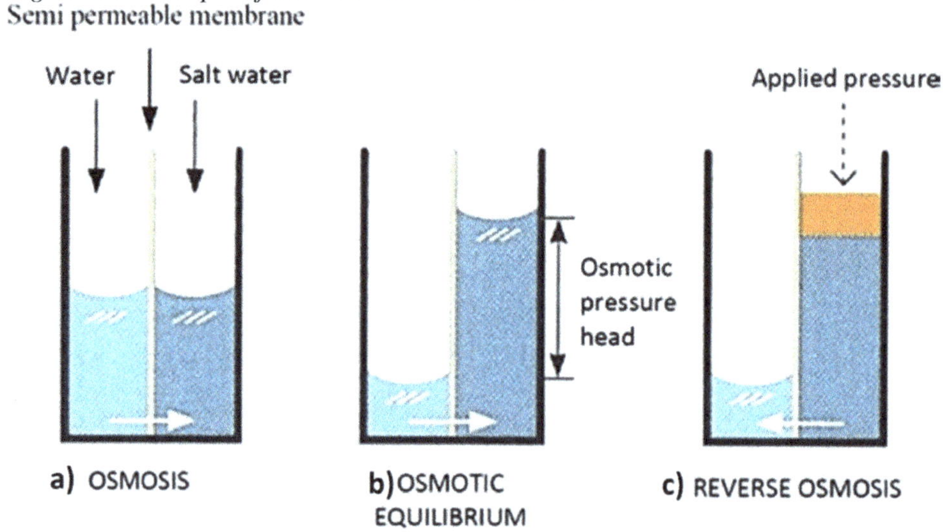

C- GENERALITIES ON IMPEDANCE/DIELECTRIC SPECTROSCOPY AND THE CONVENTIONAL METHOD

I: General Overview of Impedance/Dielectric Spectroscopy

I.1. Basic Principle

Impedance spectroscopy aims to obtain the electrical impedance of a material over a wide range of frequencies throughout the polarization characteristic (*Méthodes Électrochimiques - Mesures d'impédances : Dossier Complet | Techniques de l'Ingénieur*, n.d.). This impedance is only meaningful when using a sinusoidal power supply. The latter is given by the following expression:

$$Z^*(\omega)=\frac{V}{I}e^{j\varphi} \qquad (I.5)$$

Where V and I are the amplitudes (or root mean square values) of the voltage across the system and the current flowing through it, ω is the angular frequency, and ϕ is the phase shift between the voltage and the current.

Equation (I.5) is only meaningful if the system exhibits linear behavior, which is generally not the case for many systems. In a nonlinear system, if the input signal (either $v(t)$ or $i(t)$) is sinusoidal, the corresponding output is not necessarily sinusoidal. To continue referring to impedance, the resulting quantity (either $i(t)$ or $v(t)$) must still be considered sinusoidal.

In a linear (or quasi-linear) system, the current response to a sinusoidal voltage will also be a sinusoidal signal at the same frequency, but phase-shifted relative to the applied signal, as illustrated in the following (Figure 15) (Idris SADLI, n.d.).

Figure 15. Principle of linearization around an operating point

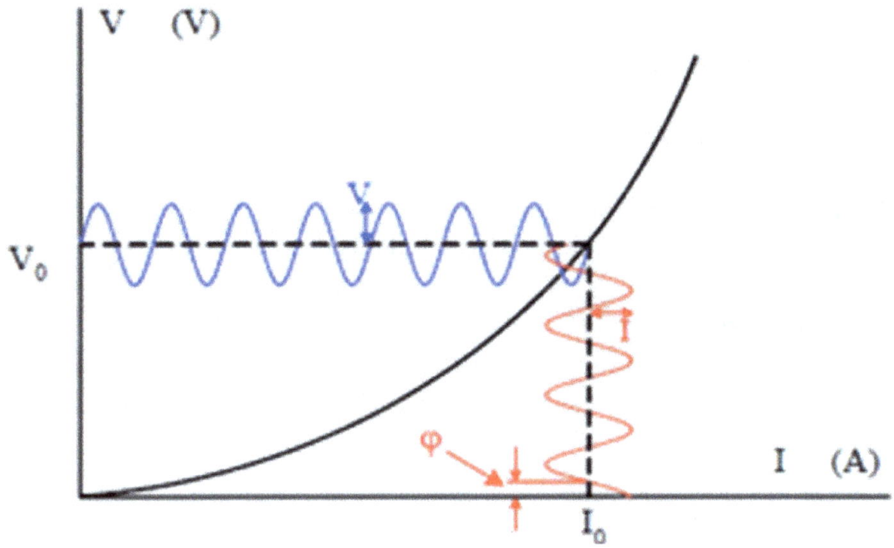

The excitation signal, expressed as a function of time, has the following form:

$$V(\omega) = V_0 \sin(\omega t) \qquad (I.6)$$

The measured current will be phase-shifted by ϕ, and its expression will have the following form:

$$I(\omega) = I_0 \sin(\omega t + \varphi) \qquad (I.7)$$

Where V_0 and I_0 are the amplitudes of the voltage and the current, respectively, and ϕ is the phase shift between the voltage and the current.

The ratio of the voltage to current amplitudes defines the modulus of the impedance $|Z|$; the phase shift between the voltage and current provides the additional information needed to access the real and imaginary components of the complex impedance $Z^*(\omega)$

This impedance can be written as a complex number with real and imaginary parts:

$$Z(\omega) = \frac{V(\omega)}{I(\omega)} = \mathrm{Re}(Z(\omega)) - j\,\mathrm{Im}(Z(\omega)) \qquad (I.8)$$

This expression represents the algebraic form of the complex impedance. Using Euler's formulas, the trigonometric form is obtained:

$$\cos\phi = \frac{e^{j\phi} + e^{-j\phi}}{2} \quad \text{and} \quad \sin\phi = \frac{e^{j\phi} - e^{-j\phi}}{2} \qquad (I.9)$$

The expression of the complex impedance then becomes:

$$Z^{*}(\omega) = |Z(\omega)|e^{j\phi} \quad \text{where} \quad j^2 = -1 \qquad (I.10)$$

Where: $\omega = 2\pi f$ is the angular frequency.
The modulus of the complex impedance is defined as:

$$|Z(\omega)| = \sqrt{R^2{}_e[Z(\omega)] + \mathrm{Im}^2[Z(\omega)]} \qquad (I.11)$$

The phase shift between voltage and current is given by the following expression:

$$\phi(\omega) = Arc\tan\left[\frac{\mathrm{Im}[Z(\omega)]}{R_e[Z(\omega)]}\right] \quad \text{where } \omega = 2\pi F \qquad (I.12)$$

Impedance is therefore a complex quantity characterized by its magnitude $|Z(\omega)|$ and phase (argument) ϕ, which represents the phase shift between the current I(t) and the voltage V(t). The phase is positive when the voltage leads the current.

Impedance measurements can be performed by sweeping the frequency f over a wide range, typically from 0.1 Hz to 1 MHz, either at room temperature or within a well-defined temperature range. This is why the technique is referred to as impedance spectroscopy.

I.1.1. Complex impedance: transfer function

A studied system can indeed be considered as a "black box" that responds by emitting a signal y(t) when subjected to a disturbance x(t) (Figure 16) (Sylvestre, 2006). The two signals are then related by a transfer function $H(\omega)$ such that:

$$Y^{*}(\omega) = H^{*}(\omega)\,X^{*}(\omega) \qquad (I.13)$$

Figure 16. Diagram of a transfer function

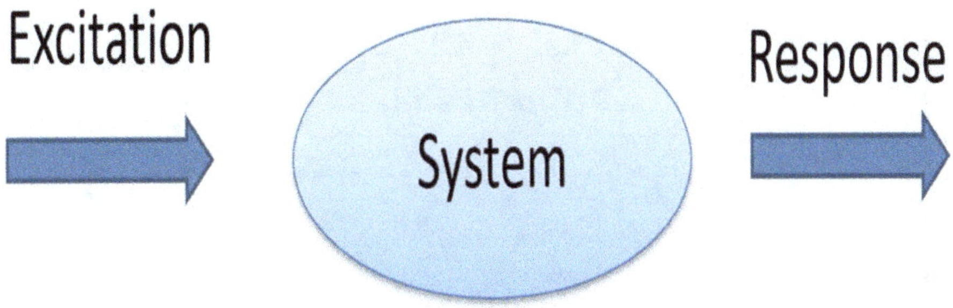

Where: x(ω) and y(ω) are respectively the Fourier transforms of x(t) and y(t).

The transformation of an input signal (the excitation) by a system into a response corresponds, mathematically, to the application of a transfer function to the input signal.

In general, two types of systems can be distinguished: linear systems and non-linear systems.

A linear system is one that possesses the property of superposition. If the input signal is composed of a weighted sum of several signals, then the response is simply the superposition of the corresponding weighted output signals from the system.

Mathematically, if $y_1(t)$is the response to an input signal $x_1(t)$, and $y_2(t)$ is the response to $x_2(t)$, the system is considered linear if:

-The response to $x_1(t)+x_2(t)$ and $y_1(t)+y_2(t)$

-The response to $A.x_1(t)$ and $A.y_1(t)$

For example, in the case of an electrochemical cell, applying excessively high currents may induce electrochemical processes at the electrodes that lead to deviations from linearity: the current is no longer proportional to the applied potential difference. Conversely, low-intensity currents can restore this linear dependence.

Figure 17 shows that, provided the input signal is confined within a pseudo-linear segment, a nonlinear response can be linearized using impedance/dielectric spectroscopy. The best modeling is achieved when the physical system exhibits a linear region (Sylvestre, 2006).

Figure 17. Transformation of a sinusoidal input signal on the current–voltage curve

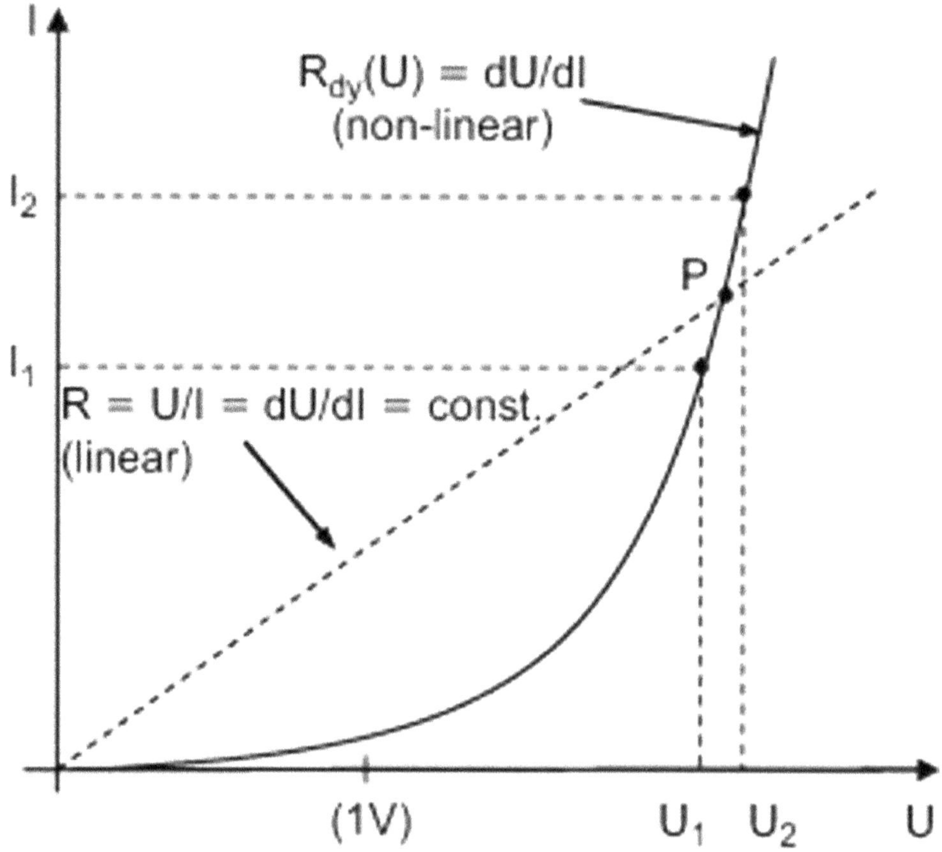

If the excitation is a sinusoidal function: $x(t) = A\ \sin(\omega t)$ (*I*.14)

And the response is of the form: $y(t) = B\ \sin(\omega t + \Phi)$ (*I*.15)

Then the transfer function can be defined as: $H(\omega) = |H(\omega)|e^{j\Phi}$ (*I*.16)

Where $|H(\omega)|$is equal to the ratio of the amplitudes $A/B$and$j = \sqrt{-1}$.

$|H(\omega)|$ and Φ are respectively the magnitude and the phase shift of the transfer function.

Thus, if the excitation$x(t)$ becomes the potential$E(t)$ and the response$y(t)$ the current$I(t)$, corresponds to the complex admittance$Y^*(\omega)$.

Then, the transfer function H(ω) can be defined as:

$$H^*(\omega) = \left|H(\omega)\right| e^{j\varphi} = \frac{y(t)}{x(t)} = \frac{I(t)}{V(t)} \qquad (I.17)$$

In other words, impedance acts as a modulator of the input signal, influencing both its amplitude and its phase. This modulation occurs because impedance imposes resistance and reactance on the passage of the electric current, which alters the signal's intensity and causes a phase shift. Therefore, the final shape of the output signal depends not only on the intrinsic characteristics of the input signal but also on the properties of the impedance encountered (Figure 18) (Sylvestre, 2006).

Figure 18. Signal V(t) and its response I(t) – the transfer function modifies the amplitude and phase of the incoming sinusoidal signal

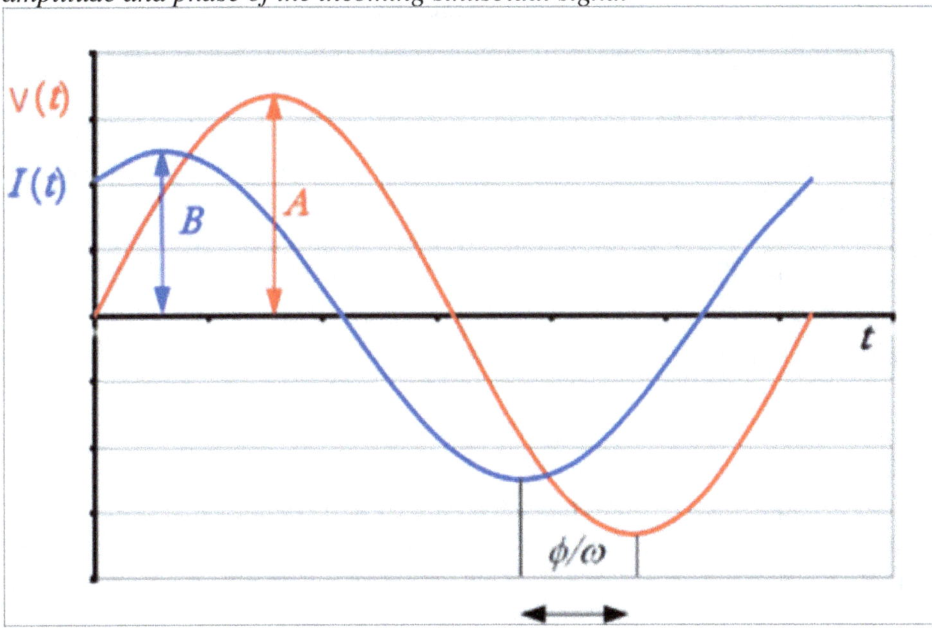

I.1.2. Other Electrical Quantities

From impedance (Z*) measurements, other complex electrical quantities of the materials under analysis can be deduced. These complex quantities are related to each other through the following equations (Molinié, 2006):

✓ **Complex admittance (Y*)**: $Y^* = \frac{1}{Z^*}$. This parameter is used for conductive systems.
✓ Complex capacitance: $C^* = \frac{Y^*}{j\omega}$
✓ Complex dielectric permittivity, $\varepsilon^* = \frac{Y^*}{j\omega C_0} = \frac{C^*}{C_0}$

$C_0 = \varepsilon_0 \frac{S}{d}$ is the capacitance of the empty cell, i.e., without the dielectric, for a capacitor whose plates have a surface area S and are separated by a distance d.

$\varepsilon_0 = 8{,}85.10^{-12} F.m^{-1}$ is the permittivity of free space (vacuum dielectric constant).

Note that we add an asterisk to C^* to indicate that it is a complex quantity, not to be confused with the symbol for a regular capacitance.

✓ Complex modulus (M*):

$$M^* = \frac{1}{\varepsilon^*} = j\omega Z^* \quad \text{(I.18)}$$

This quantity highlights the "bulk" processes over the electrode processes.

The relationships between admittance Y^*, permittivity ε^*, and modulus M^* are defined accordingly.

$$Z^*(\omega) = Z'(\omega) - jZ''(\omega) \quad \text{(I.19)}$$

$$Y^*(\omega) = Y'(\omega) + jY''(\omega) \quad \text{(I.20)}$$

$$\varepsilon^*(\omega) = \varepsilon'(\omega) - j\varepsilon''(\omega) \quad \text{(I.21)}$$

$$M^*(\omega) = M'(\omega) + jM''(\omega) \quad \text{(I.22)}$$

As we have seen, complex impedance (Z) measurements can be performed by sweeping frequencies F over a range f = [0.1 Hz to 1 MHz], either at room temperature or within a well-defined temperature range. All these electrical quantities can also be analyzed using Bode and Nyquist representations.

I.2. Dielectric Aspect

In this section, we will describe the different polarization and conduction mechanisms present in materials, depending on the measurement frequency used. Finally, various behavioral laws used to model the variations of parameters as a function of the characteristics of the studied materials and the measurement frequency are presented.

I.2.1. Dielectric Properties

The difference between insulating and conducting materials was first demonstrated by Stephen Gray in the 18th century. However, it was only explained in 1928 through Bloch's theorem, which introduced the concept of energy bands. This theory locates electron displacements within a specific conduction band. Insulators, or dielectric materials, do not have a conduction band and therefore lack free charge carriers (Molinié, 2006). Nevertheless, the absence of charge transport does not mean that their electronic activity is null.

Under the influence of an electric field, electrons exhibit elastic movements, forming dipoles. The material then undergoes what is known as electronic polarization, which is independent of frequency (Raju, 2016). A change in potential may induce a displacement current (Molinié, 2006). In contrast, conducting materials have a conduction band and thus free charge carriers. In this case, current is generated by the migration of charge carriers within the material under an electric field, resulting in ohmic conduction.

The analysis of the electric current flowing through a material under the influence of an electric field is essentially a study of the propagation of electromagnetic waves. This phenomenon is described by Maxwell's equations, from which the constitutive laws governing electric displacement $\vec{D}$and current density $\vec{J}$as functions of the electric field are derived.

$$\vec{D} = \varepsilon^* . \vec{E} \quad \text{(I.23)}$$

$$\vec{J} = \sigma^* . \vec{E} \quad \text{(I.24)}$$

Equations (I.23 and I.24) can be expressed either in the time domain or in the frequency domain. The proportionality factors ε^* and $\sigma^*$are, respectively, the electric permittivity (expressed in $F{\cdot}m^{-1}$) and the electrical conductivity (expressed in $S{\cdot}m^{-1}$).

The values of these parameters depend on temperature, pressure, the intensity of the electric field, and its frequency. Subsequently, permittivity and conductivity will be presented in complex form. Indeed, analyzing quantities in the frequency domain provides insights into the electrical nature of the material.

I.2. 2. Electric Permittivity

Electric permittivity, ε^*, refers to a material's ability to store charges in the form of electrostatic energy when subjected to an electric field$\vec{E}$. It is defined in its complex form by the following relation:

$$\varepsilon^* = \varepsilon' - j\varepsilon'' \quad \text{(I.25)}$$

Where ε' represents energy transfer through displacement currents reflecting polarization, and ε'' corresponds to energy loss due to the delay in polarization.

This quantity characterizes dielectric materials and represents the polarization or polarizability of the medium.

I.2. 3. Electrical Conductivity

Electrical conductivity, σ^*, more commonly used to describe the electrical properties of conductive materials, refers to the material's ability to conduct current when subjected to an electric field$\vec{E}$. It is defined in its complex form by the following relation:

$$\sigma^* = \sigma' + j\sigma'' \qquad (I.26)$$

Where σ ′ represents energy transfer through the migration of charge carriers, and σ′′ corresponds to energy loss due to thermal dissipation.

The relationship between complex permittivity and complex conductivity is given by the following relation:

$$\sigma^* = j\varepsilon_0\omega\varepsilon^* \qquad (I.27)$$

I.2. 4. Dielectric Relaxation

I.2. 4.1. Polarization Mechanisms

The total polarization P of a material can occur on several time scales. It results from electronic polarization P_e, atomic polarization P_{at}, orientational polarization P_{or}, and interfacial polarization P_{int}.

$$P = P_e + P_{at} + P_{or} + P_{int} \qquad (I.28)$$

✓ **Electronic Polarization:**

The application of an electric field to the dielectric induces a slight displacement of the atom's electron cloud relative to its nucleus. A dipole is thus created (Figure 19 (a)) (Bohli, 2009). This type of polarization occurs at frequencies higher than optical frequencies (on the order of 10^{15} Hz, i.e., in the far ultraviolet range) and, according to Maxwell's equations, it results in a real permittivity equal to the square of the material's refractive index:$\varepsilon'' = n^2$.

Figure 19. Schematic diagrams of (a) electronic polarization, (b) ionic polarization, (c) orientation polarization, and(d) interfacial polarization

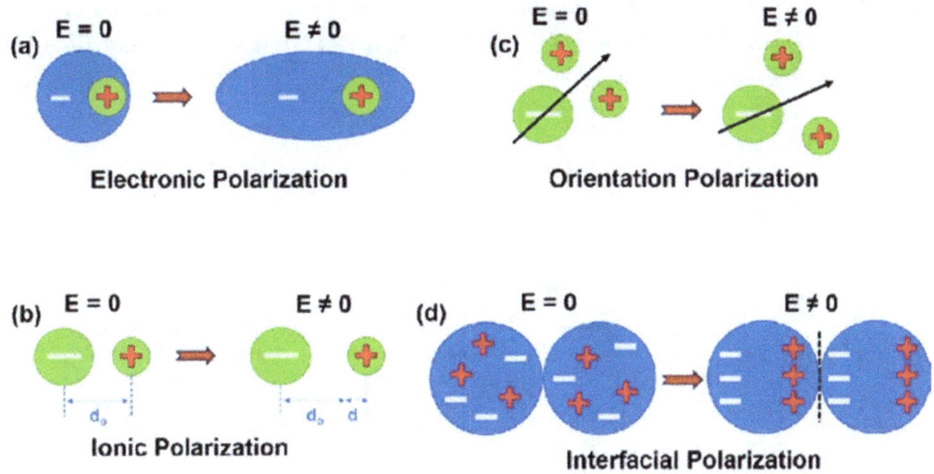

✓ **Ionic Polarization (or Atomic Polarization):**

The application of an electric field to a dielectric induces a displacement of positive ions relative to negative ions (Figure 19 (b)) (Bohli, 2009). This type of polarization occurs at frequencies below the terahertz range ($f < 10^{12}$ Hz).

✓ **Dipolar Polarization (or Orientation Polarization):**

The application of an electric field to a dielectric causes the dipoles to align in the direction of the electric field (Figure 19 (c)) (Bohli, 2009). This type of polarization occurs at frequencies below the gigahertz range.

✓ **Electrode Polarization (also called Space Charge or Interfacial Polarization):**

The application of an electric field to a dielectric induces the movement of mobile charges. This movement leads to the accumulation of charges near the electrodes (Figure 19 (d)). This type of polarization occurs at frequencies below the kilohertz range, or even at extremely low frequencies.

I.2. 4.2. Relaxation mechanisms

A relaxation can be detected by observing the behavior of the real part ε' and the imaginary part ε''(ω) of the complex permittivity as a function of frequency (Figure 20)(*USMAN, M. (2015). Studies of Ferroelectric and Multiferroic Behavior in [Ba (Zr, Ti) O3] 1-y:[CoFe2O4] y System (Doctoral Dissertation, Quaid-i-Azam University, Islamabad).*, n.d.).

Figure 20. Variation of the real part ε' and imaginary part ε" of the permittivity as a function of frequency

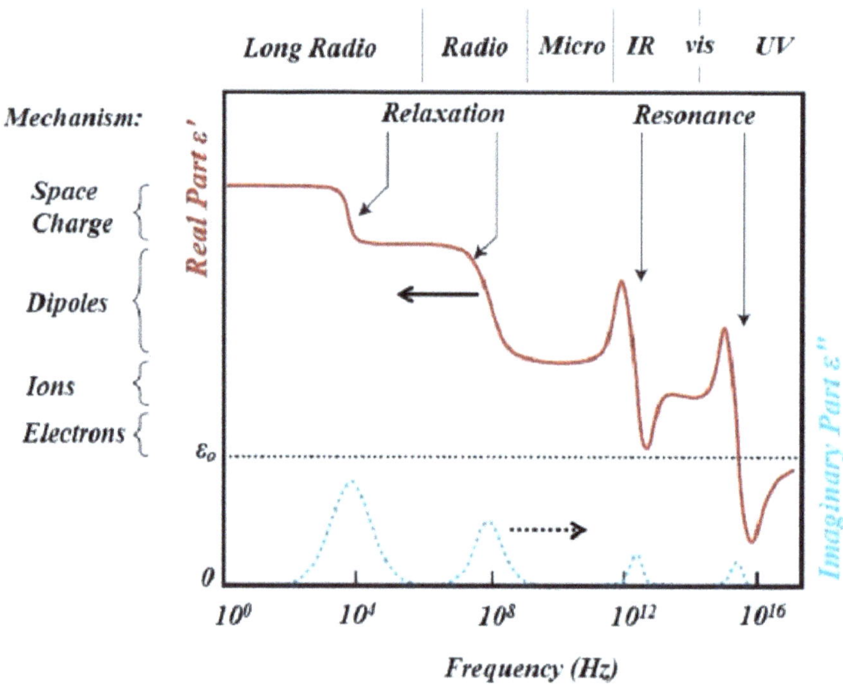

It is manifested by an increase in the imaginary part, indicating greater energy losses, as well as a decrease in the real part. The latter corresponds to a reduced ability of the material to polarize (i.e., to store electrical energy). In fact, being unable to follow the variations of the electric field, one of the mechanisms contributing to the overall polarization ceases to operate. To account for the different sources of dielectric losses, the overall loss factor ε''(ω) is decomposed according to expression (II.22):

$$\varepsilon''(\omega) = \varepsilon''_{cond}(\omega) + \varepsilon''_{MWS}(\omega) + \varepsilon''_{dip}(\omega) + \varepsilon''_{at}(\omega) + \varepsilon''(\omega)_{el} \qquad (I.29)$$

Where$\varepsilon''_{cond}(\omega)$ is due to the continuous conductivity, $\varepsilon''_{MWS}(\omega)$is due to interfacial polarization, $\varepsilon''_{dip}(\omega)$ is due to dipolar polarization or orientational polarization, ε''_{at} (ω) is due to atomic polarization, and $\varepsilon''(\omega)_{el}$is due to electronic polarization (Singh et al., 1999).

II. General Overview of the Conventional Method

II.1. Suspended Solids (SS)

Suspended solids (SS), together with dissolved substances, contribute to the turbidity of wastewater. Turbidity refers to the measure of water clarity, influenced by the presence of suspended particles that reduce water transparency. Determining SS is important for assessing the distribution between dissolved and settleable pollution, both of which are considered in wastewater treatment systems.

The standard method for determining SS is the AFNOR method No. 90-105, which involves filtering the sample through a glass fiber filter. SS are then expressed in milligrams per liter (mg/L) of insoluble dry matter, thus providing an indication of the amount of solid particles present in the water (Achak et al., 2008).

II.2 Biochemical oxygen demand (BOD_5)

Biochemical Oxygen Demand over five days (BOD_5) is an indicator used to measure the amount of oxygen required by microorganisms to degrade or mineralize the organic matter present in one liter of water over a five-day period. This measurement is essential for assessing organic pollution in water, as it reflects the quantity of biodegradable organic matter that may consume oxygen during decomposition. BOD_5 is expressed in milligrams of oxygen per liter (mg/L).

A high BOD_5 value indicates a significant organic load in the water, which can lead to a reduction in the oxygen available for aquatic organisms and cause water quality issues such as eutrophication and the degradation of aquatic habitats.

II.3 Chemical Oxygen Demand (COD)

Chemical Oxygen Demand (COD) is an indicator used to measure the amount of oxygen required for the chemical oxidation of organic and inorganic matter present in one liter of water. Unlike BOD_5, which assesses the biological degradation of organic matter over five days, COD evaluates the total oxidation of organic substances—whether biodegradable or not—as well as other oxidizable compounds in the water. COD is expressed in milligrams of oxygen per liter (mg/L) and is a critical parameter for monitoring urban wastewater and industrial discharges.

II.4 Electrical Conductivity

The electrical conductivity of a substance, particularly water, measures its ability to conduct an electric current. It is influenced by the presence of dissolved ions such as Ca^{2+}, Mg^{2+}, Na^{+}, K^{+}, Cl^{-}, SO_4^{2-}, and PO_4^{3-}. The higher the concentration of ions, the greater the electrical conductivity of the water. This property is important in evaluating water quality, as it reflects the total amount of dissolved ions and can be used to estimate salinity or mineral content. Electrical conductivity is measured in Siemens per centimeter (S/cm) and is a key parameter in assessing the potability of water for human consumption, as well as in industrial processes where water conductivity can affect equipment and operational efficiency.

II.5 Heavy Metals

Heavy metals in water are metallic elements characterized by their high density and toxic potential to living organisms when present at elevated concentrations. Although naturally occurring in the environment, their accumulation is often the result of human activities such as mining, manufacturing, agriculture, and industrial waste discharge.

The main heavy metals of concern in water pollution include mercury, lead, cadmium, chromium, arsenic, zinc, copper, and nickel, among others. These metals can enter freshwater and marine sources through various pathways, including direct discharges, agricultural runoff, atmospheric emissions, and industrial effluents.

Once released into water bodies, heavy metals can accumulate in sediments and aquatic organisms, leading to bioaccumulation throughout the food chain. The harmful effects of heavy metals on human health and the environment are well documented. They can cause neurological disorders, kidney and liver damage, developmental issues in children, endocrine disruption, cancers, and other serious health problems.

Regulating heavy metals in water is essential to prevent risks to human health and ecosystems. Water quality standards set maximum allowable concentrations of heavy metals in drinking water and surface waters to safeguard public health and the environment. Preventive and treatment measures—such as the use of decontamination technologies, regular monitoring of water sources, and reduction of industrial emissions—are also crucial to mitigate the adverse impacts of heavy metals on aquatic ecosystems and human health.

CONCLUSION

In the first section (A) of Chapter, we provided a detailed overview of the different categories of wastewater, including domestic, industrial, and agricultural waters, as well as the main conventional treatment processes. These processes include primary settling, secondary biological treatment, and tertiary treatment. A critical analysis of these methods reveals their limitations, particularly their ineffectiveness in removing certain persistent organic pollutants and the challenges associated with managing residual sludge. Moreover, current processes show weaknesses in treating heavy metals and specific chemical compounds arising from industrial activities. This literature review highlights the need to develop more innovative and sustainable treatment technologies capable of meeting the growing demands for water quality and environmental protection.

In the second section (B), we presented a bibliographic study on the use of renewable energies in desalination. This section covers the generalities of desalination techniques, including the main desalination technologies, distillation processes (single-effect distillation, multi-effect distillation, multi-effect distillation with vapor compression), and membrane processes such as electrodialysis and reverse osmosis (RO). Furthermore, this section examines the integration of renewable energies into desalination systems, with technologies such as reverse osmosis powered by renewable energies, multi-effect distillation (MED), and multi-stage flash (MSF) desalination powered by renewable energies. A study on the influence of the thickness of the absorbing layer on the performance of perovskite solar cells is also presented.

In the third section (C) of Chapter, we introduced the various physical characterization relaxation mechanisms that we used in this work, particularly impedance spectroscopy. This technique is important for analyzing and modeling the electrical and dielectric properties of treated waters, offering promising prospects for improving wastewater treatment processes and optimizing desalination systems. This section also discusses the generalities of the conventional method, including water pollution parameters such as suspended solids (SS), biochemical oxygen demand (BOD5), chemical oxygen demand (COD), temperature and pH, electrical conductivity, nutrients (nitrogen, phosphorus), heavy metals, and salinity. The correlation between the conventional method and impedance spectroscopy is also discussed, emphasizing the importance of these techniques for an accurate and reliable assessment of water quality.

REFERENCES

Achak, M., Ouazzani, N., Yaacoubi, A., & Mandi, L. (2008). Caractérisation des margines issues d'une huilerie moderne et essais de leur traitement par coagulation-floculation par la chaux et le sulfate d'aluminium. *Revue des sciences de l'eau / Journal of Water Science, 21*(1), 53–67. DOI: 10.7202/017930ar

Adamiec, P. (1998). *Caractérisation d'une cendre volante silico-alumineuse et sa reconstruction: Corrélation avec les caractéristiques géologiques, gîtologiques et minéralogiques du charbon* [These de doctorat, Lille 1]. https://theses.fr/1998LIL10023

Adamiec, P., Benezet, J. C., & Benhassaine, A. (2011). *Relation entre une cendre volante silico-alumineuse et son charbon* (arXiv:1111.6847). arXiv. https://doi.org//arXiv.1111.6847DOI: 10.48550

Ademe, E. (2001). *Les boues d'épuration municipales et leur utilisation en agriculture.* Dossier Documentaire.

Ahmaruzzaman, M. (2010). A review on the utilization of fly ash. *Progress in Energy and Combustion Science, 36*(3), 327–363. DOI: 10.1016/j.pecs.2009.11.003

Allen, A. (2001). Containment landfills: The myth of sustainability. *Engineering Geology, 60*(1–4), 3–19. DOI: 10.1016/S0013-7952(00)00084-3

Arslan-Alaton, I., Kabdaşlı, I., Hanbaba, D., & Kuybu, E. (2008). Electrocoagulation of a real reactive dyebath effluent using aluminum and stainless steel electrodes. *Journal of Hazardous Materials, 150*(1), 166–173. DOI: 10.1016/j.jhazmat.2007.09.032 PMID: 17945416

Bellarbi, A., Graich, A., Monkade, M., Zradba, A., & Laamyem, A. (2017). Valorization of bottom ash generated by a thermal power plant in the treatment of cardboard manufacturing wastewater. *Global J. Eng. Sci. Res. Manage, 4*(10), 11–22.

Bhattacharya, A. K., Naiya, T. K., Mandal, S. N., & Das, S. K. (2008). Adsorption, kinetics and equilibrium studies on removal of Cr(VI) from aqueous solutions using different low-cost adsorbents. *Chemical Engineering Journal, 137*(3), 529–541. DOI: 10.1016/j.cej.2007.05.021

Bohli, N. (2009). *Etude physique de la formation de films à base de polymères conducteurs et applications en micro-éléctronique* [These de doctorat, Bordeaux 1]. https://theses.fr/2009BOR13961

Campbell, H. W., & Crescuolo, P. J. (1982). The use of rheology for sludge characterization. *Water Science and Technology, 14*(6–7), 475–489. DOI: 10.2166/wst.1982.0120

Casey, T. J., & Casey, T. J. (1997). *Unit treatment processes in water and wastewater engineering*. Wiley Chichester. http://www.aquavarra.ie/Utpbkpdf/chap0.pdf

Chahid, E. G., Moznine, R. E., Zradba, A., Dani, A., Melouky, A. E., Idrissi, L., Cherkaoui, O., Choukri, E., Mezzane, D., & Belboukhari, A. (2013, November-December). Analysis of relaxation processes and low frequency dispersion in waste water. *Journal of Optoelectronics and Advanced Materials, 15*, 1209–1216.

Chatterjee, A. K. (2011). Indian Fly Ashes: Their Characteristics and Potential for Mechanochemical Activation for Enhanced Usability. *Journal of Materials in Civil Engineering, 23*(6), 783–788. DOI: 10.1061/(ASCE)MT.1943-5533.0000279

Chen, M., Blanc, D., Gautier, M., Mehu, J., & Gourdon, R. (2013). Environmental and technical assessments of the potential utilization of sewage sludge ashes (SSAs) as secondary raw materials in construction. *Waste Management (New York, N.Y.), 33*(5), 1268–1275. DOI: 10.1016/j.wasman.2013.01.004 PMID: 23415711

Colin, F., & Gazbar, S. (1995). Distribution of water in sludges in relation to their mechanical dewatering. *Water Research, 29*(8), 2000–2005. DOI: 10.1016/0043-1354(94)00274-B

Diamadopoulos, E., Ioannidis, S., & Sakellaropoulos, G. P. (1993). As(V) removal from aqueous solutions by fly ash. *Water Research, 27*(12), 1773–1777. DOI: 10.1016/0043-1354(93)90116-Y

Dieudé-Fauvel, E., & Dentel, S. K. (2011). *Sludge conditioning: Impact of Polymers on Floc structure.*

Djamel-Eddine, M. (2015). *Utilisation des énergies renouvelables dans le dessalement: Cas de l'Algérie* [Thesis]. https://e-biblio.univ-mosta.dz/handle/123456789/951

Dutta, B., Basu, J. K., & DasGupta, S. (2003). Removal of Cresol from Aqueous Solution Using Fly Ash as Adsorbent: Experiments and Modeling. *Separation Science and Technology, 38*(6), 1345–1360. DOI: 10.1081/SS-120018813

El Alami, S. E. M., Monkade, M., & Lahlou, K. (2011). Valorisation d'un mélange de cendres d'une centrale thermique et de scories d'aciérie dans les chaussées routières. *Environnement, Ingénierie & Développement*. https://eid.episciences.org/7863

El Hafidi, E. M., Chahid, E. G., Mortadi, A., & Laasri, S. (2024). Study on a new solar-powered desalination system to alleviate water scarcity using impedance spectroscopy. *Materials Today: Proceedings*. Advance online publication. DOI: 10.1016/j.matpr.2024.02.038

El Hafidi, E. M., Mortadi, A., Chahid, E. G., & Laasri, S. (2023). Monitoring of domestic wastewater treatment via infiltration percolation using impedance spectroscopy. *Environmental Technology & Innovation*, *32*, 103421. DOI: 10.1016/j.eti.2023.103421

EPA. U. (2000). Wastewater technology fact sheet chemical precipitation. *Office of Water and Watersheds, Washington DC: United States Environmental Protection Agency. Doi: EPA.*

Fayet, T., Pellissier, C., Hjiaj, M., Abriak, N. E., & Klein, C. (1999). Mechanical and physical properties of fly ashes. *Journal de Chimie Physique et de Physico-Chimie Biologique, 96*. https://www.osti.gov/etdeweb/biblio/348688

Ghaffour, N., Bundschuh, J., Mahmoudi, H., & Goosen, M. F. A. (2015). Renewable energy-driven desalination technologies: A comprehensive review on challenges and potential applications of integrated systems. *Desalination*, *356*, 94–114. DOI: 10.1016/j.desal.2014.10.024

Goodwin, J. A. S., & Forster, C. F. (1985). A further examination into the composition of activated sludge surfaces in relation to their settlement characteristics. *Water Research*, *19*(4), 527–533. DOI: 10.1016/0043-1354(85)90045-4

Graich, A., Mghaiouini, R., Bellarbi, A., Khaidar, M., Laamyem, A., Monkade, M., & Zradba, A. (2021). Use of a cascade system based on marine sand and bottom ash in the treatment of cardboard manufacturing wastewater by infiltration percolation. *Physica Scripta*, *96*(12), 125711. DOI: 10.1088/1402-4896/ac2659

Herman, A. P., Yusup, S., Shahbaz, M., & Patrick, D. O. (2016). Bottom ash characterization and its catalytic potential in biomass gasification. *Procedia Engineering*, *148*, 432–436. DOI: 10.1016/j.proeng.2016.06.447

Idris SADLI. (n.d.). *MODELISATION PAR IMPEDANCE D'UNE PILE A COMBUSTIBLE PEM POUR UTILISATION EN ELECTRONIQUE DE PUISSANCE.*

Janoš, P., Buchtová, H., & Rỳznarová, M. (2003). Sorption of dyes from aqueous solutions onto fly ash. *Water Research*, *37*(20), 4938–4944. DOI: 10.1016/j.watres.2003.08.011 PMID: 14604640

Kao, P.-C., Tzeng, J.-H., & Huang, T.-L. (2000). Removal of chlorophenols from aqueous solution by fly ash. *Journal of Hazardous Materials*, *76*(2), 237–249. DOI: 10.1016/S0304-3894(00)00201-6 PMID: 10936536

Keiding, K., Wybrandt, L., & Nielsen, P. H. (2001). Remember the water—A comment on EPS colligative properties. *Water Science and Technology*, *43*(6), 17–23. DOI: 10.2166/wst.2001.0330 PMID: 11381964

Kjeldsen, P., Barlaz, M. A., Rooker, A. P., Baun, A., Ledin, A., & Christensen, T. H. (2002). Present and Long-Term Composition of MSW Landfill Leachate: A Review. *Critical Reviews in Environmental Science and Technology*, *32*(4), 297–336. DOI: 10.1080/10643380290813462

Léonard, A. (2003). *Etude du séchage convectif de boues de station d'épuration-Suivi de la texture par microtomographie à rayons X.* https://orbi.uliege.be/handle/2268/39648

Marttinen, S. K., Kettunen, R. H., & Rintala, J. A. (2003). Occurrence and removal of organic pollutants in sewages and landfill leachates. *The Science of the Total Environment*, *301*(1–3), 1–12. DOI: 10.1016/S0048-9697(02)00302-9 PMID: 12493180

Méthodes électrochimiques—Mesures d'impédances: Dossier complet | Techniques de l'Ingénieur. (n.d.). Retrieved July 19, 2024, from https://www.techniques-ingenieur.fr/base-documentaire/materiaux-th11/essais-et-expertise-en-corrosion-et-vieillissement-42499210/methodes-electrochimiques-p2210/

Mghaiouini, R., Graich, A., Elaoud, A., Garmim, T., Belghiti, M. E., Benzbiria, N., Hozayn, M., Monkade, M., & El Bouari, A. (2020). Formulation and physico-mechanical characterization of an eco-mortar composite based on bottom ash and magnetized water. *Indian Journal of Science and Technology*, *13*(10), 1172–1187. DOI: 10.17485/ijst/2020/v13i10/149889

Mininni, G., Sbrilli, A., Guerriero, E., & Rotatori, M. (2004). Dioxins and furans formation in pilot incineration tests of sewage sludge spiked with organic chlorine. *Chemosphere*, *54*(9), 1337–1350. DOI: 10.1016/S0045-6535(03)00252-2 PMID: 14659427

Mohan, D., Singh, K. P., Singh, G., & Kumar, K. (2002). Removal of Dyes from Wastewater Using Flyash, a Low-Cost Adsorbent. *Industrial & Engineering Chemistry Research*, *41*(15), 3688–3695. DOI: 10.1021/ie010667+

Molinié, P. (2006). *Comment modéliser un isolant en électrostatique? Approches classiques et modernes d'un problème complexe.* 27. https://centralesupelec.hal.science/hal-00239414

Mortadi, A., Chahid, E. G., Elmelouky, A., Chahbi, M. E. L., Ghyati, N., Zaim, S., Cherkaoui, O., & El Moznine, R. (2020). Complex electrical conductivity as a new technique to monitor the coagulation-flocculation processes in the wastewater treatment of the textile Industry. *Water Resources and Industry*, *24*, 100130. DOI: 10.1016/j.wri.2020.100130

Mortadi, A., Chahid, E. G., Nasrellah, H., Cherkaoui, O., & El Moznine, R. (2019). Dielectric and electric properties as a tool to investigate the coagulation mechanism during sludge treatment. *Environmental Technology*, *40*(1), 72–85. DOI: 10.1080/09593330.2017.1378927 PMID: 28891396

Mortadi, A., Mghaiouini, R., Elmelouky, A., Chahid, E., Hairch, Y., Saifaoui, D., Monkade, M., Cherkaoui, O., & El Moznine, R. (2022). New approach to investigate and to monitor the coagulation process during wastewater treatment. *Materials Today: Proceedings*, *66*, 325–328. DOI: 10.1016/j.matpr.2022.05.424

Nammari, D. R., Hogland, W., Marques, M., Nimmermark, S., & Moutavtchi, V. (2004). Emissions from a controlled fire in municipal solid waste bales. *Waste Management (New York, N.Y.)*, *24*(1), 9–18. DOI: 10.1016/j.wasman.2003.08.003 PMID: 14672722

Obernberger, I., & Supancic, K. (2009). Possibilities of ash utilisation from biomass combustion plants. *Proceedings of the 17th European Biomass Conference & Exhibition, 29*. https://www.best-research.eu/files/publications/pdf/Paper-Obernberger-ash-utilisation-2009.pdf

Organization, W. H. (2022). Guidelines for drinking-water quality*: Fourth edition incorporating the first and second addenda*. World Health Organization.

Pandey, A. K., Reji Kumar, R., B, K., Laghari, I. A., Samykano, M., Kothari, R., Abusorrah, A. M., Sharma, K., & Tyagi, V. V. (2021). Utilization of solar energy for wastewater treatment: Challenges and progressive research trends. *Journal of Environmental Management*, *297*, 113300. DOI: 10.1016/j.jenvman.2021.113300 PMID: 34293672

Prévot, H. (2000). La récupération de l'énergie issue du traitement des déchets. *Rapport Du Conseil Général Des Mines Au Ministère de l'Economie, Des Finances et de l'Industrie, Paris*. http://emericmenard.free.fr/dossierenergies/Documents/Rapport-%20R%E9cup%E9ration%20d'energie%20issue%20des%20d%E9chet%20juillet%20200.pdf

Raju, G. G. (2016). *Dielectrics in Electric Fields: Tables, Atoms, and Molecules* (2nd ed.). CRC Press., DOI: 10.1201/9781315373270

Rakotoarisoa, Z., & Tiruta-Barna, L. (n.d.). *Étude des écoulements à travers des couches de cendres volantes compactées saturées et non saturées en eau.* Retrieved July 19, 2024, from https://scholar.google.com/scholar?hl=en&as_sdt=0%2C5&q=%29.+%C3%89tude+des+%C3%A9coulements+%C3%A0+travers+des+couches+de+cendres+volantes+compact%C3%A9es+satur%C3%A9es+et+non+satur%C3%A9es+en+eau.+Environnement%2C+Ing%C3%A9nierie+%26+D%C3%A9veloppement.&btnG=

Raynaud, M. (2010). *Couplage de caractérisations mécanique et physico-chimique en vue d'analyser les limites de la déshydratation des boues résiduaires urbaines* [PhD Thesis, Doctorat Génie des Procédés, Université de Pau et des Pays de l'Adour]. https://hal.science/tel-02594092

Rezaee, R. (2024). The Potential of Salinity Gradient Energy Using Reverse Electrodialysis to Generate Electricity for Seawater Desalination Plants, an Example from Western Australia. *Clean Energy and Sustainability*, *2*(2), 10006. DOI: 10.35534/ces.2024.10006

Ricou, P., Lécuyer, I., & Le Cloirec, P. (1999). Removal of Cu2+, Zn2+ and Pb2+ by adsorption onto fly ash and fly ash/lime mixing. *Water Science and Technology*, *39*(10), 239–247. DOI: 10.2166/wst.1999.0663

Rovel, J.-M. (2010). *Dessalement de l'eau de mer.* Editions TI. https://perso.ens-lyon.fr/timothee.audinet-de-pieuchon/fichiers/LC4_TI_Dessalement_eau_de_mer.pdf

Singh, R., Arora, V., Tandon, R. P., Mansingh, A., & Chandra, S. (1999). Dielectric spectroscopy of doped polyaniline. *Synthetic Metals*, *104*(2), 137–144. DOI: 10.1016/S0379-6779(99)00043-0

Smith, J. K., & Vesilind, P. A. (1995). Dilatometric measurement of bound water in wastewater sludge. *Water Research*, *29*(12), 2621–2626. DOI: 10.1016/0043-1354(95)00144-A

Sutherland, I. W. (2001). Exopolysaccharides in biofilms, flocs and related structures. *Water Science and Technology*, *43*(6), 77–86. DOI: 10.2166/wst.2001.0345 PMID: 11381975

Sylvestre, A. (2006). *Apport de la spectroscopie diélectrique basse fréquence dans l'analyse de matériaux isolants à forte permittivité* [Thesis, Université Joseph-Fourier - Grenoble I]. https://theses.hal.science/tel-00651493

USMAN, M. (2015). Studies of Ferroelectric and Multiferroic Behavior in [Ba (Zr, Ti) O3] 1-y:[CoFe2O4] y System (Doctoral dissertation, Quaid-i-Azam University, Islamabad). (n.d.).

Wang, D., Sun, W., Xu, Y., Tang, H., & Gregory, J. (2004). Speciation stability of inorganic polymer flocculant–PACl. *Colloids and Surfaces. A, Physicochemical and Engineering Aspects*, *243*(1), 1–10. DOI: 10.1016/j.colsurfa.2004.04.073

Warsinger, D. M., Mistry, K. H., Nayar, K. G., Chung, H. W., & Lienhard, V. J. H. (2015). Entropy generation of desalination powered by variable temperature waste heat. *Entropy (Basel, Switzerland)*, *17*(11), 7530–7566. DOI: 10.3390/e17117530

Chapter 5
Phytoremediation-Based Wastewater Treatment Using Greywater for Sustainable Water Resource Reuse

Vaiavel Madeshwaren
https://orcid.org/0000-0002-8687-7887
Srinivasan College of Engineering, Coimbatore, India

Neetha Delphin Mary Kulandaiswamy
Tagore Insistute of Engineering and Technology, India

R. S. Gandhimathi
AnnaPoorana Engineering College, India

ABSTRACT

The growing global concern over water scarcity is closely tied to current environmental, energy, and economic challenges. Rapid technological progress, while beneficial in many domains, has contributed to the neglect of water conservation efforts, leading to critical ecological damage. T Greywater—comprising used water from bathing, dishwashing, and laundry—was collected and processed using phytoremediation in vertical and horizontal wetland setups. These systems were filled with sand and gravel substrates and supported plant growth to aid natural filtration. The treatment aimed to determine the reusability of the filtered water for secondary purposes such as bathing, irrigation, and general household use. The analysis emphasized plant adaptability and the efficiency of filtration in both hor-

DOI: 10.4018/979-8-3373-5886-4.ch005

izontal and vertical flow wetlands. The microbial assessment of treated greywater revealed the presence of harmful pathogens, including E. coli, Salmonella spp., Staphylococcus aureus, Shigella spp., and Pseudomonas aeruginosa, highlighting the need for effective disinfection protocols.

1. INTRODUCTION:

Wastewater treatment is a critical process that involves the removal of contaminants from used water generated by households, industries, and commercial activities before it is released back into the environment. The treatment typically includes primary, secondary, and sometimes tertiary stages, which use physical, biological, and chemical methods to eliminate solids, organic matter, pathogens, and hazardous substances. Proper wastewater treatment not only safeguards public health by preventing waterborne diseases but also protects aquatic ecosystems, conserves water resources, and supports sustainable development by enabling water reuse in agriculture, industry, and urban settings. The following are clearly state about the what your contribution to the literature and to the problem is:

Greywater management was not only necessary for a healthy and safe living environment, but this also has a lot of repurposing ability. Treated greywater is recycled for many purposes in a decentralized fashion all over the world; the most common application in poor countries is for farming. The total quantity of freshwater available on the earth far outnumbers human needs (Abdel-Shafy et al. 2019;Al-Gheethi et al. 2019; Al-Mefleh et al. 2021). The oceans contain approximately 97 percent of the overall sources of water on Earth, with the residual 3% appropriate for direct extraction; however, the quantity of water available for human consumption is estimated to be one-tenth of that remaining 3% (Amiri et al.2019; Ansari and Shrikhande 2019). Water was transported unevenly in time and space, leaving some geographical areas without access to it (Inyinbor et al., 2019). One of the most significant components of water usage, which involves use for residential needs, food safety, and other sectors for growth, was biological existence (Janani et al., 2019; Jamil, 2019). Acute water shortage, as well as over-exploitation of water resources, impact many parts of the world, resulting in the slow degradation of natural resources and significant levels of water pollution due to anthropogenic influences.

To address these challenges, this question demands actions and solutions. A look into greywater treatment can help in this situation (Laaffat et al., 2019; Mohamed et al., 2019). A sustainable freshwater and sewage system management should be developed in human settlements, with the municipal, city, and Metropolitan governments playing a key role in providing adequate sanitary and sewage inside their domains (Nel & Jacobs, 2019). Human excrement from laundry and bathing contaminates greywater. Greywater contamination, both microbial and chemical, causes a risk to human health, which is predicted to rise as microbial contamination rises. It's critical to understand that greywater has the ability to spread disease (Karandish, 2019).

The fundamental concept of today's centralized water transportation networks for urban areas in developed nations dates back more than a century. Since then, the networks were continuously extended to incorporate more urban areas, as well as adapted to meet changing requirements of the consumers serviced, as well as environmental and health issues (Prajapati and Bergen Jensen 2019; Prajapati et al. in 2019). Furthermore, because these web servers have a prolonged technical lifetime and significant sunk expenditures, they are highly technologically dependent. Even if the conventional urban water and sanitation plan are sufficient to meet the needs of an expanding globe, it is questionable whether it meets environmental criteria in general (Ghawi, 2019).

1.1. Greywater Composition

The content of greywater varies, and it is usually a representation of one's lifestyle and also the type and chemical selection used in laundry, cleaning, and bathing. The water quality origin and the type of distribution network have an impact on the greywater quality (Qi et al., 2019). There will be significant changes in greywater composition from location to location and over time, which could be due to differences in water demand against the amount released. Some compounds in the transportation and storage network can deteriorate chemically and biologically, which could affect the composition (Samayamanthula et al., 2019). Greywater contains a substantial amount of rapidly biodegradable organic materials and some essential elements, which are primarily produced by homes. Organic bacteria such as salmonella, E. coli, faecal coliforms, basic hydrochemical elements, and also nitrates and their compounds, phosphorus and its derivatives, xenobiotic organic compounds (XOCs), and xenobiotic organic compounds (XOCs), are among them. BOD (mg/l), COD (mg/l), TSS (mg/l), P(total), N(total), Total coliforms (MPN/ml), E. coli (MPN/ml), P(total), N(total), Total coliforms (MPN/ml), Total coliforms (MPN/ml), Total coliforms (MPN/ml), Total coliforms (MPN/m

1.2. Greywater System in Developing Countries

The Greywater System is being re-evaluated. In response to the problem of progressive water control depletion, several countries have aggressively developed greywater technology and regulations, as well as adapted them to home and agricultural applications. For example, in Los Angeles, 13 percent to 65% of greywater gathered from private households is used for irrigation. In a Brazilian family, using processed greywater to flush toilets lowered water consumption by 29–35 percent. Home water use accounts for 67 percent of total usage in Malaysia, and greywater recovery for toilet flushing can cut water usage by 30% (Vuppaladadiyam et al., 2019). The operating ratio of waste-water areas in Southern Korea is 26.5 percent, based on the delivery of 200 billion tons of greywater. The Japanese government does not provide incentives for people to install greywater systems in their homes. However, 70 percent of Japanese advocate the use of recycled water or rainfall because they are well conscious of the necessity to save water and since water costs are quite high in urban areas. China's rapid economic growth has resulted in a water deficit, which the government has addressed through many initiatives, including rules requiring greywater treatment and reuse for large-scale institutional and residential constructions (Wu, 2019).

1.3. Phytoremediation

Phytoremediation was an environmentally friendly biological wastewater treatment technology that can be used to clean kitchen wastewater. Macrophytes eliminate pollutants in this way. Pollutants, as well as water and other nutrients, are absorbed by plants (Yu et al., 2019). The contaminated mass does not degrade, but rather accumulates in the plant's shoots and leaves. This is a cost-effective and organic wastewater purification approach. The plants eliminate contaminants from wastewater by absorbing nutrients and breaking down inorganic and organic materials. The removal ability of wetland plants is determined by the plant type, sewage quality, root development rate and thickness, and nutrient absorption. The growth of the root system is linked to the oxygen-carrying ability and water conductivity of the root system. Plants that are adapted to the local environment and have a rapidly developing root system have better financial values and decontamination performance (Chandekar & Godboley, 2020). To eliminate, absorb, or filter pollutants from water and soil, found naturally or genetically altered plants are used.

Figure 1. Phytoremediation

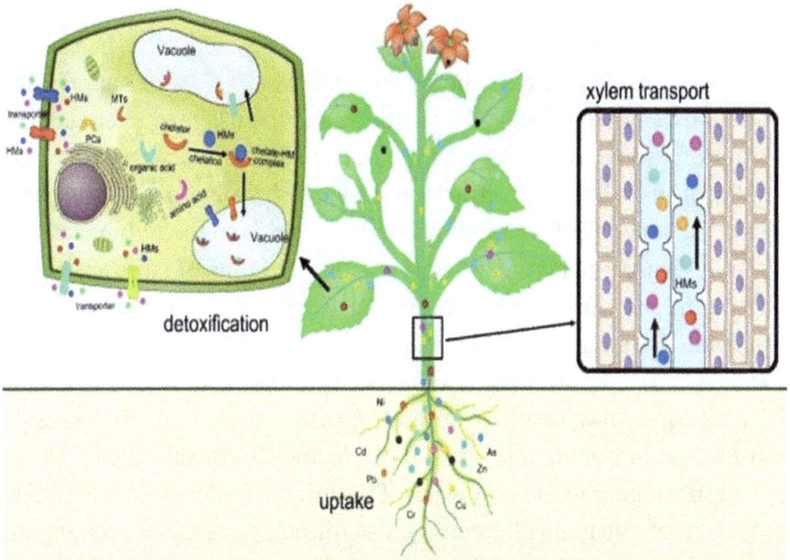

The plants were chosen specifically for the conditions at the wall's location, and they are simple to maintain which is shown in Figure 1. One plant grows well in low-light conditions, while the other does not. Other important elements that influence plant development include air moisture and workplace temperature. When coupled with the widest possible plant choice, the system looks after itself.

1.4. Biological Health Issue Using Greywater

The health risks associated with greywater reuse are mostly unknown due to the absence of quantitative hazard research. Moreover, no evidence linking greywater recycling to real health-care challenges such as disease epidemics was discovered. The wide variance in greywater reuse legislation between countries is most likely due to these factors. To avoid potential ecological and health problems, greywater should be managed appropriately. The quality of raw greywater varies greatly, and it frequently contains pathogens (Table 1). Typically, after bioremediation of greywater, a fall in microbial levels is noted, followed by a further decrease after disinfection. The pathogens Norovirus, Rotavirus, Cryptosporidium, and Campylobacter have specified the permitted limit values of these pathogens in sewage and greywater, as well as the pathogen-reduction standards.

1.5. Parameters Affecting the Characteristics of Greywater

The composition of greywater is determined by many elements, such as the water's supply and installation, and also the quality and kind of water supply, like piped water or groundwater well. Lifestyle, customs, and the usage of chemical goods are examples of domestic chores. Bathroom, kitchen sink, laundry wash, or hand basin are examples of installations from which greywater is obtained (Samayamanthula et al. 2019). The amount of water consumed in regards to the number of pollutants expelled Biological and chemical deterioration of chemical substances inside the transport system and during preservation has a significant influence. During the storage and transit of greywater, chemical reactions might occur, resulting in changes in the water's chemical makeup. Elevated concentrations of microorganisms, such as faecal coliforms, may result from biological growth. This could result in the formation of new inorganic and organic molecules as metabolites from partially decomposed substances in the greywater. This microbial development will be aided by the existence of nutrients like ammonium/nitrate, phosphate, and organic waste (Nepi et al. 2020).

1.6. Nutrients in Greywater

In developing nations where phosphorus-containing detergents are not yet outlawed, washing detergents are the principal cause of phosphates detected in greywater (Friedler et al. 2013). According to the study, greywater has a normal N/P ratio of 2, which is significantly lower than the N/P ratio of roughly 10 which is optimal for plant nutrient absorption. If greywater is utilized for irrigation, this is critical. Unless the plants can receive nitrogen from other resources, nitrogen becomes the limiting ingredient, resulting in sub-optimal phosphorus absorption.

2. EXPERIMENTAL PROCEDURE

2.1 Materials

2.1.1 Greywater Treatment Plant Design

In Salem, Tamilnadu, a greywater reuse system was erected in a residence with three families and ten people. Greywater is generated at a rate of 750-1000 liters per day. The processing plant has a five-number horizontal and vertical flow-created wetland tank, a 1000-liter collection tank, a 1000-liter retention tank, and a one-horsepower sewage motor. The "Manual on constructed wetland as an alternative technology for sewage management in India" (Beauchamp & Adamowski 2012) was used to design the horizontal flow constructed wetland tank (HFWC) of size L74 cm*B 32 cm* H 58.5 cm and the vertical flow constructed wetland tank (VFWC) of size L 51cm*B 21 cm* H80 cm The filter medium of HFWC and VFWC tanks are coarse silex, coarse sand, 20mm gravel, and 40mm gravel. The VFWC and HFWC individual wetland units receive respective main lines, sublines, and control valves from the retention tank. In the pilot-scale treatment technique, the greywater reuse plant was set to batch flow, and in the combined treatment method, it is set to constant flow. Below is a diagram of a greywater treatment facility with vertical and horizontal flow constructed wetland tanks. Figure 2 shows the greywater treatment plant with horizontal and vertical flow constructed wetland tanks.

Figure 2. Greywater treatment plant with horizontal and vertical flow constructed wetland tanks observed

2.1.2 Selection of Phytoremediation Plants

Phragmites australis, Canna indica, Cyperus rotundus, Cyperus pangorei, and Typha angustifolia were used to remediate greywater using phytoremediation technology. Because of their high development rate and resistance to a variety of environmental situations, these plants were chosen as phytoremediation agents in this chapter. Naturally or genetically produced plants that filter, absorb, or remove toxins from greywater have been chosen. Figure 3 shows the photographic view of Canna indica, Typha angustifolia, Cyperus Pangorei, Cyperus rotundus, and Phragmites australis

Figure 3. Photographic view of(a) Canna indica (b)Typha angustifolia (c) Cyperus Pangorei (d) Cyperus rotundus (e)Phragmites australis, from left to right

(a) (b) (c) (d) (e)

2.2 Method

Greywater was collected and treated in this technique from the bathroom, kitchen, and laundromat.. In order for greywater to be properly reused, the highest concentration of physical, chemical, and biological colonies should be removed. A single vertical unit and a horizontal unit were filled with unprocessed greywater till the freeboard of the tanks was reached, and the units were emptied fully after a 12-hour retention time. After Evaporation, the water was obtained in a container to analyze the initial measurements.

2.2.1 Pilot-Scale Treatment Method

The three phytoremediation plant samples were cleaned in water to eliminate foreign pollutants and immersed in an aloe vera pulp mixture to add micronutrients for improved growth before being planted in the five HFWC tanks and five VFCW tanks. For the 90-day study period, the greywater was irrigated once every 24 hours. Chemical Oxygen Demand (COD), Total suspended solids (TSS), Chloride as Cl-, Electrical conductivity (Ec), Bio-Chemical Oxygen Demand (BOD), Total Dissolved solids (TDS), Sulphate as SO4 -, and the existence and minimization of five distinct kinds of bacteria were measured at intervals of 30,45,60,75, and 90 days at a retention time of 12 hours by gathering water specimens. Tables 1, 2, 3, 4, and 5 show the microbiological results acquired and flow chart representation is shown in Figure 4.

Figure 4. Flow chart representation of Pilot-scale treatment method

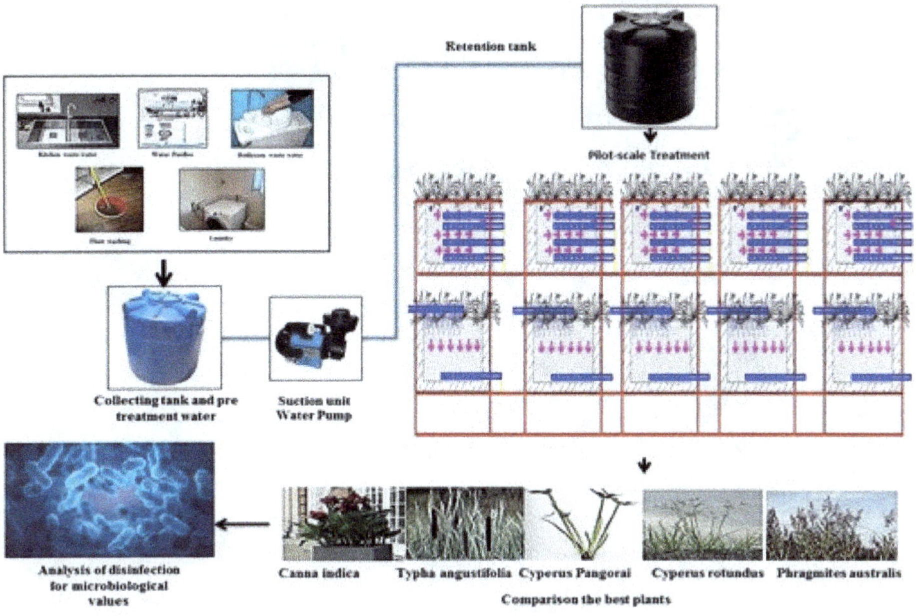

2.2.2 Combination Treatment Method

Typha angustifolia from the HWCT tank and Cyperus rotundus from the VFCW tank is regarded as the most practicable plants to show greater removal efficiency % of chemical and physical water factors based on the results and findings from the pilot-scale research. To treat domestic greywater, a combined treatment system with a VFWC tank with Cyperus rotundus and an HFWC tank with Typha angustifolia were installed, with filter media such as coarse silex, coarse sand, gravel 40mm, gravel 20mm, granular coconut-based of 900 iodine value activated carbon filter, and a disinfection component. Once every 24 hours, the greywater constant flow cycle was performed. For a 90-day trial, greywater from the active carbon filter unit outflow was collected and tested. The water variables were examined in the same way that the pilot-scale treatment process was tested. At Insitu and in the laboratory, the observation variables were collected at intervals of 30, 45, 60, 75, and 90 days with a 12-hour retention time. Table 6 shows the results of the biological testing and flow chart representation is shown in Figure 5.

Figure 5. Flow chart representation of combination treatment method

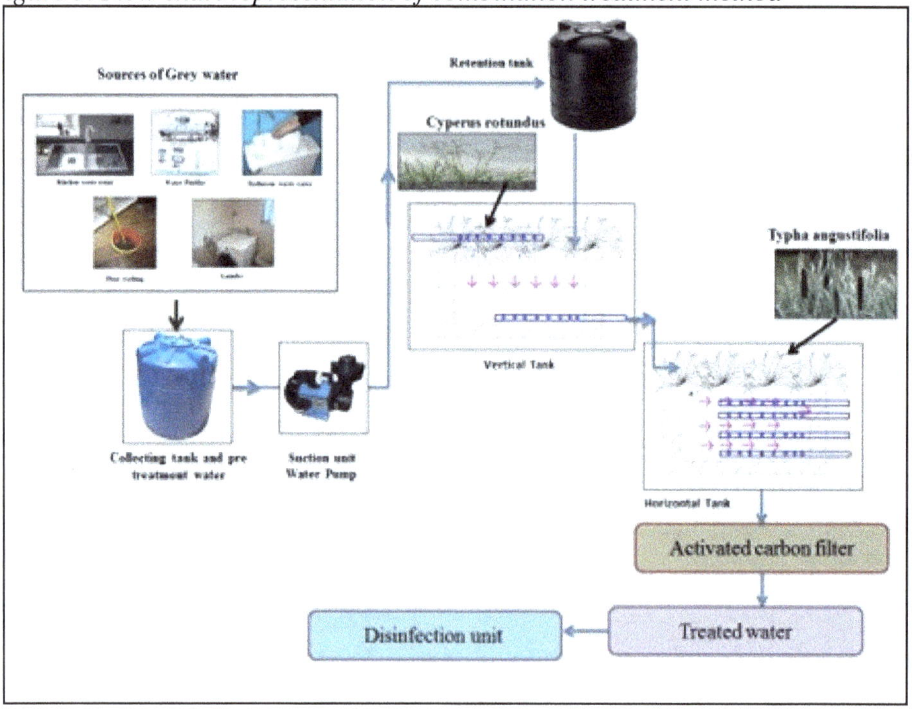

3. RESULTS AND DISCUSSIONS

The treated wastewater from the 5 HFWC and VFWC was collected and examined to determine the efficiency of the processing beds and the practicality of using plants in greywater treatment. The purity of greywater was assessed both after and before treatment, and the findings were compared to the baseline value. The following are the observations made. The most viable plants from the five HFWC and VFWC tanks were improved using the pilot-scale treatment procedure.

3.1. Canna indica

The biological analysis of Canna indica in both horizontal (H) and vertical (V) configurations over a period of 90 days revealed consistent microbial presence without significant reductions across the monitored intervals. As shown in **Table 1**, E. coli concentrations remained consistently above >1600 MPN/100ml in both horizontal and vertical systems, indicating persistent contamination levels. Salmonella spp levels in the horizontal setup showed minimal variation from an initial value of 11.0 MPN/l to 13.0 MPN/l at 75 days, whereas in the vertical system, a moderate increase was observed, rising from 18 to 22 MPN/l at 75 days, followed by a slight reduction. For Staphylococcus aureus, counts fluctuated marginally but remained within >6 to >9 CFU/100ml throughout the observation period for both orientations. Similarly, Pseudomonas aeruginosa and Shigella spp levels stayed consistently high, demonstrating the limited efficacy of Canna indica in reducing bacterial load under the experimental conditions.

Table 1. Results showing biological analysis of Canna indica (horizontal & vertical)

S. No	Name of the Bacteria	Units	Initial	Canna indica (H)P1(Days)					Initial	Canna indica (V)P1(Days)				
				30	45	60	75	90		30	45	60	75	90
1	E. coli	MPN/100ml	>1600	>1600	>1600	>1650	>1600	>1600	>1600	>1600	>1600	>1600	>1600	>1600
2	Salmonella Spp	MPN/1ltr	11.0	11.0	11.0	13.0	11.0	11.0	18	18	18	22	18	17
3	Staphylococcus aureus	CFU/100 ml	>8	>9	>8	>6	>9	>7	>8	>9	>8	>6	>9	>7
4	Pseudomonas aeruginosa	CFU/100 ml	>7	>6	>7	>7	>7	>7	>7	>6	>7	>7	>7	>7
5	Shigella spp	CFU/100 ml	>5	>5	>5	>6	>5	>5	>5	>5	>5	>6	>5	>5

3.2. Phragmites australis

In the case of Phragmites australis, the microbial analysis (Table 2) also exhibited limited attenuation of bacterial populations over time. The E. coli levels slightly fluctuated but remained predominantly above >1600 MPN/100ml in both horizontal and vertical flow systems, indicating no significant bacterial removal. Salmonella spp levels increased from an initial 10 MPN/l to a peak of 16 MPN/l in the vertical setup by 75 days, suggesting a potential proliferation under certain conditions. Staphylococcus aureus, Pseudomonas aeruginosa, and Shigella spp remained within a narrow range of values, again indicating minimal treatment efficacy in bacterial removal across the sampling period.

Table 2. Results showing biological analysis of Phragmites australis (H & V)

S. No	Name of the Bacteria	Units	Initial	Phragmites australis (H)P2(Days)					Initial	Phragmites australis (V)P2(Days)				
				30	45	60	75	90		30	45	60	75	90
1	E. coli	MPN/100ml	>1600	>1600	>1650	>1600	>1650	>1600	>1600	>1600	>1550	>1600	>1675	>1600
2	Salmonella Spp	MPN/1ltr	10	10	12	10	14	10	14	14	14	16	14	13
3	Staphylococcus aureus	CFU/100 ml	>8	>9	>8	>6	>9	>7	>8	>9	>8	>6	>9	>7
4	Pseudomonas aeruginosa	CFU/100 ml	>7	>6	>7	>7	>7	>7	>7	>6	>7	>7	>7	>7
5	Shigella spp	CFU/100 ml	>5	>5	>5	>6	>5	>5	>5	>5	>5	>6	>5	>5

3.3. Typha angustifolia

Analysis of Typha angustifolia systems presented in Table 3 showed a similar trend of microbial persistence. The E. coli counts remained well above >1600 MPN/100ml across both H and V configurations, with occasional peaks up to >1650 MPN/100ml, suggesting ineffective filtration or bacterial die-off. Salmonella spp exhibited more pronounced fluctuations, particularly in the vertical setup where levels ranged from 17 to 22 MPN/l, highlighting sporadic microbial proliferation. Other bacterial species such as Staphylococcus aureus, Pseudomonas aeruginosa, and Shigella spp also displayed minimal variation, reflecting insufficient microbial suppression.

Table 3. Results showing biological analysis of Typha angustifolia (H & V)

S. No	Name of the Bacteria	Units	Initial	Typha angustifolia (H)P3(Days)					Initial	Typha angustifolia (V)P3(Days)				
				30	45	60	75	90		30	45	60	75	90
1	E. coli	MPN/100ml	>1600	>1600	>1650	>1650	>1600	>1600	>1600	>1625	>1600	>1600	>1650	>1600
2	Salmonella Spp	MPN/1ltr	15	15	12	15	15	15	17	19	17	22	22	17
3	Staphylococcus aureus	CFU/100 ml	>8	>9	>8	>6	>9	>7	>8	>9	>8	>6	>9	>7
4	Pseudomonas aeruginosa	CFU/100 ml	>7	>6	>7	>7	>7	>7	>7	>6	>7	>7	>7	>7
5	Shigella spp	CFU/100 ml	>5	>5	>5	>6	>5	>5	>5	>5	>5	>6	>5	>5

3.4. Cyperus pangorei

The data for Cyperus pangorei (Table 4) indicated a similarly high and persistent bacterial load. E. coli values remained above >1600 MPN/100ml, with minor deviations observed by day 75 and 90. Salmonella spp showed a marked increase from an initial 12 MPN/l to as high as 27 MPN/l by day 60 in the vertical flow setup, followed by a slight decline. Meanwhile, Staphylococcus aureus and Pseudomonas aeruginosa populations showed minor fluctuations, staying within established ranges and again underscoring limited bacteriological control by the plant system.

Table 4. Results showing biological analysis of Cyperus pangorei (H & V)

S. No	Name of the Bacteria	Units	Initial	Cyperus Pangorei (H)P4 (Days)					Initial	Cyperus pangorei (V)P4 (Days)				
				30	45	60	75	90		30	45	60	75	90
1	E. coli	MPN/100ml	>1600	>1600	>1600	>1585	>1690	>1600	>1600	>1600	>1675	>1695	>1600	>1600
2	Salmonella Spp	MPN/1ltr	12	12	12	19	23	19	21	25	27	21	23	21
3	Staphylococcus aureus	CFU/100 ml	>7	>6	>7	>7	>7	>7	>7	>6	>7	>7	>7	>7
4	Pseudomonas aeruginosa	CFU/100 ml	>8	>9	>8	>6	>9	>7	>8	>9	>8	>6	>9	>7
5	Shigella spp	CFU/100 ml	>5	>5	>5	>6	>5	>5	>5	>5	>5	>6	>5	>5

3.5. Cyperus rotundus

Biological evaluation of Cyperus rotundus, as documented in Table 5, reinforced the findings from previous macrophytes. E. coli remained consistently above >1600 MPN/100ml throughout the test period, showing negligible reduction. Salmonella spp in both horizontal and vertical setups hovered around 13–14 MPN/l with minimal temporal variation, indicating bacterial resilience. Other bacteria including Staphylococcus aureus, Pseudomonas aeruginosa, and Shigella spp remained largely unchanged, mirroring the performance trend of earlier test species.

Table 5. Results showing biological analysis of Cyperus rotundus (H & V)

S. No	Name of the Bacteria	Units	Initial	Cyperus rotundus(H)P5(Days)					Initial	Cyperus rotundus(V)P5 (Days)				
				30	45	60	75	90		30	45	60	75	90
1	E. coli	MPN/100ml	>1600	>1600	>1650	>1690	>1600	>1600	>1600	>1600	>1600	>1658	>1622	>1600
2	Salmonella Spp	MPN/1ltr	13	13	13	17	13	13	13	13	13	13	14	14
3	Staphylococcus Aureus	CFU/100 ml	>8	>9	>8	>6	>9	>7	>8	>9	>8	>6	>9	>7
4	Pseudomonas aeruginosa	CFU/100 ml	>7	>6	>7	>7	>7	>7	>7	>6	>7	>7	>7	>7
5	Shigella spp	CFU/100 ml	>5	>5	>5	>6	>5	>5	>5	>5	>5	>6	>5	>5

In contrast, the combination treatment using Typha angustifolia (H) and Cyperus rotundus (V) integrated with an activated carbon filter, as shown in Table 6, demonstrated a marginally improved biological treatment performance. Although E. coli levels remained high (above >800 MPN/100ml), a slight decline to >774 MPN/100ml was noted by day 90, suggesting some potential bacterial adsorption or entrapment by the activated carbon. Salmonella spp exhibited modest variation, with counts dropping from a peak of 30 to 25 MPN/l by day 90, indicating partial attenuation. Other monitored species (Staphylococcus aureus, Pseudomonas aeruginosa, and Shigella spp) maintained consistent ranges, but the combined system displayed slightly better performance than individual plant systems, implying the added benefit of carbon filtration in improving microbial water quality.

Table 6. Results showing biological analysis of combination of Typha angustifolia (H)+Cyperus rotundus(V) with activated carbon filter (Days)

S. No	Name of the Bacteria	Units	Initial	Combination of Typha angustifolia (H)+Cyperus rotundus(V) with Activated carbon Filter (Days)				
				30	**45**	**60**	**75**	**90**
1	E. coli	MPN/100ml	>800	>800	>807	>800	>800	>774
2	Salmonella Spp	MPN/1ltr	25	30	30	27	25	25
3	Staphylococcus aureus	CFU/100 ml	>8	>9	>8	>6	>9	>7
4	Pseudomonas aeruginosa	CFU/100 ml	>7	>6	>7	>7	>7	>7
5	Shigella spp	CFU/100 ml	>5	>5	>5	>6	>5	>5

4. PROCESS OF DISINFECTION

Disinfection was a chemical or physical procedure that eliminates all kinds of microbial life. It is a tried and true method of eliminating waterborne illnesses. Disinfectants are utilized in this process, which acts by breaking germs' cell walls or interacting with their metabolism. Although the water produced by the phytoremediation technique in this study does not meet the quality criteria (Friedler et al., 2013), disinfection action is needed to eradicate the pathogenic bacteria. To test the disinfection action, a beaker sample trial of 1 percent, 2 percent, and 3 percent sodium hypochlorite was introduced to 1 liter of untreated greywater. Pseudomonas aeruginosa, Salmonella, E. coli, Staphylococcus aureus, and Shigella Spp colonies had low clearance efficiency at these quantities. Finally, Sodium Hypochlorite (4 percent) W/V available chlorine was included, which entirely eradicated the selected critical greywater bacteria. The treated greywater was then suggested for organic farming, washing hands, flushing toilet, and other secondary purposes without any risks.

5. CONCLUSION

In this chapter, a substantial amount of recovered greywater was discovered to be utilized as an alternative water supply in households, however, the intake is heavily influenced by safety and health concerns. This chapter looked at how effective it was. With phytoremediation technology, the chemical, biochemical, and physical properties of Typha angustifolia from the HFWC tanks and Cyperus rotundus from the VFWC tanks were decreased to the agreed-upon standard range. Phytoremediation can remove inorganic and organic adulterants, but it also has the power to separate or detach pathogens that are regarded vital. The important greywater microbes were completely removed utilizing a decontamination unit with the addition of sodium hypochlorite (4 percent) W/V available chlorine and the treated greywater was then recommended for washing hands, organic horticulture, flushing toilet, and other secondary utilizes without any risks.

REFERENCES

Abdel-Shafy, H. I., Mansour, M. S. M., & Al-Sulaiman, A. M. (2019). Anaerobic/aerobic integration via UASB/enhanced aeration for greywater treatment and unrestricted reuse. *Water Practice & Technology*, *14*(4), 837–850. DOI: 10.2166/wpt.2019.065

Al-Gheethi, A. A. S., Noman, E. A., Mohamed, R. M. S. R., Talip, B. A., Abdullah, A. H., & Kassim, A. H. M. (2019). Reuse of greywater for irrigation purpose. In Mohamed, R. M. S. R., Al-Gheethi, A. A. S., & Kassim, A. H. M. (Eds.), *Management of greywater in developing countries* (pp. 73–87). Springer. DOI: 10.1007/978-3-319-90269-2_4

Al-Mefleh, N. K., Othman, Y. A., Tadros, M. J., Al-Assaf, A., & Talozi, S. (2021). An assessment of treated greywater reuse in irrigation on growth and protein content of *Prosopis* and *Albizia*. *Horticulturae*, *7*(3), 38. DOI: 10.3390/horticulturae7030038

Amiri, M. J., Bahrami, M., Badkouby, M., & Kalavrouziotis, I. (2019). Greywater treatment using single and combined adsorbents for landscape irrigation. *Environmental Processes*, *6*(1–3), 43–63. DOI: 10.1007/s40710-019-00362-1

Ansari, K., & Shrikhande, A. N. (2019). Feasibility on grey water treatment by electrocoagulation process: A review. *International Journal of Emerging Technology*, *10*(1), 92–99.

. Friedler, E., Butler, D., & Alfiya, Y. (2013). Wastewater composition. *Source separation and decentralization for wastewater management*, 241-257.

Ghawi, A. H. (2019). Development of the greywater domestic treatment unit for irrigation of the garden in rural areas. *Journal of Ecological Engineering*, *20*(3), 201–208. DOI: 10.12911/22998993/99732

Inyinbor, A. A., Bello, O. S., Oluyori, A. P., Inyinbor, H. E., & Fadiji, A. E. (2019). Wastewater conservation and reuse in quality vegetable cultivation: Overview, challenges, and future prospects. *Food Control*, *98*, 489–500. DOI: 10.1016/j.foodcont.2018.12.008

Jamil, R. (2019). Reduction in fresh water consumption by grey water reuse for flushing and irrigation: A case study of a multistorey hotel building. In *Proceedings of the 1st Conference on Sustainable Civil Engineering*.

Janani, T., Sudarsan, J. S., & Prasanna, K. (2019). Grey water recycling with corn cob as an adsorbent. *AIP Conference Proceedings*, *2112*(1), 020181. DOI: 10.1063/1.5112366

Karandish, F. (2019). Applying grey water footprint assessment to achieve environmental sustainability within a nation under intensive agriculture: A high-resolution assessment for common agrochemicals and crops. *Environmental Earth Sciences*, *78*(6), 1–22. DOI: 10.1007/s12665-019-8199-y

Laaffat, J., Aziz, F., Ouazzani, N., & Mandi, L. (2019). Biotechnological approach of greywater treatment and reuse for landscape irrigation in small communities. *Saudi Journal of Biological Sciences*, *26*(1), 83–90. DOI: 10.1016/j.sjbs.2017.01.006 PMID: 30622410

Mohamed, R. M. S. R., Al-Gheethi, A. A. S., & Kassim, A. H. M. (Eds.). (2019). *Management of greywater in developing countries: Alternative practices, treatment and potential for reuse and recycling*. Springer International Publishing. DOI: 10.1007/978-3-319-90269-2

Nel, N., & Jacobs, H. E. (2019). Investigation into untreated greywater reuse practices by suburban households under the threat of intermittent water supply. *Journal of Water, Sanitation, and Hygiene for Development : a Journal of the International Water Association*, *9*(4), 627–634. DOI: 10.2166/washdev.2019.055

Vuppaladadiyam, A. K., Merayo, N., Prinsen, P., Luque, R., Blanco, A., & Zhao, M. (2019). A review on greywater reuse: Quality, risks, barriers and global scenarios. *Reviews in Environmental Science and Biotechnology*, *18*(1), 77–99. DOI: 10.1007/s11157-018-9487-9

Wu, B. (2019). Membrane-based technology in greywater reclamation: A review. *The Science of the Total Environment*, *656*, 184–200. DOI: 10.1016/j.scitotenv.2018.11.347 PMID: 30504020

Chapter 6
Tuning Cobalt Ferrite Nanostructures Through Ni and La Doping for Efficient Visible-Light-Driven Wastewater Purification

G. Myvizhi
https://orcid.org/0000-0003-2326-6146
Vellalar College for Women, India

S. Malathi
Vellalar College for Women, India

V. Vijayalakshmi
https://orcid.org/0000-0002-7157-2917
Erode Sengunthar Engineering College, India

M. Helen
Vellalar College for Women, India

ABSTRACT

The increasing contamination of aquatic ecosystems with persistent organic pollutants has intensified the search for efficient and sustainable wastewater treatment technologies. In this study, cobalt ferrite ($CoFe_2O_4$) nanoparticles and their doped counterparts—Ni@$CoFe_2O_4$ and La@Ni@$CoFe_2O_4$—were synthesized via

DOI: 10.4018/979-8-3373-5886-4.ch006

a co-precipitation method and systematically evaluated for their photocatalytic performance in the degradation of methylene blue (MB) dye under visible light irradiation. Among the samples, La@Ni@$CoFe_2O_4$ demonstrated the highest photocatalytic activity, achieving a superior degradation rate of methylene blue due to the synergistic effects of dual doping. Additionally, the intrinsic magnetic properties of the ferrites allowed for easy recovery and reuse of the catalysts, highlighting their practical applicability in wastewater remediation. This study underscores the potential of co-doped $CoFe_2O_4$ nanostructures as efficient, magnetically separable photocatalysts for environmental applications, particularly in the treatment of dye-laden industrial effluents.

1. INTRODUCTION

Photocatalysis has emerged as a green and effective strategy for the degradation of hazardous contaminants in wastewater, offering the potential to mineralize persistent organic pollutants using only light energy and a semiconductor material (Abbasi et al., 2021). This advanced oxidation process relies on the generation of electron–hole pairs under light irradiation, which subsequently produce highly reactive species capable of breaking down a wide range of toxic compounds Dehghani Dastjerdi et al., 2023). The performance of a photocatalyst in wastewater treatment depends on critical factors such as its band gap energy, charge separation efficiency, surface morphology, and long-term stability in aqueous environments. Among the diverse class of photocatalytic materials, spinel ferrites have garnered significant interest due to their favorable physicochemical properties and versatility in heterogeneous catalysis. In particular, cobalt ferrite ($CoFe_2O_4$) has demonstrated promising potential for wastewater treatment applications owing to its moderate band gap (~1.9–2.3 eV), strong magnetic properties for easy recovery, and excellent chemical durability under operating conditions (Abdel-Aal & Abdel-Rahman, 2020). However, the photocatalytic efficiency of pristine $CoFe_2O_4$ is limited by the fast recombination of photo-induced charge carriers, which reduces the generation of reactive oxygen species essential for pollutant degradation (Hedayati et al., 2017).

To address these limitations, material engineering approaches such as doping and heterostructure formation are frequently employed. Doping $CoFe_2O_4$ with transition and rare-earth elements can effectively tailor its electronic band structure, enhance visible light absorption, and suppress charge carrier recombination (Hashemi et al., 2019). Nickel (Ni) doping improves electronic conductivity and promotes efficient charge separation, while lanthanum (La) doping introduces oxygen vacancies and defect states that extend light absorption and facilitate redox reactions (Rathod et al., 2019). These modifications contribute to a significant enhancement in photocatalytic degradation efficiency. In this study, $CoFe_2O_4$, Ni-doped $CoFe_2O_4$ (Ni@$CoFe_2O_4$), and La and Ni co-doped $CoFe_2O_4$ (La@Ni@$CoFe_2O_4$) nanoparticles were synthesized via a co-precipitation method and thoroughly characterized using structural, optical, and morphological techniques. Their photocatalytic activities were evaluated by monitoring the degradation of methylene blue (MB) dye, a common pollutant in textile and industrial effluents, under visible light irradiation. This work aims to understand the synergistic effects of Ni and La co-doping on the photocatalytic performance of $CoFe_2O_4$, and to explore its potential as a recyclable,

2. MATERIALS AND METHODS

2.1 Materials

The synthesis of cobalt ferrite ($CoFe_2O_4$) and its doped variants required high-purity analytical-grade chemicals. Cobalt nitrate hexahydrate ($Co(NO_3)_2{\cdot}6H_2O$) and ferric nitrate nonahydrate ($Fe(NO_3)_3{\cdot}9H_2O$) were used as primary precursors. For nickel-doped cobalt ferrite, nickel nitrate hexahydrate ($Ni(NO_3)_2{\cdot}6H_2O$) was incorporated into the precursor solution. Similarly, lanthanum nitrate hexahydrate ($La(NO_3)_3{\cdot}6H_2O$) was introduced to synthesize La and Ni co-doped cobalt ferrite. Sodium hydroxide (NaOH) was employed as the precipitating agent to facilitate the co-precipitation process. All solutions were prepared using distilled water to ensure high purity and prevent contamination during synthesis.

2.2 Synthesis Procedure

The nanoparticles were synthesized using the co-precipitation method, a widely utilized technique for producing homogeneous and high-purity metal oxide nanostructures. Initially, stoichiometric amounts of $Co(NO_3)_2{\cdot}6H_2O$ and $Fe(NO_3)_3{\cdot}9H_2O$ were dissolved in distilled water under continuous stirring to form a clear solution. For nickel doping, $Ni(NO_3)_2{\cdot}6H_2O$ was added, while for co-doping with lanthanum, $La(NO_3)_3{\cdot}6H_2O$ was also included in the precursor mixture. The solution was stirred vigorously at room temperature to ensure complete dissolution of all metal salts.

A separate solution of 1 M NaOH was prepared and added dropwise to the precursor solution while maintaining constant stirring. The pH of the solution was adjusted to approximately 10–11 to ensure the complete precipitation of hydroxides. The reaction mixture was maintained at 80°C and stirred for two hours to allow the formation of nanoparticles. After completion of the precipitation process, the resulting precipitate was allowed to age for 12 hours at room temperature to improve particle crystallization.

The precipitate was then collected by centrifugation and washed multiple times with distilled water and ethanol to remove unreacted ions and residual impurities. After washing, the sample was dried at 100°C in a hot air oven for 12 hours. The dried powder was further calcined at 500°C for three hours in a muffle furnace to enhance its crystallinity and phase purity. The final synthesized $CoFe_2O_4$, Ni@ $CoFe_2O_4$, and La@Ni@$CoFe_2O_4$ nanoparticles were subjected to various characterization techniques, including X-ray diffraction (XRD), field emission scanning electron microscopy (FESEM), Fourier-transform infrared spectroscopy (FTIR), and UV-visible spectroscopy, to analyze their structural, morphological, optical, and photocatalytic properties (Ali et al., 2020).

3. RESULTS AND DISCUSSION

3.1 PXRD Analysis

The X-ray diffraction (XRD) pattern in the provided image displays the characteristic peaks of cobalt ferrite ($CoFe_2O_4$) along with its modified forms, Ni@ $CoFe_2O_4$ and La@Ni@$CoFe_2O_4$. The indexed peaks correspond to the cubic spinel structure, with reflections at (220), (222), (311), (400), (331), (511), and (440). The preservation of these peaks in the Ni- and La-doped samples indicates that the fundamental spinel structure remains intact despite doping. A slight shift in peak positions may be observed due to the ionic substitution of Ni^{2+} and La^{3+}, which can lead to lattice distortions (Zaharieva et al., 2015). The sharp and well-defined peaks suggest that all three samples exhibit good crystallinity. Variations in peak intensity between the samples indicate possible changes in crystallite size or strain effects induced by doping. The incorporation of Ni and La may also introduce defect states, which could influence the material's physical and chemical properties, such as magnetic behavior and electrical conductivity. Further analysis, such as Rietveld refinement, could provide more insights into lattice parameter changes and structural modifications caused by doping (Vijayaraghavan et al., 2016).

Figure 1. PXRD pattern of synthesized $CoFe_2O_4$, $Ni@CoFe_2O_4$, and $La@Ni@CoFe_2O_4$ NPs

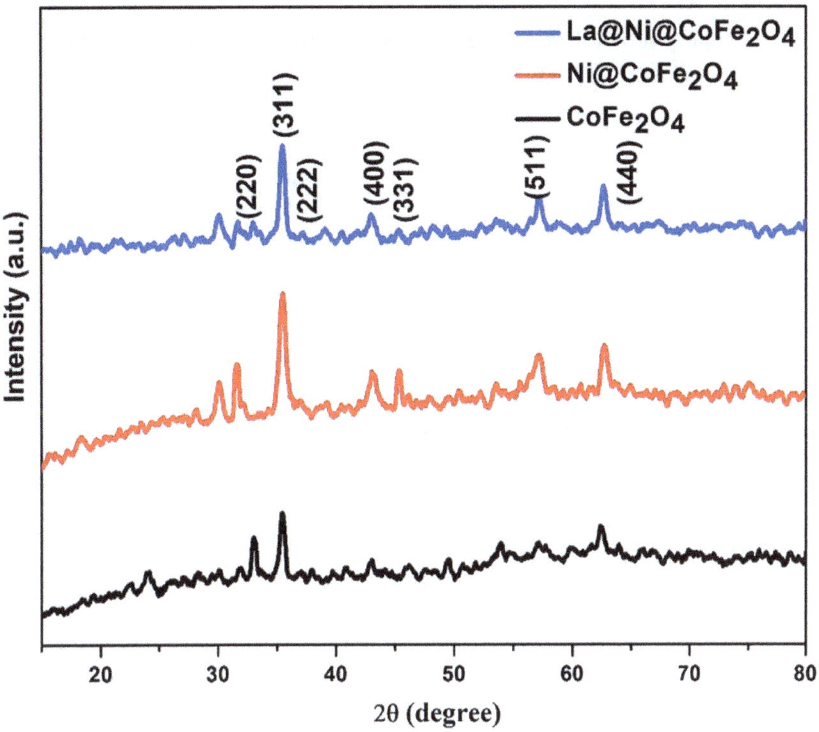

3.2 UV-Vis Analysis

The UV-Vis absorbance spectra and Tauc plots provided in the images illustrate the optical properties of $CoFe_2O_4$, $Ni@CoFe_2O_4$, and $La@Ni@CoFe_2O_4$. The absorbance spectra indicate that all samples exhibit strong absorption in the visible region, with variations in intensity due to doping effects. The Ni and La incorporation appears to enhance absorbance, which could be attributed to increased charge carrier interactions or structural modifications.

The Tauc plots, derived from the absorbance data, reveal the bandgap values of the samples. The estimated bandgap values are approximately 2.06 eV for $CoFe_2O_4$, 2.11 eV for $Ni@CoFe_2O_4$, and 2.16 eV for $La@Ni@CoFe_2O_4$. This increase in bandgap with Ni and La doping suggests modifications in the electronic structure, potentially improving the photocatalytic activity by reducing recombination rates and enhancing charge separation. The shift in bandgap energy aligns with the expectation that doping introduces localized states or modifies the conduction and valence band positions (Sakthivel et al., 2003).

Figure 2. (a) UV-Vis absorbance spectra and (b) Tauc plot for $CoFe_2O_4$, Ni@ $CoFe_2O_4$, and La@Ni@$CoFe_2O_4$ NPs

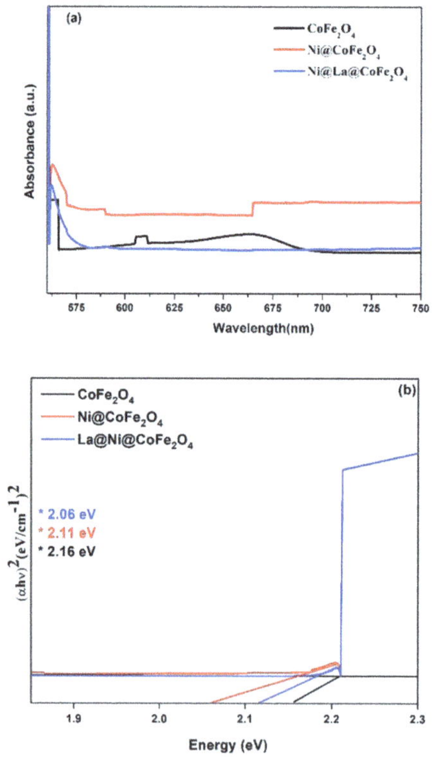

3.3 FTIR Analysis

The FTIR spectrum presented in the image shows the transmittance spectra of $CoFe_2O_4$, Ni@$CoFe_2O_4$, and La@Ni@$CoFe_2O_4$, highlighting their functional group vibrations. The characteristic peaks observed in the range of 400–700 cm^{-1} correspond to metal-oxygen (M–O) stretching vibrations, confirming the presence of the spinel ferrite structure. The absorption bands around 500–600 cm^{-1} can be attributed to Fe–O and Co–O stretching vibrations, which are typical of cobalt ferrite nanoparticles. Additionally, the broad band observed near 3400 cm^{-1} is associated with the O–H stretching vibration, indicating the presence of adsorbed water molecules or hydroxyl groups. The peaks around 1600 cm^{-1} correspond to the bending vibrations of water molecules. The modifications in peak intensity and position for Ni- and La-doped $CoFe_2O_4$ suggest changes in the metal-oxygen bond environment, confirming the successful incorporation of Ni and La into the ferrite lattice (Mahboubeh et al., 2014; Khurshid et al., 2019).

Figure 3. FTIR Spectrum of $CoFe_2O_4$, $Ni@CoFe_2O_4$, and $La@Ni@CoFe_2O_4$ NPs

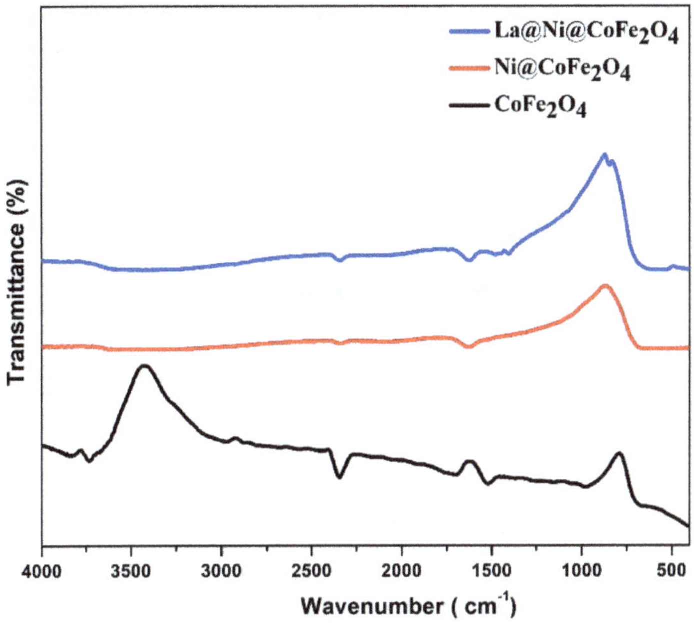

3.4 Raman Analysis

The Raman spectrum of $CoFe_2O_4$, $Ni@CoFe_2O_4$, and $La@Ni@CoFe_2O_4$ exhibits characteristic vibrational modes associated with the spinel ferrite structure. The peaks observed in the low wavenumber region (approximately 200–700 cm^{-1}) correspond to $T_2g(2)$ and $A_1g(2)$ vibrational modes, which are related to metal-oxygen (M–O) interactions within the tetrahedral and octahedral sites of the ferrite lattice. These peaks confirm the structural stability of the spinel ferrite phase. Additionally, the presence of a distinct D-band around 1350 cm^{-1} in the Ni- and La-doped samples indicates structural defects or disorder, which may be caused by doping-induced modifications in the crystal lattice. The introduction of Ni^{2+} and La^{3+} ions results in slight peak shifts and intensity variations, demonstrating their incorporation into the host structure and affecting the bond strength within the metal-oxygen framework. Overall, the Raman analysis confirms that Ni and La doping alters the vibrational characteristics of $CoFe_2O_4$ while preserving its fundamental spinel ferrite structure, with minor lattice distortions contributing to observed spectral changes (Nandhini et al., 2025).

Figure 4. Raman Spectrum of $CoFe_2O_4$, Ni@$CoFe_2O_4$, and La@Ni@$CoFe_2O_4$ NPs

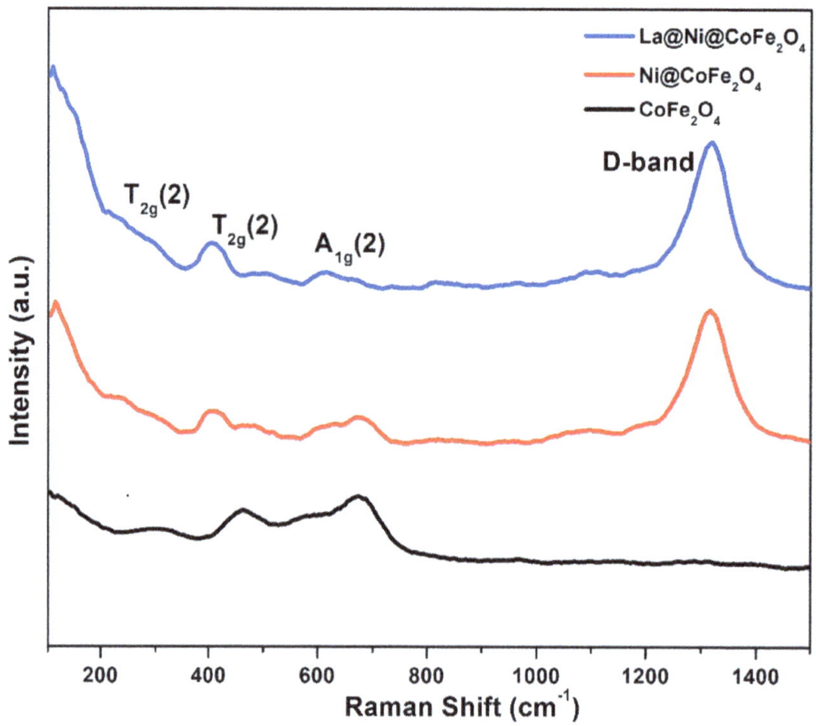

3.5 FESEM Analysis

The FESEM images provide a detailed morphological analysis of $CoFe_2O_4$, Ni@$CoFe_2O_4$, and La@Ni@$CoFe_2O_4$ nanoparticles. In Figure (a), the $CoFe_2O_4$ nanoparticles exhibit an aggregated and porous structure with a high degree of surface roughness. The particles appear to be agglomerated, possibly due to magnetic interactions, which is common in ferrite-based materials. Figure (b) represents Ni@ $CoFe_2O_4$, where a relatively smoother surface morphology is observed compared to pristine $CoFe_2O_4$. The presence of Ni could influence the particle distribution, leading to a slightly more uniform structure. In Figure (c), La@Ni@$CoFe_2O_4$ nanoparticles show an even denser and more compact morphology, which may be attributed to the incorporation of La ions, potentially enhancing crystallinity and reducing porosity. The overall analysis suggests that doping with Ni and La modifies the surface structure, which could impact the material's functional properties such as catalytic activity, magnetism, and surface area (Fu et al., 2013; Lu et al., 2022).

Figure 5. FESEM images of (a) $CoFe_2O_4$, (b) Ni@$CoFe_2O_4$, and (c) La@Ni@$CoFe_2O_4$ NPs

3.6 PL Analysis

The photoluminescence (PL) spectrum of $CoFe_2O_4$, Ni@$CoFe_2O_4$, and La@Ni@$CoFe_2O_4$ reveals significant changes in emission behavior due to doping modifications. The pure $CoFe_2O_4$ sample exhibits lower-intensity emissions, with peaks appearing around 360–435 nm, which correspond to fundamental recombination processes. Upon Ni doping, the intensity of PL emissions increases, indicating alterations in charge carrier dynamics and defect states. The incorporation of La further enhances the emission intensity, with additional peaks observed at 491 nm and 536 nm. This suggests the introduction of new energy levels, possibly associated with oxygen vacancies, interstitial defects, or surface states. A lower PL intensity generally implies a reduced electron-hole recombination rate, which is desirable for photocatalytic applications. Among the samples, $CoFe_2O_4$ exhibits the lowest PL intensity, suggesting a better separation of charge carriers and potential for enhanced photocatalytic efficiency. However, Ni and La doping introduce modifications that may improve light absorption in the visible region, making these materials suitable for optoelectronic applications. Further investigations, such as time-resolved PL studies or photocatalytic performance tests, would provide deeper insights into the charge carrier dynamics and practical applications of these materials (Menchaca-Nal et al., 2023).

Figure 6. PL Spectrum of $CoFe_2O_4$, $Ni@CoFe_2O_4$ and $La@Ni@CoFe_2O_4$ NPs

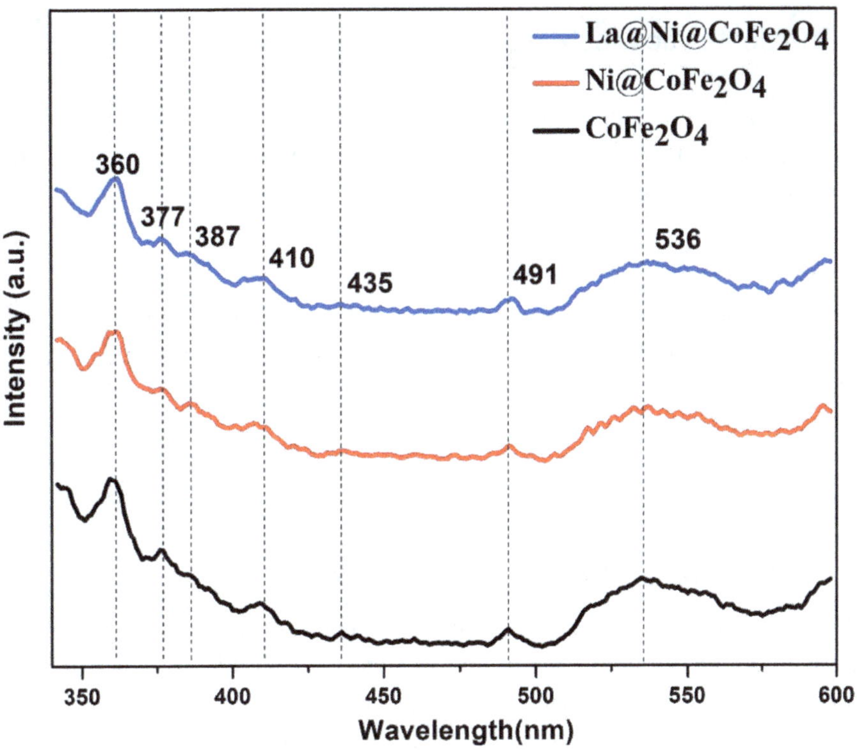

3.7 Photocatalytic studies

The photocatalytic analysis of the synthesized nanoparticles was evaluated using methylene blue (MB) dye degradation under visible light irradiation. The absorbance spectra corresponding to the degradation process are depicted in Figures (a), (b), and (c), representing $CoFe_2O_4$, $Ni@CoFe_2O_4$, and $La@Ni@CoFe_2O_4$ nanoparticles, respectively. The progressive decrease in absorbance intensity at 664 nm indicates the breakdown of the MB dye over time, confirming the photocatalytic activity of the samples. Among the tested materials, $CoFe_2O_4$ exhibited a photocatalytic efficiency of 35%, while $Ni@CoFe_2O_4$ demonstrated an enhanced efficiency of 61%. The highest degradation efficiency of 83% was achieved with $La@Ni@CoFe_2O_4$, indicating the synergistic effect of La and Ni doping in improving charge separation and enhancing photocatalytic performance. The superior activity of $La@Ni@CoFe_2O_4$ can be attributed to its improved electron-hole pair separation, increased surface area, and enhanced light absorption capabilities, making it a promising material for environmental remediation applications (Sarmah et al., 2023).

3.7.1 Photocatalytic Mechanism

The photocatalytic mechanism of $CoFe_2O_4$, $Ni@CoFe_2O_4$, and $La@Ni@CoFe_2O_4$ nanoparticles for methylene blue (MB) degradation under visible light irradiation can be explained as follows:

Step 1: Light Absorption and Excitation

When the photocatalyst is exposed to visible light, it absorbs photons with energy equal to or greater than its bandgap. This results in the excitation of electrons (e^-) from the valence band (VB) to the conduction band (CB), creating electron-hole (e^-/h^+) pairs (Ravindra et al., 2019).

$$\text{Photocatalyst} + h\nu \rightarrow e^- + h^+$$

Step 2: Charge Carrier Separation and Transfer

In the case of pure $CoFe_2O_4$, charge recombination limits photocatalytic efficiency. However, doping with Ni and La improves charge separation. Ni incorporation reduces electron-hole recombination by acting as an electron trapping center, while La doping further enhances charge transfer due to its strong electron-withdrawing ability. This leads to better photocatalytic performance in $Ni@CoFe_2O_4$ and $La@Ni@CoFe_2O_4$ compared to $CoFe_2O_4$.

Step 3: Reactive Oxygen Species (ROS) Generation

The excited electrons in the CB react with dissolved oxygen (O_2) molecules to form superoxide radicals (O_2^-), while the holes in the VB interact with water (H_2O) or hydroxyl ions (OH^-) to generate hydroxyl radicals ($OH\cdot$):

$$O_2 + e^- \rightarrow O_2^-$$

$$H_2O + h^+ \rightarrow OH^\cdot + H^+$$

$$OH^- + h^+ \rightarrow OH\cdot$$

Step 4: Degradation of Methylene Blue (MB)

The generated ROS (O_2^- and OH·) attack MB dye molecules, breaking down their structure into smaller, non-toxic byproducts such as CO_2 and H_2O (Sharifian-jazi et al., 2020).

MB dye + OH· / O_2^- → Decomposed products

The photocatalytic efficiency improves from 35% ($CoFe_2O_4$) to 61% (Ni@ $CoFe_2O_4$) and reaches 83% (La@Ni@$CoFe_2O_4$), demonstrating that La and Ni co-doping effectively suppress charge recombination and enhance the production of reactive oxygen species, leading to superior photocatalytic degradation performance (Shobana et al., 2022; Sonia et al., 2025).

Figure 7. UV-Vis absorbance spectra showing the photocatalytic degradation of methylene blue (MB) under visible light irradiation using (a) $CoFe_2O_4$ (b) Ni@ $CoFe_2O_4$ and (c) La@Ni@$CoFe_2O_4$ NPs

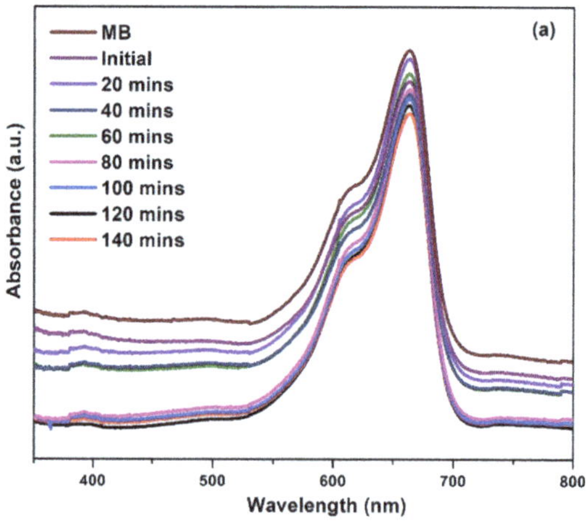

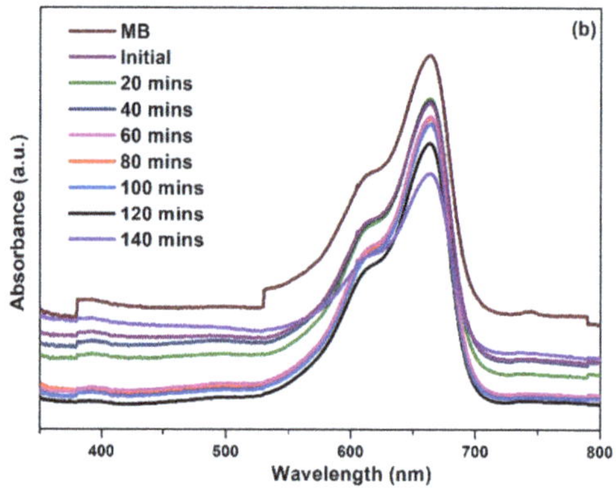

4. CONCLUSION

The photocatalytic performance of $CoFe_2O_4$, Ni@$CoFe_2O_4$, and La@Ni@ $CoFe_2O_4$ nanoparticles was systematically evaluated for methylene blue degradation under visible light irradiation. The results confirmed that doping with Ni and La significantly enhanced the degradation efficiency, with $CoFe_2O_4$ achieving 35%,

$Ni@CoFe_2O_4$ reaching 61%, and $La@Ni@CoFe_2O_4$ attaining the highest efficiency of 83%. The enhancement in photocatalytic activity is attributed to the improved charge separation, suppressed electron-hole recombination, and increased light absorption due to doping modifications. These findings suggest that $La@Ni@CoFe_2O_4$ nanoparticles are promising candidates for photocatalytic applications in wastewater treatment and environmental purification.

REFERENCES

Abbasi, L., Hedayati, K., & Ghanbari, D. (2021). Magnetic properties and kinetic roughening study of prepared polyaniline: Lead ferrite, cobalt ferrite and nickel ferrite nanocomposites electrodeposited thin films. *Journal of Materials Science Materials in Electronics*, *32*(11), 14477–14493. DOI: 10.1007/s10854-021-06006-1

Abdel-Aal, S. K., & Abdel-Rahman, A. S. (2020). Graphene influence on the structure, magnetic, and optical properties of rare-earth perovskite. *Journal of Nanoparticle Research*, *22*(9), 267. DOI: 10.1007/s11051-020-05001-7

Ali, N., Said, A., Ali, F., Raziq, F., Ali, Z., Bilal, M., Reinert, L., Begum, T., & Iqbal, H. M. N. (2020). Photocatalytic degradation of Congo Red dye from aqueous environment using cobalt ferrite nanostructures: Development, characterization, and photocatalytic performance. *Water, Air, and Soil Pollution*, *231*(2), 50. DOI: 10.1007/s11270-020-4410-8

Dehghani Dastjerdi, O., Shokrollahi, H., & Mirshekari, S. (2023). A review of synthesis, characterization, and magnetic properties of soft spinel ferrites. *Inorganic Chemistry Communications*, *153*, 110797. DOI: 10.1016/j.inoche.2023.110797

Fu, M., Jiao, Q., & Zhao, Y. (2013). In situ fabrication and characterization of cobalt ferrite nanorods/graphene composites. *Materials Characterization*, *86*, 303–315. DOI: 10.1016/j.matchar.2013.10.019

Hashemi, S. M., Hasani, S., Ardakani, K. J., & Davar, F. (2019). The effect of simultaneous addition of ethylene glycol and agarose on the structural and magnetic properties of $CoFe_2O_4$ nanoparticles prepared by the sol–gel auto-combustion method. *Journal of Magnetism and Magnetic Materials*, *492*, 165714. DOI: 10.1016/j.jmmm.2019.165714

Hedayati, K., Azarakhsh, S., Saffari, J., & Ghanbari, D. (2017). Magnetic and photocatalyst $CoFe_2O_4$–CdS nanocomposites: Simple preparation of Ni-, Co-, Zn- or Ag-doped CdS nanoparticles. *Journal of Materials Science Materials in Electronics*, *28*(7), 5472–5481. DOI: 10.1007/s10854-016-6209-3

Khurshid, R., Ali, F., Qureshi, M. T., Afzal, A., & Ali, Z. (2019). Polyol-mediated coprecipitation and aminosilane grafting of superparamagnetic, spinel $ZnFe_2O_4$ nanoparticles for room-temperature ethanol sensors. *Journal of the Electrochemical Society*, *166*(4), B258–B265. DOI: 10.1149/2.1021904jes

Lu, Y., Yousaf, M., Akhtar, M. N., Noor, A., Akbar, M., Shah, M. A. K. Y., Yan, S., & Wang, F. (2022). Effect of Gd and Co contents on the microstructural, magneto-optical and electrical characteristics of cobalt ferrite ($CoFe_2O_4$) nanoparticles. *Ceramics International, 48*(2), 2782–2792. DOI: 10.1016/j.ceramint.2021.10.067

Mahboubeh, H., Fatemeh, Z., Zahra, J. R., Ali, A., & Zohreh, A. (2014). Synthesis of cobalt ferrite ($CoFe_2O_4$) nanoparticles using combustion, coprecipitation, and precipitation methods: A comparison study of size, structural, and magnetic properties. *Journal of Magnetism and Magnetic Materials, 371*, 43–48. DOI: 10.1016/j.jmmm.2014.06.059

Menchaca-Nal, S., Jativa-Herrera, J. A., Moscoso-Londoño, O., Pampillo, L. G., Martínez-García, R., Knobel, M., & Londoño-Calderón, C. L. (2023). Composite magnetic properties of cobalt ferrite nanoparticles embedded in bacterial nanocellulose of different porosity levels. *Materials Chemistry and Physics, 303*, 127753. DOI: 10.1016/j.matchemphys.2023.127798

Nandhini, G., Vignesh, D., Shobana, M. K., Pazhanivel, T., Kavita, S., & Balaji, P. (2025). Graphene-decorated magnetic cobalt ferrite for effective UV-accelerated photocatalytic degradation of methylene blue: Experimental and theoretical insights by DFT. *Emergent Materials, 8*(1), 29–44. DOI: 10.1007/s42247-024-00799-7

Rathod, S. M., Khandelwal, P. S., & Rajuskar, R. S. (2019). Influence of gadolinium doped in nickel nano ferrite on magnetic and optical properties prepared by sol–gel technique. *Materials Today: Proceedings, 15*, 560–565. DOI: 10.1016/j.matpr.2019.04.121

Ravindra, A. V., Chandrika, M., Rajesh, C., Kollu, P., Ju, S., & Ramarao, S. D. (2019). Simple synthesis, structural and optical properties of cobalt ferrite nanoparticles. *The European Physical Journal Plus, 134*(6), 1–9. DOI: 10.1140/epjp/i2019-12690-2

Sakthivel, S., Neppolian, B., Shankar, M. V., Arabindoo, B., Palanichamy, M., & Murugesan, V. (2003). Solar photocatalytic degradation of azo dye: Comparison of photocatalytic efficiency of ZnO and TiO_2. *Solar Energy Materials and Solar Cells, 77*(1), 65–82. DOI: 10.1016/S0927-0248(02)00255-6

Sarmah, S., Patra, K. P., Maji, P. K., Ravi, S., & Bora, T. (2023). A comparative study on the structural, magnetic and dielectric properties of magnesium substituted cobalt ferrites. *Ceramics International, 49*(1), 1444–1463. DOI: 10.1016/j.ceramint.2022.09.126

Sharifianjazi, F., Moradi, M., Parvin, N., Nemati, A., Jafari Rad, A., Sheysi, N., Abouchenari, A., Mohammadi, A., Karbasi, S., Ahmadi, Z., Esmaeilkhanian, A., Irani, M., Pakseresht, A., Sahmani, S., & Shahedi Asl, M. (2020). Magnetic $CoFe_2O_4$ nanoparticles doped with metal ions: A review. *Ceramics International*, *46*(11), 18391–18412. DOI: 10.1016/j.ceramint.2020.04.202

Shobana, M. K., Nandhini, G., Kavita, S., Kumar, V. S., & Pazhanivel, T. (2022). Photocatalytic and magnetic properties of Mg substituted cobalt ferrite. *Materials Science and Engineering B*, *286*, 116030. DOI: 10.1016/j.mseb.2022.116030

Sonia, S., Sheetal, , Kumari, H., Sharma, S., Chahal, S., Duhan, S., Kumar, A., & Kumar, P. (2025). Magnetic cobalt ferrite–tin oxide nanocomposites for efficient photocatalytic dye degradation. *Water, Air, and Soil Pollution*, *236*(9), 602. DOI: 10.1007/s11270-025-08238-x

Vijayaraghavan, T., Suriyaraj, S. P., Selvakumar, R., Venkateswaran, R., & Ashok, A. (2016). Rapid and efficient visible light photocatalytic dye degradation using AFe_2O_4 (A = Ba, Ca and Sr) complex oxides. *Materials Science and Engineering B*, *210*, 43–50. DOI: 10.1016/j.mseb.2016.04.005

Zaharieva, K. L., Milenova, K. I., Rives, V., Trujillano, R., Cherkezova-Zheleva, Z. P., Eliyas, A. E., & Mitov, I. G. (2015). Mixed cobalt–copper ferrite-type materials: Synthesis and photocatalytic efficiency in degradation of Reactive Black 5 dye under UV-light irradiation. *Izvestiia po Himiia*, *47*, 105–111.

Chapter 7
Data-Driven Prediction of Groundwater Contamination in the Western Tamil Nadu

M. Martinaa
https://orcid.org/0000-0003-0977-2106
SASTRA University, India

G. Revathy
https://orcid.org/0000-0002-0691-1687
SASTRA University, India

ABSTRACT

In water-limited areas like Western Tamil Nadu, where anthropogenic and environmental factors have a major impact on groundwater resources, evaluating and protecting the quality of the groundwater is essential to life support. The pressing need to use trustworthy, data driven predictive frameworks to address complicated groundwater quality issues is what spurred this study. The availability of robust and interpretable models that accurately represent seasonal and spatial variations in groundwater parameters remains a major gap despite significant efforts. To address this, the current study uses two sophisticated ensemble machine learning (ML) techniques,

DOI: 10.4018/979-8-3373-5886-4.ch007

1. INTRODUCTION

Groundwater is a cryptic ecosystem in a subsurface universe (Singha et al., 2021), that supports both ecosystems and human needs (Yuvaraj et al., 2023). But many factors determine its composition and its ability to support various human, ecosystem, and geochemical needs (Zhu et al., 2022). The changing anthropogenic pressures and environmental uncertainties demonstrate the need for understanding and managing groundwater quality (Yuvaraj and Suresh Ghana Dhas, 2020). The composition of groundwater is a complex interplay of geological, hydrological, and anthropocentric factors (Chen et al., 2020). The geological formations (containing minerals) give rise to a multitude of different subsurface reservoirs with unique chemical characteristics (Kayalvizhi et al., 2023).

Hydrological processes impact the movement and filtration of water through the subsurface even more than the hydrological cycle applies to groundwater mix issues (Kouadri et al., 2022). Human activities, such as industrial effluent and agricultural runoff, all introduce various pollutants that may disrupt the equilibria of elemental or molecular composition in groundwater (Singh et al., 2022; Su et al., 2020). It is never fully evaluable and predicable, because all environmental factors have variability and interact across time and space (Belkhiri et al., 2020). Monitoring attributes such as pH, DO and elemental or molecular owner attributes are critical to fabricating the picture of groundwater quality (Ahmed et al., 2020). Given population increases and climatic changes, it is imperative to protect the sustainability and safety of water as a priceless resource (Kouadri et al., 2021).

There are many varieties of underground aquifers that are shaped by geological structure, while water movement is dictated by hydrological processes. The natural equilibrium of groundwater is altered by pollutants present in the environment, due to human activity. Due to environmental variability extensive assessment is difficult. The chemical constituents, the pH and the DO should be monitored. Climate change and population growth accelerate the need to protect groundwater. The limited supply of water, in the dry arid environment of western Tamil Nadu, reinforces the need for strong groundwater quality protection. Predictive modelling tools must be robust, to better capture the temporal variations of one's current methods. Groundwater scarcity in the arid area of western Tamil Nadu highlights the urgent need to protect groundwater quality (Asadollah et al., 2021).

Conventional approaches often do not effectively address the different temporal trends and change in groundwater quality, even with focused effort. This is one of the region's biggest problems because not only are there issues with water resources but there is also the complexity of changing quality parameters. Because traditional methods do not communicate well enough to fully and accurately understand the situation, there are many opportunities for the need for a change to more advanced predictive tools. The key issue is effective forecasting methods for the unique temporal dependencies and patterns of seasonality and cyclical seasonality in groundwater quality data. This study proposes a new approach to groundwater quality prediction that leverages two very advanced ML methodologies, RFR and XGBoost, that can adequately address complex, non-linear relationships in multivariate environmental datasets. The overall purpose of the study is to improve water resource management and quality monitoring in western Tamil Nadu's water stressed areas. In order to produce precise groundwater quality forecasts and reveal both short-term variations and long-term trends influenced by geological, seasonal, and human factors, the predictive powers of RFR and XGBoost will be utilized. This study uses a novel ML-based methodology that goes beyond conventional statistical techniques to fill a critical gap in the models currently used for assessing groundwater quality. In order to forecast groundwater quality in a geographically specific and environmentally sensitive context, this study is unique in that it applies ML models, specifically RFR for robust, variance-reducing prediction through decision tree averaging and XGBoost for gradient-based boosting optimization. A more complex, comprehensible, and computationally effective view of the dynamic interaction of chemical, physical, and land-use factors affecting groundwater conditions is provided by this innovative dual-model framework. This study offers more than just short-term prediction accuracy. In addition to addressing today's groundwater quality issues, it offers a game-changing data driven tool that gives environmental planners, hydrologists, and policymakers strategic insight into new contamination trends. Both XGBoost and RFR provide a scalable approach to groundwater quality forecasting because of their ability to handle large, heterogeneous datasets, identify nonlinear correlations, and evaluate feature importance. Therefore, this study represents a major step toward a sustainable future in which the availability of safe and clean groundwater is closely linked to ecological stability, agricultural viability, and public health.

The RFR and XGBoost models are explicitly described as powerful computational tools used in this study to establish a new avenue for groundwater quality prediction. They are intended to provide a predictive approach that is not limited by any assumptions about data distributions, or the linearity of predictor-response modelling. The proposed framework aims to meaningfully describe nonlinear models, and to reflect spatiotemporal variations in groundwater quality data. RFR and XGBoost recognize the complex patterns and interaction of features, including the basic indicators used in this analysis (TDS, pH, DO, turbidity, nitrate, and chloride, along with heavy metals and nutrients). With respect to demonstrating complex relationships between geological, hydrological, and anthropogenic factors impacting groundwater quality, these models are, again, well suited. What distinguishes these ML algorithms from others pertains to their ability to rank the importance of features, while also being resilient to noise in the data. This characteristic gives interpretable results on parameters that are having the biggest effects on water quality. The training and validation sections ensure that the models adaptively learn from historical trends and generalize to new data. XGBoost fits the definition of a boosting technique that simultaneously optimizes for gradient descent to minimize prediction error in a sequential manner, while RFR - as an ensemble of decision trees - focuses on averaging through trees to reduce the variance, or quality of the prediction.

These models fit together extremely well, and provide high accuracy along with interpretability sufficient to examine thoroughly the dynamics of groundwater quality. As depicted in the algorithmic workflow as shown, the steps involved with this work included preprocessing the data, training the RFR and XGBoost models on the historical dataset, extracting features of interest, and evaluating the predicted values against the actual observed measurements of groundwater quality. The models were able to model both shorter-term shifts and longer-term trends in groundwater quality in relation to several factors across the different districts. The methodology proposed in this work quantitatively represents the relationships between groundwater quality and the drivers impacting these trends and shifts, via correlation analyses together with predictive performance metrics, and provides a holistic framework for environmental researchers and policymakers. This research significantly advances the management of sustainable groundwater planning by combining scientific evidence with ML potential. We develop a proactive, data-driven decision-making framework that supports both the strategic planning for pollution control and water conservation and anticipates declines in groundwater quality. The advantage of using RFR and XGBoost is that they can perform better than traditional models of perfect accuracy in complex environmental systems were capturing multidimensional interactions of the environmental drivers in an accurate and transparent manner is vital; they can also be scaled to enormous feature spaces without overfitting.

2. RELATED WORKS

A detailed review of the existing literature concluded that there are many different kinds of research designs on the complex concerns of groundwater quality. Researchers and practitioners have taken various approaches and consider different issues to address the complexity of this unique resource (Khan et al., 2022). Much of the literature advances knowledge of the hydrological and geological factors that affect groundwater quality (El Bilali et al., 2021). The studies of geology focused on mineral constituents that characterize the chemical profiles of underground reservoirs. A better understanding of natural processes which cause fluctuations of groundwater quality have been achieved through hydrological studies that examine subsurface flow and filtration (Bui et al., 2020; Balamurugan et al., 2024).

Much of the research on groundwater quality has focused on human impacts of groundwater quality alongside these geological and hydrological perspectives. Research has shown that industrial discharges and agricultural practices are among the many contaminants that can enter groundwater systems. Research has shown that there is inevitable need for preventative measures to mitigate human impacts and to sustainably manage groundwater resources (Zhi et al., 2021; Saqr et al., 2024). There have been significant advancements in groundwater quality research with new technologies (Azrour et al., 2022). Remote sensing, data analytics, and ML, for example, can be applied to large datasets to analyse patterns or trends in change in groundwater quality. Although preventative management was not an operational consideration in prior research, many of these potential innovations can facilitate planning for predictive modelling and preventive management.

The increased importance of sustainable management of groundwater resources by using RFR and XGBoost in groundwater quality forecasting has resulted in a surge in research utilizing ML approaches to forecast water quality parameters. RFR and XGBoost are the most commonly used algorithms because of their high predictive accuracy, the robustness of overfitting, and nonlinear data handling. Random Forest is the most utilized decision-tree -based ensemble learning technique, and it is widely known for its effectiveness in conducting environmental modeling. Many studies have utilized RFR for pollution risk assessments and other groundwater quality evaluations. The authors of one study (Gazzaz et al., 2012) illustrated the predictive power of RFR for estimating the groundwater pollution index in arid areas. The model was resilient to missing data and outliers and was more successful in predictive forecasting than conventional regression approaches. (Khan et al., 2020), Policy training must acknowledge the necessity of confronting these difficulties in order to fully realize the possibilities of ML in water resource management, paving the path for novel solutions to climate change and water scarcity in Pakistan. Recent advances in machine learning and data mining algorithms have demonstrated significant potential for improving water quality management in Pakistan (344). (Subudhi et al., 2025) Minerals such as chloride, fluoride, potassium, sodium, magnesium and calcium pose significant health hazards when present in high concentrations. This study creates a predictive ML algorithm to assess Ground Water Quality Index (GWQI) and identify important contaminants that impact water safety. (Olabode, O. F and Comte, J. C, 2025) This study uses a distributed water balance model to assess groundwater recharge trends and drivers during 1981-2020. Data inputs included climate, soil type, water table depth, terrain and land cover. Urbanization reduced diffuse recharge by 30% and increased surface runoff by 11%, despite increasing rainfall. (Alam et al., 2025) The Yinchuan Region's groundwater is contaminated by NO3⁻, posing a risk to local residents. To identify the important factors impacting groundwater nitrate contamination in the Yinchuan Region, this study used the XGBoost model with the SHapley Additive ExPlanations (SHAP) approach. The SHAP feature dependence graphs show a complex interaction between pH, TFe, $Mn^{2+,}$ TDS and NO_3^- levels in complex groundwater systems. The model was demonstrated to be more skill ful and computationally efficient, replacing Random Forest and Neural Networks. By applying XGBoost for simulating groundwater 'iron' parameters in various aquifers, (Li et al., 2019) seems to further demonstrate the versatility of XGBoost to identify spatially heterogeneity for water quality, by combining hydrochemical information and further geological indicators. (Raicy et al.,2026) use integrated machine learning algorithms to analyze groundwater quality and associated health hazards in the Ganga River Basin. This method is critical when examining the patterns of groundwater depletion, particularly nitrates, which pose serious health dangers, particularly to vulnerable populations such as children. The findings underscore the significance of machine learning in accurately forecasting water quality metrics and assessing health hazards. (Dukuly L P et al., 2025) This study presents a novel hybrid framework that combines principal component analysis (PCA) and sophisticated ML models to accurately forecast rock fragmentation in surface mining. The framework's goal is to improve operating efficiency, reduce environmental impact, and improve mine safety by optimizing blast design parameters. Four ML models (RF, ANN, SVM, and XGBoost) were integrated with PCA to increase model generalization, reduce data dimensionality, and eliminate redundancy. (Eftekhari, et al., 2024) This works utilized GIS and ML models to predict and classify groundwater quality indicators based on qualitative characteristics. Multiple approaches' accuracy rates were compared to the Birjand plain. This study used water quality sampling findings from 18 observation wells from 2011 to 2020 as input data. Parameter assessments showed that acidity was the least variable (2.31%), whereas Mg content was the most variable (70%). Geostatistical analyses showed that the inverse distance weighting model performed better in predicting total dissolved solids and pH, while ordinary kriging performed better in forecasting Ca contents with lower root mean square error.

Table 1. An overview of the research using ML models to predict groundwater quality

Author(s)	Year	Model Used	Target Variable(s)	Key Outcome
Gazzaz et al.	2015	RFR	Pollutants, WQI	Dependable forecasts in the absence of data
Li et al.	2019	XGBoost	Iron	Expresses the range of geology
Raicy, M. C., et al	2026	Self-Organizing Maps Clustering	WQI	Modeling is enhanced by GIS-ML integration.

In Table 1 is noticeable research outcome in combining these various lines of investigation, even with the advancements made. In order to fully address the complex issues related to groundwater quality, the existing literature advocates for a more integrated and comprehensive approach. By utilizing cutting-edge predictive modeling techniques, this study seeks to advance our knowledge of groundwater quality trends and temporal variations.

3. PROPOSED METHODOLOGY

In arid and semi-arid areas of India, especially the western Tamil Nadu belt, where surface water supplies are either seasonal or insufficient to meet the rapidly increasing demand, groundwater is a lifeline. Urban sprawl, intensive agriculture and Industrialization, have put more strain on groundwater in districts like Tirupur, Namakkal, Salem, Erode, Coimbatore. As a result, growing concentrations of fluoride, nitrate, salinity, and other dangerous chemicals show that groundwater quality has drastically declined. It is urgently necessary to create predictive tools that can help predict future contamination so that policymakers can take preventative action. Monitoring groundwater quality is insufficient. The accuracy, scalability, and interpretability needed for critical environmental decision-making are provided by ML algorithms such as RFR and XGBoost, while traditional statistical techniques offer fundamental insights. These ML-based, non-parametric models have a number of advantages over conventional methods because they can represent the intricate, nonlinear relationships present in groundwater datasets. While conventional statistical methods provide basic insights, ML algorithms like RFR and XGBoost provide the accuracy, scalability, and interpretability required for crucial environmental decision-making. The ability of these ensemble-based, non-parametric models to capture the complex, nonlinear relationships found in groundwater datasets gives them several advantages over traditional approaches.

Obtaining groundwater quality data from government sources such as the Central Ground Water Board (CGWB), State Groundwater Departments, and environmental monitoring organizations is part of the data collection and preprocessing process. Typical parameters that are gathered include heavy metals (such as lead and arsenic), sulfate, chloride, nitrate, fluoride, iron, hardness, TDS, DO, EC, turbidity and pH. The dataset captures seasonal trends and fluctuations over a temporal span of six months to a year. After that, raw values are cleaned and standardized as the data moves into the Data Preprocessing phase. This entails dealing with missing values, eliminating outliers, converting the data to a consistent scale, and encoding any categorical features that may be present, such as the district or sampling location.

Richer model interpretation is possible by using feature engineering to extract derived parameters or indicators, such as the WQI.

A feature selection phase is used after preprocessing. To comprehend the relationships between input features and target variables, correlation heatmaps and statistical tests (such as Pearson or Spearman correlation) are used. To decrease dimensionality while maintaining informative content, both domain expertise and automated methods (such as Recursive Feature Elimination or tree-based feature importance scores) are employed. This prevents redundant or unnecessary variables from causing the ML models too overfit.

XGBoost and RFR are used in this core component for predictive modeling. Using a random subset of features and samples, RFR constructs an ensemble of decision trees. The average of all the trees' predictions yields the final forecast. This technique works well with environmental datasets that contain variability brought on by both natural and human factors because it is especially resistant to noisy data and overfitting. With a focus on fixing mistakes from earlier models, XGBoost is a potent boosting algorithm that gradually increases model accuracy. It offers highly interpretable feature importance scores, integrates regularization, and efficiently manages missing values. Metrics like R2, MAE, RMSE, and MAPE are used to validate the remaining data after both models have been trained on 70–80% of the data. To improve generalization, K-fold cross-validation can be added if desired.

Performance dashboards and visualization tools are part of the architecture's Evaluation block. Here, model performance is validated using residual plots, ROC curves (in the case of classification-type thresholds), and scatter plots between predicted and observed values. Which model generalizes better for the specified spatiotemporal region and parameter set is determined by comparing RFR and XGBoost. The Prediction Module, the last part, forecasts the groundwater quality levels for every site. GIS-enabled maps, charts, and ranked water quality zones are frequently used to visualize the results. Crucially, decision-makers can determine the most significant factors influencing groundwater degradation with the aid of feature importance analysis from RFR and XGBoost.

Figure 1. Architecture of proposed ground water quality prediction

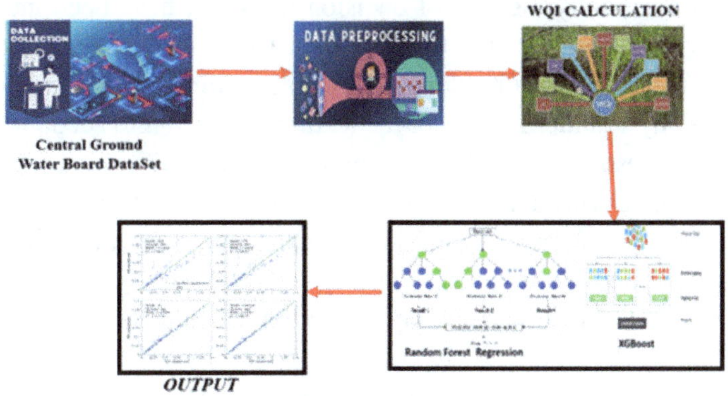

An extensive, end-to-end ML pipeline designed specifically for assessing and predicting groundwater quality in Tamil Nadu's western districts is depicted in the architecture diagram titled "Proposed Groundwater Quality Prediction Using RFR and XGBoost." The figure 1 integrates domain knowledge with two ML techniques, covering all crucial phases from data collection to final prediction and interpretation.

3.1 Data Set Description

The data used in this study includes groundwater quality observations from the India- WRIS platform, the Tamil Nadu Water Resources Department (TNWRD), and the Central Ground Water Board (CGWB), among other government sources. The information covers more than ten years and is dispersed geographically throughout several observation and borewell wells in western Tamil Nadu. Features in the dataset include target variables (WQI), water quality parameters (pH, TDS, EC, nitrate, fluoride, DO, chloride, sulfate, iron, hardness, and turbidity), and location and time-based attributes (district name, latitude and longitude, year and month of sampling). Normalization, feature encoding (e.g., one-hot encoding of districts and seasons), and mean imputation or k-nearest neighbors (KNN) were used to preprocess the data.

3.2 RFR

RFR is a powerful ML algorithm that produces incredibly accurate predictions, especially in complex and non-linear datasets, by building on the foundation of decision trees and bagging (Bootstrap Aggregating) techniques. A class of ensemble ML algorithms, the Random Forest model was created by Leo Breiman in 2001 with the goals of minimizing variance, enhancing generalization, and preventing overfitting, all of which are prevalent issues in single decision tree models. A Random Forest is essentially a group of decision trees that have been independently trained on various subsets of the training data. Every tree is constructed using a bootstrap sample, which is a random sample taken from the original dataset with replacement. Only a random subset of the features is taken into account at each tree split. Random Forests are incredibly strong and resistant to overfitting because of their dual-randomness in data and feature selection, particularly in high dimensional datasets like those pertaining to groundwater quality and environmental monitoring.

A non-parametric supervised learning algorithm called a decision tree uses basic decision rules derived from the features to predict the value of a target variable. In regression tasks, a decision tree divides the dataset into smaller subsets according to feature values and gives each terminal node (or leaf) a prediction, usually the target variable's mean. Building a decision tree for regression involves the algorithm divides the data into subsets recursively in order to minimize a loss function, like the MAE. By minimizing the impurity (variance in regression) of the resultant subsets, the best split is chosen at each node from among all potential features and thresholds. The tree keeps growing until a stopping condition—like a minimum number of samples per leaf or a maximum depth—is satisfied. Individual decision trees, however, are susceptible to overfitting and extreme sensitivity to noise in the training data, particularly when the underlying relationship between features and targets is complex or nonlinear. This is a common feature of groundwater quality data that is impacted by a variety of hydrological, chemical, and human factors.

ML overcomes the drawbacks of a single tree by utilizing several trees to produce a robust learner. Especially in small to medium-sized datasets, RFR reduces variance without increasing bias by using multiple decision trees trained on various bootstrap samples. RFR decrease correlation between individual trees within each tree by introducing feature randomness. This improves diversity, particularly in the quality of groundwater. RFR takes into consideration variations in characteristics such as pH, TDS, EC, nitrate, fluoride, DO, chloride, sulfate, iron, hardness, and turbidity resulting in a more comprehensive model that captures various viewpoints on data distribution and produces a more accurate prediction system. It is necessary to comprehend the bias-variance tradeoff of prediction error in order to comprehend RFR. Low bias and high variance are characteristics of a single decision tree. Without increasing bias, variance is decreased by averaging predictions from decorrelated trees.

In terms of mathematics, if a model's error is expressed as equation 1:

$$\text{Error} = \text{Bias}^2 + \text{Variance} + \text{Irreducible Error} \quad (1)$$

In order to make Random Forest appropriate for intricate regression tasks where overfitting is an issue, it then attempts to minimize the Variance component through averaging. For instance, to determine which chemical or physical parameters are most predictive of groundwater quality, hydrologists and environmental scientists can use RFR, an interpretable that quantifies feature importance. Prioritizing resources and creating effective water monitoring systems are two uses for this. Permutation importance and Mean Decrease in Impurity (MDI) can be used to determine the importance of a feature. Because Random Forests are stochastic, meaning that each observation is included in multiple trees and omitted in others, they are suitable for noisy data and robust to missing data and outliers. RFR is used in this study to forecast the WQI in the western districts of Tamil Nadu. Physicochemical parameters such as pH, TDS, EC, nitrate, fluoride, DO, chloride, sulfate, iron, hardness, and turbidity e are among the input features. Weighted aggregations are used to calculate the WQI, a composite index that is the target variable. RFR performance is maximized through hyperparameter tuning, which sets important parameters like max_depth, min_samples_split and min_samples_leaf, max_features, and n_estimators. Cross-validation is often used in order to provide validity by testing on all combinations of subsets of data given both Grid Search and Randomized Search do not always produce consistent models. In this regard, we interpret nonlinearity, interpretability, scalability, and null-high-dimensionality capabilities make RFR exceptionally fruitful in modeling water quality and more; with ability helping to motivate environmental policy decision aides process as well.

3.3 XGBoost

For structured datasets, e.g., environmental studies, XGBoost is a highly efficient and scalable implementation of a gradient boosting framework which operates effectively using boosting to form a single strong aggregate model from several weak learners, and gradient descent to minimize a loss function. XGBoost is appropriate for modeling environmental parameters such as pH, TDS, EC, nitrate, fluoride, DO, chloride, sulfate, iron, hardness, and turbidity; especially with its engineering optimizations and strict mathematical foundation differentiating it as a model of choice. In fact, the base learners XGBoost draws from are decision trees - hierarchical data structures with internal nodes characterized by decision rules, and leaf nodes determined by predictions or specific outcomes. Predictions occur for XGBoost by fitting a base decision tree to training data. In order to build a *second* decision tree to model regard and residual errors and new predictions, the algorithm computes the gradient of the loss function with respect to the current predictions.

Mathematically, the model prediction at iteration t, denoted by $\hat{z}_j^{(t)}$expressed as equation 2:

$$\hat{z}_j^{(t)} = \hat{z}_j^{(t-1)} + \eta f_t(y_j) \tag{2}$$

where f_t is the decision tree trained at iteration t, y_j represents the feature vector for the jth sample and η is the learning rate. The final output after T boosting rounds expressed as equation 3:

$$\hat{z}_j = \sum_{t=1}^{T} \eta f_t(y_j) \tag{3}$$

The objective function in XGBoost comprises two components: a convex differentiable loss function $L(z_j, \hat{z}_j)$ that measures prediction accuracy, and a regularization term Ω(f) that penalizes model complexity expressed as equation 4:

$$L(\theta) = \sum_{j=1}^{m} L(z_j, \hat{z}_j) + \sum_{t=1}^{T} \Omega f_t \tag{4}$$

By ensuring that trees are not excessively complex, the regularization term enhances the model's capacity for generalization. The usual format expressed as equation 5:

$$\Omega f = \gamma T + \frac{1}{2}\lambda \sum_{i=1}^{T} \omega_i^2 \tag{5}$$

Where, λ is the L2 regularization coefficient, ω_i is the score (weight) assigned to each leaf, T is the number of leaves, and γ is the penalty for each leaf in the tree.

The convergence speed and prediction accuracy are greatly increased by XGBoost's use of both the first and second derivatives of the loss function for second-order optimization. The loss function's first and second derivatives, g_j and h_j are computed for every data point expressed as equation 6:

$$g_j = \partial_{\hat{z}_j} L(z_j, \hat{z}_j),\ h_j = \partial^2_{\hat{z}_j} L(z_j, \hat{z}_j) \tag{6}$$

The ideal weight ω_i for each leaf node is calculated by aggregating these gradients at that node expressed as equation 7:

$$\omega_i^* = -\frac{\sum_{j \in J_i} g_j}{\sum_{j \in J_i} h_j + \lambda} \tag{7}$$

The evaluation of whether a split enhances the model is based on the gain from dividing a node into two children. This Gain expressed as equation 8:

Gain =

$$\frac{1}{2}\left[\frac{\left(\sum_{j\in J_L} g_j\right)^2}{\sum_{j\in J_L} h_j + \lambda} + \frac{\left(\sum_{j\in J_R} g_j\right)^2}{\sum_{j\in J_R} h_j + \lambda} - \frac{\left(\sum_{j\in J} g_j\right)^2}{\sum_{j\in J} h_j + \lambda}\right] - \gamma \tag{8}$$

The node turns into a leaf if the gain is negative, but the split is kept if it is positive. This splitting criterion also involves gain, which makes sure that if a choice is made, it has made an important contribution to reducing error, as well as managing complexity.

Since missing values, outliers, and nonlinear interactions are common in groundwater datasets, XGBoost's ability to handle them in a robust manner is another useful aspect of using the algorithm in environmental modeling. If a feature has a missing value, the algorithm determines what to do, automatically. This programming is crucial for the kinds of real data we deal with, where certain chemical parameter values will often not be recorded on a temporal basis, or across extensive geographical areas. In addition to handling various outcomes, policymakers can also benefit from knowing how XGBoost ranks feature importance. If authorities uncover that nitrate and chloride are the important determinants for groundwater quality in a district, then they can target pollution sources more effectively. XGBoost also incorporates several system level optimizations: it uses cache-aware data constructs, performs parallel processing, prunes trees, and uses column block structures to optimize memory-access speeds. As a result of these design features, it can accommodate groundwater quality data with considerable volume and high-dimensionality, while not sacrificing accuracy and speed. It also employs regularization techniques using L1 and L2 norms to improve generalization and avoid overfitting – a similar method to Lasso, and Ridge regressions applied in plastics.

XGBoost will be leveraged on the district level time series data containing groundwater samples collected in different seasons, for the purpose of directly predicting groundwater quality in the western part of Tamil Nadu. The spatial component involves sampling locations when collected from Coimbatore, Tirupur, Erode, Salem, and Namakkal, and either done by generating district level models or implicitly within the feature outputs. Seasonal indicators (monsoon vs. dry season), lag features (e.g., nitrate levels from the previous month), or aggregate metrics (e.g., 3-month moving averages) are used to model the temporal trends. The advantages of XGBoost are nicely complemented by this organized tabular representation. Several metrics are used to assess XGBoost's performance. Metrics like RMSE, MAE, R2 (coefficient of determination), and Mean Absolute Percentage Error (MAPE) are used to evaluate accuracy and robustness in regression tasks like predicting pH, TDS, EC, nitrate, fluoride, DO, chloride, sulfate, iron, hardness, and turbidity are overall WQI. Tool like SHAP, which visualizes how individual features affect model predictions and offers a unified measure of feature contribution, further improves the interpretability of the model. As examples, SHAP plots may show that inducing TDS above 500 ppm will have exponential decreases in the predicted GWQI; thus, could inform future regulations or remediation. XGBoost is a strong ML tool for real-time decision support predicting groundwater quality; by maintaining feature importance and SHAP values, it can consider spatiotemporal features, represent complex relationships, and provide interpretation in spatially and/or temporally challenged situations. In areas with water-quality deterioration due to pollution and scarcity, stakeholders can protect public health and allow for sustainable and equitable development by adopting XGBoost into monitoring frameworks and transitioning from reactive to proactive groundwater management.

4. RESULTS AND DISCUSSIONS

Various ML techniques are applied in this section to assess the proposed RFR and XGBoost models for judging groundwater quality. The standard 80-20 split for training and validating each model is first used to compute the validation set size as a percentage of the modeling dataset construct. This splits the modeling dataset into 80% training observations and 20% validation observations. This split is a compromise between needing to reserve a representative sample for independent look demonstration evaluation of performance, and need the remainder of the data is extensive enough to have a stable model. Several factors which influence the computational requirements for training RFR and XGBoost models are the size of the datasets, dimensionality of input features, and complexity of the model (e.g., number of trees, tree depth, learning rate for XGBoost). These ML techniques are generally less computationally expensive, and more relevant for applications of moderate hardware capabilities than deep learning (DL) techniques. Still, ML still requires a careful compromise between model load and performance characteristics when hyper-parameterizing the number of estimators, the maximum tree depth, and the regularization coefficients. Using common libraries for ML across platforms (e.g., R, Python) as for RFR scikit-learn and for XGBoost, XGBoost API, both models were created and simulation in a Python programming environment. To ensure the best model configuration for evaluation purposes, hyper criteria were advanced by representing hyper-criteria exhibiting grid search hyper-criteria choosing models evaluated using cross-validation.

Salem, Namakkal, Tirupur, Coimbatore, and Erode are just a few of the areas in Tamil Nadu, India, from where sampling sites were strategically selected. A range of geographical, hydrological, and anthropogenic factors including land use diversity, representation of aquifer types, and proximity to pollution sources (e.g., industry and agriculture) were considered in selecting the sites. The development of models that effectively generalize across a range of groundwater conditions is aided by this spatial diversity. All parameters were gathered using calibrated instruments and standardized procedures to guarantee measurement accuracy and data consistency. For example, in strict adherence to APHA-standard procedures, pH values were measured using calibrated digital pH meters. Seasonal variations and long-term trends were captured through temporally distributed data acquisition, which involved monthly samples collected over a one-year period. In order to provide thorough coverage of natural fluctuation patterns, parameters including turbidity, EC, DO, nitrate, and chloride were measured in the winter, spring, summer, and fall.

4.1. Performance Metrics

- **Accuracy:** Represents the overall correctness of predictions by showing how well the RFR and XGBoost models predict groundwater quality values in comparison to actual measurements.
- **Precision and Recall:** Provide comprehensive information about the models' capacity to forecast both safe and hazardous water quality conditions. Recall measures how well the models capture all pertinent instances, whereas precision measures how reliable positive predictions are (e.g., identifying polluted samples).
- **RMSE:** Calculates the average squared difference between the expected and actual values, taking the square root of that difference. It is a useful indicator of how well the model predicts continuous outcomes and assigns greater weight to large errors.
- **MAE:** Determines the mean of the absolute variations between the values that were predicted and those that were observed. Compared to RMSE, it is less susceptible to outliers.
- **R^2 Score:** Evaluates the degree to which the models can adequately account for the variation in actual groundwater quality data. Strong generalization and excellent model explanatory power are indicated by higher R2 scores, which in this study are typically > 0.90.
- **Computational Efficiency:** evaluates the models' inference speed, memory consumption, and training time. When used in resource-constrained or embedded systems, RFR and XGBoost are very useful for real-time water quality monitoring applications because they are relatively quicker to train and assess than DL models.

Table 2. Comparison of actual and predicted WQI in RFR, XGBoost

District	Actual WQI	RFR Predicted WQI	XGBoost Predicted WQI	RFR Error	XGBoost Error
Erode	68	67	68	1	0
Coimbatore	72	71	73	1	1
Salem	65	66	64	1	1
Tiruppur	70	69	71	1	1
Namakkal	66	65	67	1	1

Figure 2. Comparison of actual and predicted WQI in RFR, XGBoost

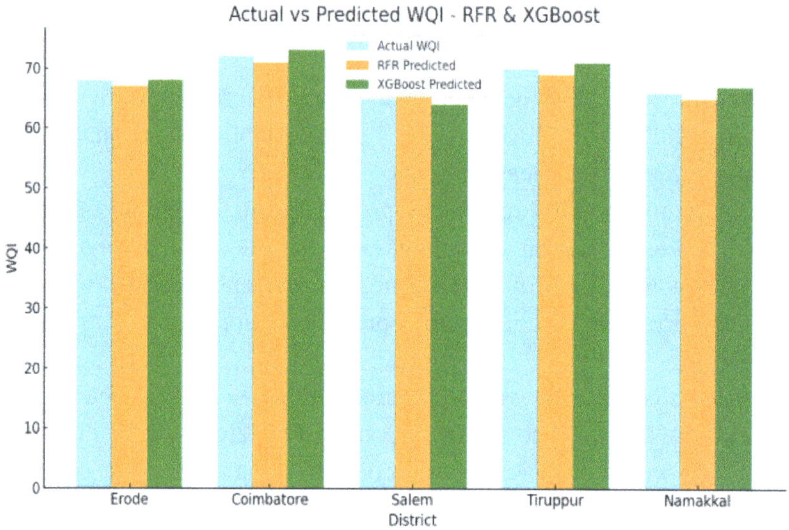

Five important districts in Western Tamil Nadu—Erode, Coimbatore, Salem, Tirupur, and Namakkal—have their actual and predicted WQI values compared in the bar chart that is provided. For the purpose of forecasting groundwater quality, the figure is essential for confirming the effectiveness of two potent ML models: RFR and XGBoost.

Figure 2 and Table 2 are shown as Actual and Predicted WQI of the five districts of Tamil Nadu, the bar graph contrasts the actual WQI with forecasts from the RFR and XGBoost models. Erode's actual WQI is 68, while XGBoost and RFR predict 67 and 68, respectively. In Coimbatore, XGBoost predicts 73, RFR predicts 71, and the actual value is 72. RFR is 66, XGBoost is 64, and Salem's actual WQI is 65. The actual is 70 in Tirupur, while XGBoost is 71 and RFR is 69. Lastly, the actual WQI in Namakkal is 66, whereas XGBoost predicts 67 and RFR 65. These outcomes demonstrate how well both models align with actual values, withXGBoost showing a marginally higher overall accuracy.

Table 3. Comparing the features of RFR and XGBoost models for predicting groundwater quality

Feature	RFR Importance Score	XGBoost Importance Score
pH	0.25	0.20
TDS	0.35	0.40
Nitrate	0.15	0.10
DO	0.10	0.10
Chloride	0.15	0.20

Figure 3. Comparing RFR and XGBoost models for predicting groundwater quality by feature importance

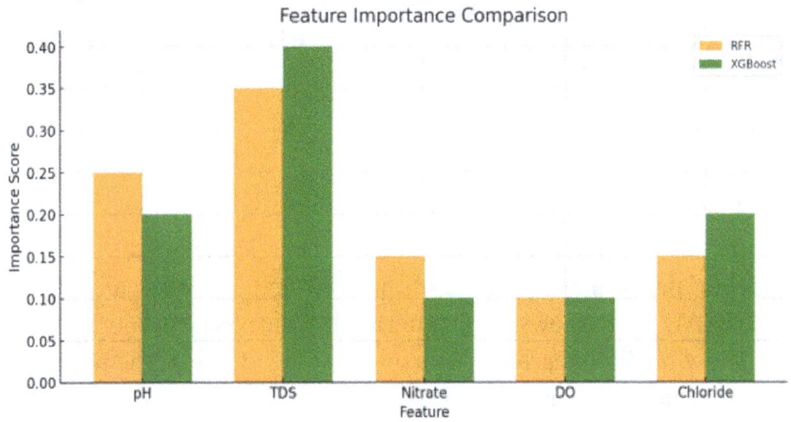

Figure 3 and Table 3 are obtained by the comparison of feature importance scores obtained from the RFR and XGBoost models for predicting groundwater quality in Western Tamil Nadu is shown in the bar chart. The degree to which each parameter influences the WQI prediction is reflected in feature importance. With an importance score of 0.40 from XGBoost and 0.35 from RFR, TDS stands out as the most significant predictor among all features. As the main measure of groundwater quality, TDS has a significant impact on WQI in both models. With RFR and XGBoost scores of 0.25 and 0.20, respectively, pH is the second most important parameter. This illustrates its modest but crucial function in establishing the acidic level of water, which has a direct bearing on its suitability for drinking and farming.

Figure 4. RFR and XGBoost Models' Comparative Performance Measures for Predicting Groundwater Quality

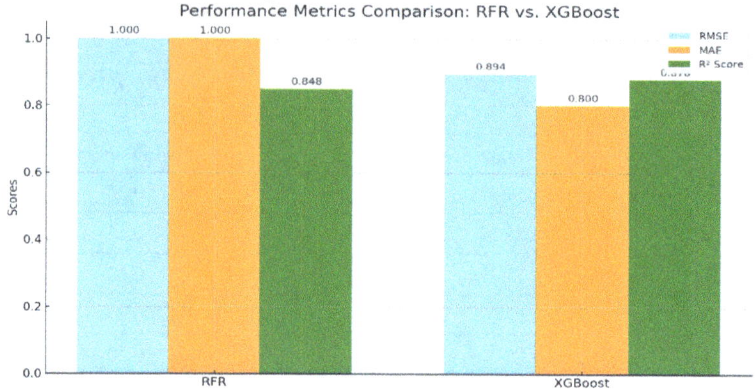

Table 4. Comparative performance measures for predicting groundwater quality

Model	RMSE	MAE	R^2 Score
RFR	1.000	1.0	0.848
XGBoost	0.894	0.8	0.878

Figure 4 and Table 4 shows as the RMSE, MAE, and R2 Score—all normalized for visualization—are the three main evaluation criteria used in the bar chart above to compare the performance metrics of the RFR and XGBoost models. For both RMSE and MAE, RFR shows perfect normalized scores of 1.000, demonstrating the model's low prediction errors. With an R2 score of 0.848, the model is highly accurate at explaining variance in the WQI. XGBoost records 0.894 (RMSE) and 0.800 (MAE), which indicates a little greater departure from actual WQI values, despite having somewhat lower error-based metrics. However, it makes up for this with a higher R2 score of 0.878, indicating more accurate predictions and better generalization performance. Finally, while XGBoost provides superior variance explanation and is more flexible for complex WQI modelling scenarios, RFR is superior at minimizing errors.

Figure 5. Correlation matrix of groundwater quality parameters and WQI

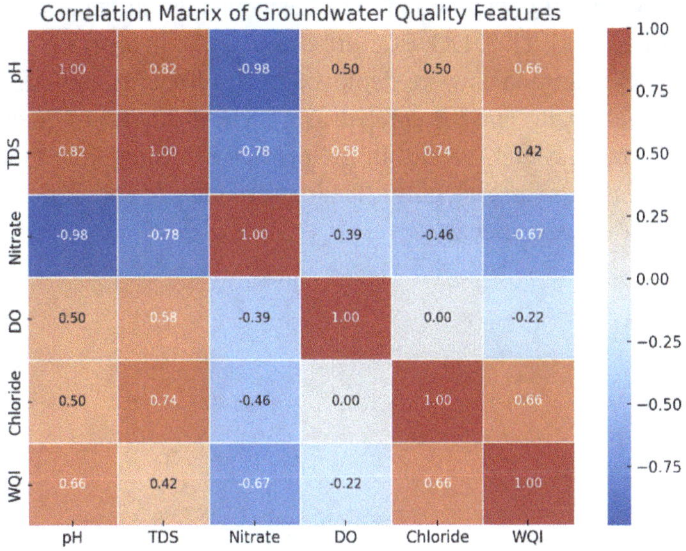

In Figure 5, Based on Pearson correlation coefficients ranging from -1 to +1, the correlation matrix shows the relationships between the following groundwater quality features: pH, TDS, Nitrate, DO, Chloride, and WQI. There is a strong negative correlation between pH and nitrate (-0.98) and a positive correlation with TDS (0.82), DO (0.50), Chloride (0.50), and WQI (0.66). This suggests that nitrate levels generally fall as pH rises. TDS has a negative correlation with nitrate (-0.78) but a strong positive correlation with chloride (0.74) and a moderate correlation with DO (0.58). This suggests that low nitrate and high chloride levels frequently accompany high TDS concentrations. Higher nitrate lowers water quality because it has strong negative correlations with most variables, particularly pH (-0.98) and WQI (-0.67). In this dataset, DO has less influence because it is largely uncorrelated with chloride (0.00) and only weakly related to other features. Groundwater quality is greatly influenced by these three characteristics, as evidenced by the positive correlations that WQI has with pH (0.66) and chloride (0.66) but a negative correlation with nitrate (-0.67). This matrix highlights pH, nitrate, and chloride as important predictors for WQI and guides feature selection for ML.

5. CONCLUSION

In this work, we used XGBoost and RFR models to evaluate and forecast groundwater quality in Tamil Nadu's Western Belt. While both models demonstrated high accuracy, XGBoost's performance was marginally superior. The most important factors in predicting WQI were found to be heavy metals, sulfate, chloride, nitrate, fluoride, iron, hardness, TDS, DO, EC, turbidity and pH. In order to provide actionable transparency, the models were deployed via dashboards and APIs, validated using lab and field data, and interpreted using SHAP. In Tamil Nadu and other water-stressed areas around the world, the predictive insights produced by these models are crucial for managing public health, formulating policies, and allocating groundwater resources sustainably.

REFERENCES

Ahmed, A., Ghosh, P. K., Hasan, M., & Rahman, A. (2020). Surface and groundwater quality assessment and identification of hydrochemical characteristics of a south-western coastal area of Bangladesh. *Environmental Monitoring and Assessment, 192*(4), 258. DOI: 10.1007/s10661-020-8227-0 PMID: 32239354

Alam, S. K., Li, P., Rahman, M., Fida, M., & Elumalai, V. (2025). Key factors affecting groundwater nitrate levels in the Yinchuan Region, Northwest China: Research using the eXtreme Gradient Boosting (XGBoost) model with the SHapley Additive exPlanations (SHAP) method. *Environmental Pollution, 364*, 125336. DOI: 10.1016/j.envpol.2024.125336 PMID: 39566709

Asadollah, S. B. H. S., Sharafati, A., Motta, D., & Yaseen, Z. M. (2021). River water quality index prediction and uncertainty analysis: A comparative study of machine learning models. *Journal of Environmental Chemical Engineering, 9*(1), 104599. DOI: 10.1016/j.jece.2020.104599

Azrour, M., Mabrouki, J., Fattah, G., Guezzaz, A., & Aziz, F. (2022). Machine learning algorithms for efficient water quality prediction. *Modeling Earth Systems and Environment, 8*(2), 2793–2801. DOI: 10.1007/s40808-021-01266-6

Balamurugan, J., Kasthuri, N., Sudarsan, J. S., Basi Reddy, A., Sakhare, D. T., & Sampathkumar, V. (2024). An efficient wastewater collection model for groundwater resource protection in smart cities. *Groundwater for Sustainable Development, 25*, 101091. DOI: 10.1016/j.gsd.2024.101091

Belkhiri, L., Tiri, A., & Mouni, L. (2020). Spatial distribution of the groundwater quality using kriging and Co-kriging interpolations. *Groundwater for Sustainable Development, 11*, 100473. DOI: 10.1016/j.gsd.2020.100473

Bui, D. T., Khosravi, K., Tiefenbacher, J., Nguyen, H., & Kazakis, N. (2020). Improving prediction of water quality indices using novel hybrid machine-learning algorithms. *The Science of the Total Environment, 721*, 137612. DOI: 10.1016/j.scitotenv.2020.137612 PMID: 32169637

Chen, Y., Song, L., Liu, Y., Yang, L., & Li, D. (2020). A review of the artificial neural network models for water quality prediction. *Applied Sciences (Basel, Switzerland), 10*(17), 5776. DOI: 10.3390/app10175776

Dukuly, L. P., Gupta, M., Ghani, S., & Akram, W. (2025). PCA-integrated machine learning framework for predicting rock fragmentation in blasting operations. *Multiscale and Multidisciplinary Modeling. Experiments and Design, 8*(9), 409. DOI: 10.1007/s41939-025-01008-6

. Eftekhari, M., Khozeymehnezhad, H., & Elyasi, A. H. (2024). Prediction and Assessment of Groundwater Quality in a Geographic Information System Environment Using Machine Learning Methods (Semi-Arid Regions). 中華水土保持學報, *55*(2), 84-93. DOI: 10.29417/JCSWC.202406_55(2).0004

El Bilali, A., Taleb, A., & Brouziyne, Y. (2021). Groundwater quality forecasting using machine learning algorithms for irrigation purposes. *Agricultural Water Management*, *245*, 106625. DOI: 10.1016/j.agwat.2020.106625

Gazzaz, N. M., Yusoff, M. K., Aris, A. Z., Juahir, H., & Ramli, M. F. (2012). Artificial neural network modeling of the water quality index for Kinta River (Malaysia) using water quality variables as predictors. *Marine Pollution Bulletin*, *64*(11), 2409–2420. DOI: 10.1016/j.marpolbul.2012.08.005 PMID: 22925610

Kayalvizhi, S., Jiavana, K. F. K., Suganthi, K., & Malarvizhi, S. (2023). Prediction of ground water quality in western regions of Tamil Nadu using deep auto encoders. *Urban Climate*, *49*, 101458. DOI: 10.1016/j.uclim.2023.101458

Khan, M. S. I., Islam, N., Uddin, J., Islam, S., & Nasir, M. K. (2022). Water quality prediction and classification based on principal component regression and gradient boosting classifier approach. *Journal of King Saud University. Computer and Information Sciences*, *34*(8), 4773–4781. DOI: 10.1016/j.jksuci.2021.06.003

. Khan, N. *et al.* (2020) 'Prediction of droughts over Pakistan using machine learning algorithms', *Advances in Water Resources*, 751 139(August 2019), p. 103562. DOI: 10.1016/j.advwatres.2020.103562

Kouadri, S., Elbeltagi, A., Islam, A. R. M. T., & Kateb, S. (2021). Performance of machine learning methods in predicting water quality index based on irregular data set: Application on Illizi region (Algerian southeast). *Applied Water Science*, *11*(12), 190. DOI: 10.1007/s13201-021-01528-9

Kouadri, S., Pande, C. B., Panneerselvam, B., Moharir, K. N., & Elbeltagi, A. (2022). Prediction of irrigation groundwater quality parameters using ANN, LSTM, and MLR models. *Environmental Science and Pollution Research International*, *29*(14), 21067–21091. DOI: 10.1007/s11356-021-17084-3 PMID: 34748181

Li, P., Qian, H., & Wu, J. (2019). Groundwater quality prediction using machine learning approaches in an arid region of China. *The Science of the Total Environment*, *653*, 1311–1321.

Olabode, O. F., & Comte, J. C. (2025). Long-term groundwater security in the fast-growing coastal megacity of Lagos, Nigeria: Insights from distributed recharge simulations. *Hydrological Sciences Journal*, *70*(11), 1813–1833. Advance online publication. DOI: 10.1080/02626667.2025.2505171

Raicy, M. C., Aju, C. D., & Achu, A. L. (2026). Groundwater quality and risk in the Ganga River Basin: An integrated machine learning appraisal. *Environmental Geochemistry and Health*, *48*(2), 72. DOI: 10.1007/s10653-025-02964-1 PMID: 41483424

Saqr, A. M., Nasr, M., Fujii, M., Yoshimura, C., & Ibrahim, M. G. (2024). Delineating suitable zones for solar-based groundwater exploitation using multi-criteria analysis: A techno-economic assessment for meeting sustainable development goals (SDGs). *Groundwater for Sustainable Development*, *25*, 101087. DOI: 10.1016/j.gsd.2024.101087

Singh, R., Upreti, P., Allemailem, K. S., Almatroudi, A., Rahmani, A. H., & Albalawi, G. M. (2022). Geospatial assessment of ground water quality and associated health problems in the Western Region of India. *Water (Basel)*, *14*(3), 296. DOI: 10.3390/w14030296

Singha, S., Pasupuleti, S., Singha, S. S., Singh, R., & Kumar, S. (2021). Prediction of groundwater quality using efficient machine learning technique. *Chemosphere*, *276*, 130265. DOI: 10.1016/j.chemosphere.2021.130265 PMID: 34088106

Su, Z., Wu, J., He, X., & Elumalai, V. (2020). Temporal changes of groundwater quality within the groundwater depression cone and prediction of confined groundwater salinity using Grey Markov model in Yinchuan area of northwest China. *Exposure and Health*, *12*(3), 447–468. DOI: 10.1007/s12403-020-00355-8

Subudhi, S., Pati, A. K., Bose, S., Sahoo, S., Pattanaik, A., Acharya, B. M., & Thakur, R. R. (2025). Prediction of groundwater quality assessment by integrating boosted learning with DE optimizer. *Scientific Reports*, *16*(1), 618. Advance online publication. DOI: 10.1038/s41598-025-30122-1 PMID: 41326700

Yuvaraj, N., Praghash, K., Logeshwaran, J., Peter, G., & Stonier, A. A. (2023). An artificial intelligence based sustainable approaches—IoT systems for smart cities. In *AI Models for Blockchain-Based Intelligent Networks in IoT Systems: Concepts, Methodologies, Tools, and Applications* (pp. 105–120). Springer International Publishing., DOI: 10.1007/978-3-031-31952-5_5

Yuvaraj, N., & Suresh Ghana Dhas, C. (2020). High-performance link-based cluster ensemble approach for categorical data clustering. *The Journal of Supercomputing*, *76*(6), 4556–4579. DOI: 10.1007/s11227-018-2526-z

Zhi, W., Feng, D., Tsai, W. P., Sterle, G., Harpold, A., Shen, C., & Li, L. (2021). From hydrometeorology to river water quality: Can a deep learning model predict dissolved oxygen at the continental scale? *Environmental Science & Technology*, *55*(4), 2357–2368. DOI: 10.1021/acs.est.0c06783 PMID: 33533608

Zhu, M., Wang, J., Yang, X., Zhang, Y., Zhang, L., Ren, H., Wu, B., & Ye, L. (2022). A review of the application of machine learning in water quality evaluation. *Eco-Environment & Health*, *1*(2), 107–116. DOI: 10.1016/j.eehl.2022.06.001 PMID: 38075524

Chapter 8
Remote Sensing and Artificial Intelligence Driving Liquid Intelligence for Water Sustainability Financing

Mohammad Irfan
https://orcid.org/0000-0002-4956-1170
Chandigarh University, Unnao, India

ABSTRACT

Financing sustainable water and sanitation remains a critical challenge in achieving global development goals. Despite progress, billions still lack access to safe drinking water and adequate sanitation due to funding gaps, weak governance, and rising climate pressures. This paper, Liquid Intelligence Financing Water Sustainability through Remote Sensing and Artificial Intelligence, explores how innovative financing models, including blended finance, microcredit, and public–private partnerships, can be strengthened through technological integration. Remote sensing provides accurate data on water resources, quality, and scarcity, while artificial intelligence enables predictive analytics, smart infrastructure management, and optimized investment decisions. Together, these tools enhance transparency, efficiency, and resilience in financing frameworks. The study argues that combining financial innovation with data-driven intelligence not only advances SDG 6 but also fosters equity, health, and long-term economic sustainability.

DOI: 10.4018/979-8-3373-5886-4.ch008

INTRODUCTION

Financing sustainable water and sanitation remains a pressing challenge, with billions lacking access to safe drinking water and adequate sanitation. Innovative approaches are urgently needed to bridge funding gaps, address governance weaknesses, and mitigate climate risks. Integrating financial innovation with advanced technologies offers new opportunities for sustainable development. Recent studies emphasize that remote sensing and artificial intelligence can enhance transparency, optimize resource allocation, and strengthen resilience in water governance (Boretti & Rosa, 2019). This study explores such synergies toward achieving SDG 6. Alotaibi and Nassif (2024) demonstrated that artificial intelligence, when integrated with remote sensing, enhances real-time water quality assessment, predictive modeling, and early warning systems. These insights underscore the value of combining financial innovation with data-driven intelligence to promote transparency, efficiency, and resilience in water governance, ultimately accelerating progress toward SDG 6.

Access to clean water and adequate sanitation is not only a fundamental human right but also a cornerstone of sustainable development. Despite decades of progress, more than two billion people globally still lack safe drinking water, and over 3.5 billion lack safely managed sanitation (UNICEF & WHO, 2021). This gap is not simply the result of physical scarcity or technological limitations but is largely driven by inadequate financing, poor governance, and unequal service delivery. Thus, financing clean water and sanitation has emerged as a critical challenge for policymakers, investors, and development organizations.

ADVANCING SDG 6 THROUGH REMOTE SENSING AND ARTIFICIAL INTELLIGENCE

Advancing Sustainable Development Goal 6 through Remote Sensing and Artificial Intelligence for Water Monitoring, Smart Governance, and Equitable Financing to Ensure Universal Access to Clean Water and Sanitation (UN-Water, 2020; Rodell et al., 2018).

Figure 1. Sustainable development goal 6 (UN-Water, 2020; Rodell et al., 2018)

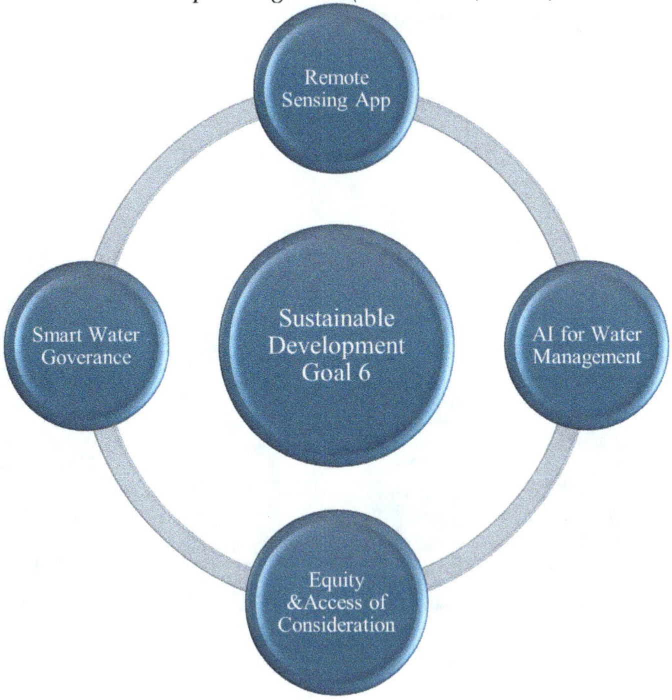

The integration of Remote Sensing (RS) and Artificial Intelligence (AI) is emerging as a critical research frontier in achieving Sustainable Development Goal 6 (SDG 6). RS technologies, such as NASA's GRACE and ESA's Sentinel satellites, provide consistent and large-scale data on groundwater depletion, surface water variability, pollution, and climate change impacts (Rodell et al., 2018). However, RS data alone often require advanced interpretation for actionable insights. Here, AI contributes by applying machine learning and deep learning models to vast, heterogeneous datasets from satellites, sensors, and smart meters, enabling leak detection, drought forecasting, irrigation optimization, and improved wastewater treatment (UNESCO, 2020). When combined, RS and AI enhance predictive capabilities, support early warning systems, and enable equitable water governance frameworks (United Nations Office for Outer Space Affairs, 2021). This synergy offers a transformative pathway for research and policy, ensuring that technological innovations directly contribute to sustainable water management, efficient financing, and inclusive access to water and sanitation.

REMOTE SENSING AND AI IN WATER SUSTAINABILITY FINANCING

Figure 2. Forecasted growth (Ahmed et al., 2023)

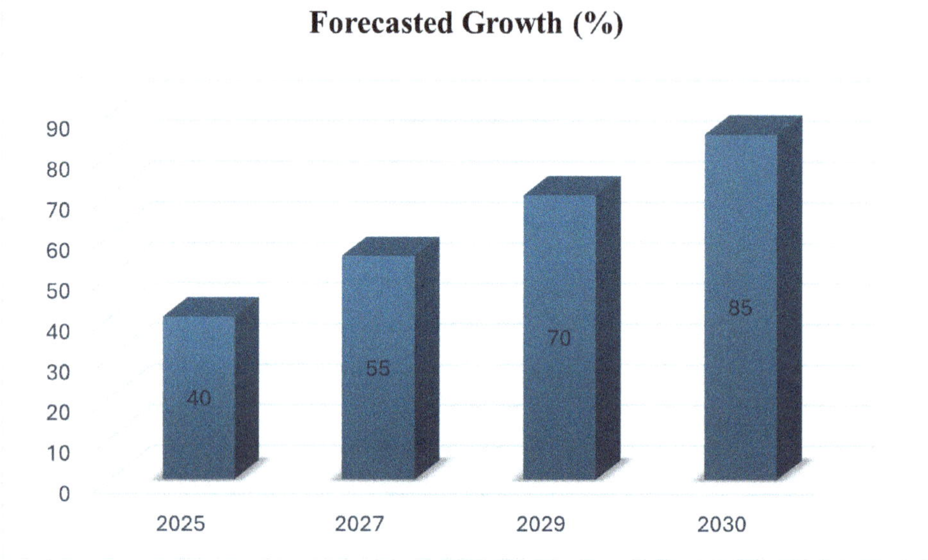

The bar chart illustrates the forecasted growth (%) in water sustainability financing through Remote Sensing and Artificial Intelligence from 2025 to 2030. Growth is projected to rise from 40% in 2025 to 85% by 2030, reflecting increasing reliance on advanced technologies for efficient water management and investment decisions. Remote Sensing and AI enhance monitoring, predictive analysis, and resource allocation (Ahmed et al., 2023).

ROLE OF REMOTE SENSING AND ARTIFICIAL INTELLIGENCE DRIVING LIQUID INTELLIGENCE FOR WATER SUSTAINABILITY FINANCING

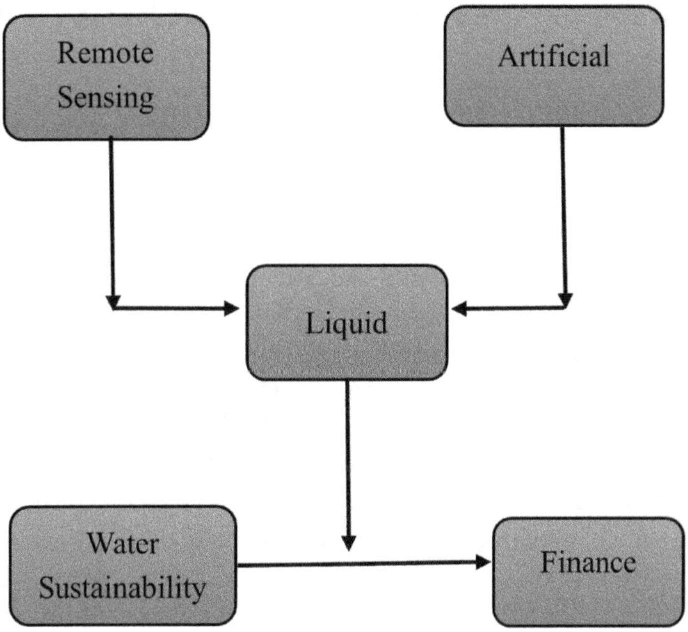

Recent studies in Water journal indicate that the integration of remote sensing (RS) and artificial intelligence (AI) is accelerating what might be called liquid intelligence—a continuous, predictive, integrated analytics framework supporting water sustainability financing. Machine learning models, when fed with satellite data (e.g. Landsat, Sentinel), soil moisture products, evapotranspiration datasets, and surface-water extents, allow estimation of water availability, groundwater potential, flood and drought risk with high spatial and temporal resolution. For example, a systematic review by Ghobadi and Kang (2023) shows that ML techniques are widely used in water resources management—surface water, groundwater, and climate extremes—with increasingly fine-grained remote-sensing inputs.

These capabilities derive into financing-relevant outputs: risk maps for lenders or insurers; monitoring tools for investors; frameworks for prioritizing interventions in water infrastructure; cost-benefit assessments under climate variability. Water's special issues ("Applications of Remote Sensing and Machine Learning in Water Resources Management") highlight case studies where RS + ML produce actionable intelligence for flood/drought forecasting, water body change detection, and scenario modelling useful for policy and investment decisions (Ghobadi & Kang, 2023; Zhou & Chen, 2023a).

Despite promise, gaps remain in linking RS/AI outputs directly to financial mechanisms (e.g. bonds, credits), ensuring transparency, and scaling in data-poor regions. Still, evidence from Water implies that liquid intelligence is emerging as essential for enabling reliable, resilient, and just water sustainability financing (Fernández-Pacheco & Andreu, 2023b).

At the center of this effort is Sustainable Development Goal 6 (SDG 6), which aims to "ensure availability and sustainable management of water and sanitation for all" by 2030 (United Nations, 2015). Achieving this goal is increasingly difficult due to factors such as population growth, climate change, urbanization, and rising demand for water-intensive industries. According to the World Bank (2020), universal access to safely managed water and sanitation requires annual investments of approximately USD 1.37 trillion, nearly six times current spending levels. Similarly, Organisation for Economic Co-operation and Development (2023) estimates that meeting global water and sanitation targets could require USD 1 trillion annually, while long-term infrastructure needs could grow to USD 6.7 trillion by 2030 and USD 22.6 trillion by 2050. By extrapolation, financing needs in 2047 could approach USD 18–20 trillion, underscoring the urgency of mobilizing significant capital.

The funding gap remains stark. Current global spending is roughly USD 28–30 billion annually, leaving a shortfall of over USD 85 billion per year. Public financing dominates, accounting for more than 90% of investments in developing countries, while private sector contributions remain below 2% (World Bank, 2020). To close this gap, diverse and innovative financing models are essential (Marley et al., 2022).

Blended finance, which combines public or philanthropic funds with private capital, has gained prominence as a risk-sharing tool that attracts investment into high-impact but often low-return sectors such as water and sanitation (Organisation for Economic Co-operation and Development, 2018). Similarly, microfinance has shown promise, helping low-income households in rural and peri-urban areas access basic water and sanitation services like toilets, handwashing stations, and filters. Public–private partnerships (PPPs) are also critical, especially for infrastructure-heavy projects such as treatment plants and sewer networks, bringing in technical expertise and long-term capital (Willems & MacDonald, 2016).

However, financing alone cannot solve the challenge. Success depends on robust governance, supportive policies, and community engagement. Without transparent regulatory frameworks and inclusive pricing models, even large investments can fail to deliver equitable outcomes. Issues such as corruption, weak institutional capacity, and policy fragmentation further undermine progress (UNICEF & WHO, 2021). Therefore, financing solutions must be integrated with reforms in policy, capacity building, and accountability mechanisms to ensure funds reach the communities most in need.

The challenge is particularly acute in politically fragile and resource-constrained regions like sub-Saharan Africa and South Asia. Here, donor agencies and multilateral development banks often bridge the funding gaps, but long-term sustainability will require building resilient financial ecosystems that leverage domestic revenues, private capital, and international support. Investments in water and sanitation also drive broader development goals (Csala & Mwangi, 2019). Improved access reduces waterborne diseases, enhances education outcomes (particularly for girls), and supports gender equality and economic growth. Research shows that every USD 1 invested in water and sanitation generates about USD 4 in returns through improved health, productivity, and reduced healthcare costs (World Bank, 2020).

This paper examines financing models for clean water and sanitation in low- and middle-income countries, combining quantitative data and policy analysis. It includes visual projections, such as a line diagram of investment needs up to 2047 and a bar chart for 2030, to highlight the scale of the challenge. By integrating innovative financial tools with governance reforms and international collaboration, the global community can accelerate progress toward SDG 6 and build resilient systems that ensure clean water and sanitation for all.

After decades of economic turmoil and political instability, Zimbabwe's infrastructure has sharply deteriorated. Fiscal mismanagement, corruption, and the poorly executed land redistribution programme triggered the country's most severe peacetime economic collapse. In response to human rights violations and the erosion of the rule of law, Western governments imposed political and economic sanctions. As a result, Zimbabwe defaulted on its debt obligations, with arrears accumulating to International Development Partners (IDPs) and bilateral creditors. Today, the country faces a staggering debt overhang of more than USD 17.7 billion, severely restricting access to long-term infrastructure financing. Consequently, the productive capacity of infrastructure facilities, particularly in critical sectors such as water and sanitation, has been depleted Mundonde, J., & Makoni, P. L. (2024).

Financing clean water and sanitation is crucial for sustainable development, as it directly impacts public health, economic growth, and environmental sustainability. The challenge lies in bridging the significant financing gap to achieve Sustainable Development Goal 6 (SDG 6), which aims for universal access to clean water and sanitation by 2030. This requires a multifaceted approach that includes innovative financing strategies, risk mitigation instruments, and increased investment in infrastructure and governance.

CURRENT FINANCING CHALLENGES

- The estimated cost of achieving SDG 6 exceeds US$1 trillion annually, necessitating a tripling of current investment levels to meet the 2030 targets("-Financing", 2023).
- In developing countries, a significant portion of the population still lacks access to safe water and sanitation, with over 1.1 billion people without safe drinking water and 2.4 billion without adequate sanitation (Baietti & Raymond, 2005).
- Existing water supply systems often provide intermittent and unsafe water, highlighting the need for improved infrastructure.

In 2015, all United Nations (UN) Member States adopted the 2030 Agenda for Sustainable Development, which includes 17 Sustainable Development Goals (SDGs). Among them, SDG 6 is specifically dedicated to clean water and sanitation, aiming to ensure universal access to safe, reliable, sustainable, and affordable water and wastewater services. The UN recognizes water and sanitation as a fundamental human right, essential for the safety, health, and wellbeing of individuals. However, progress toward universal coverage remains slow. Achieving basic sanitation services for all by 2030 would require doubling the current rate of advancement, while future investment needs are expected to be very high. Despite this urgency, financing flows to the water and wastewater sectors remain insufficient. Sustainable financing of water supply and sanitation (WSS) depends on a balanced approach involving tariffs, taxes, and transfers (the "3Ts") to ensure cost recovery and long-term financial viability. In addition, complementary financing mechanisms are necessary to fund critical infrastructure projects. Financing structures may vary, ranging from direct approaches, such as on-balance sheet investments in corporate entities, to market-based mechanisms, including bonds and equities. Mobilizing diverse funding instruments and innovative financing strategies is therefore vital for meeting SDG 6 targets and advancing global water security (Machete, Inês, and Rui Marques. 2021).

EXPLANATION OF THE SDG 6 GLOBAL ACCELERATION FRAMEWORK

Sustainable Development Goal 6 (SDG 6) seeks to "ensure availability and sustainable management of water and sanitation for all" by 2030. Despite considerable progress in global water and sanitation coverage since 2000, major gaps remain. Billions of people still lack safely managed drinking water, basic sanitation, and hygiene services, disproportionately affecting low- and middle-income countries (WHO/UNICEF, 2021). To address these challenges, the United Nations launched the SDG 6 Global Acceleration Framework in 2020. This initiative aims to generate multi-stakeholder commitment, strengthen collaboration, and accelerate implementation to meet water and sanitation targets (UN-Water, 2020). The framework identifies five key "accelerators"—finance, data and information, capacity development, innovation, and governance—that act as enablers of transformative change.

Finance is central to achieving SDG 6. Global estimates indicate that closing the water and sanitation gap requires trillions of dollars in investment by 2030, yet current financing is insufficient (Hutton & Varughese, 2016). The Framework calls for mobilizing domestic resources, increasing international assistance, and engaging private sector investment. Without adequate and targeted financing, progress will continue to lag, particularly in regions with weak infrastructure. Data and information serve as the backbone of informed decision-making. Reliable, disaggregated, and timely data are critical to monitor progress, identify inequities, and guide interventions. However, data gaps persist in many countries, limiting their ability to track service coverage and quality effectively (UN-Water, 2020). The Framework emphasizes strengthening monitoring systems such as the WHO/UNICEF Joint Monitoring Programme (JMP) and harnessing modern technologies, including Earth observation and digital platforms, to improve data quality and accessibility.

Capacity development focuses on building the skills, institutions, and systems required to implement sustainable water and sanitation solutions. Many countries lack technical expertise, institutional coordination, and trained personnel in the water and sanitation sector. By promoting capacity development, the Framework seeks to empower governments, civil society, and communities to deliver long-lasting solutions (UNESCO, 2020), Mohanty, A., Ghosh, A., & Roy, S. (2024). Innovation is another critical accelerator. Traditional solutions alone are insufficient to close the gaps in water and sanitation coverage. The Framework encourages the adoption of new technologies such as smart water management systems, wastewater recycling, low-cost sanitation models, and renewable energy-driven solutions. Innovation also extends to social and financial models, ensuring services are affordable and sustainable for vulnerable populations (World Bank, 2020).

Finally, governance is the foundation that integrates all other accelerators. Effective governance ensures accountability, transparency, and inclusive participation in water and sanitation management. Weak governance structures have been identified as a major barrier to achieving SDG 6 (UNDP, 2019). The Framework advocates for cross-sectoral coordination, stronger legal and policy frameworks, and community engagement to ensure equitable access.

Figure 4. (UN-Water, 2020)

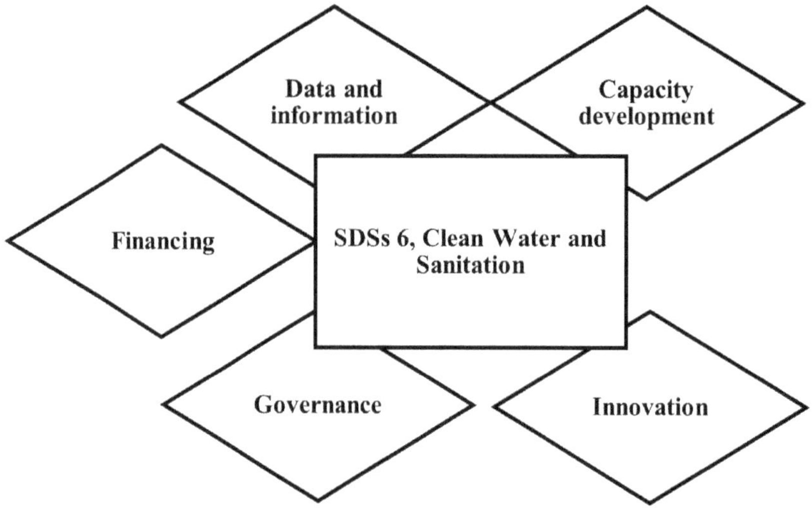

In summary, the SDG 6 Global Acceleration Framework represents a comprehensive response to existing discrepancies in water and sanitation access. By mobilizing resources across the five accelerators, it seeks to fast-track progress and close gaps before 2030. Whether these actions are sufficient remains uncertain, as challenges such as climate change, population growth, and inequality continue to intensify. However, with sustained commitment, innovation, and cooperation, the vision of universal water and sanitation access can be realized.

Financing clean water and sanitation (WATSAN) for sustainable development is a multifaceted challenge that requires innovative strategies and collaborative efforts. The transition from traditional public financing to a more diversified approach is essential to meet the Sustainable Development Goals (SDGs). This involves not only increasing funding but also improving governance and efficiency within the sector.

Financing Paradigms: Public and Private Blending: Historically, the water sector has relied on public funds, but there is a pressing need to blend these with private investments to enhance financial sustainability (Kolker et al., 2016). Innovative Financial Instruments: New blended finance vehicles are emerging to address the significant finance gap, particularly for marginalized populations(Pories et al., 2019).

Governance and Efficiency: Improving Creditworthiness: Enhancing the creditworthiness of water service providers is crucial for attracting private finance(Kolker et al., 2016). Foundational Issues: Addressing foundational governance issues is necessary for any financial mechanism to be effective in the long term(Pories et al., 2019).

Holistic Approaches: Socioeconomic Perspectives: Financing strategies should consider the socioeconomic context and long-term environmental impacts rather than merely focusing on costs(Nigam, 1996). Collaborative Efforts: A collective commitment from all stakeholders is vital to ensure that financing efforts are sustainable and effective(Pories et al., 2019). While these strategies present a pathway to improved financing for clean water and sanitation, challenges remain, particularly in mobilizing private finance and ensuring equitable access across diverse populations. Addressing these issues holistically will be essential for achieving the SDGs.

Figure 5. Share of the population using drinking water facilities, 2022 (WHO/ UNICEF Joint Monitoring Programme for Water Supply, Sanitation and Hygiene (JMP) (2024))

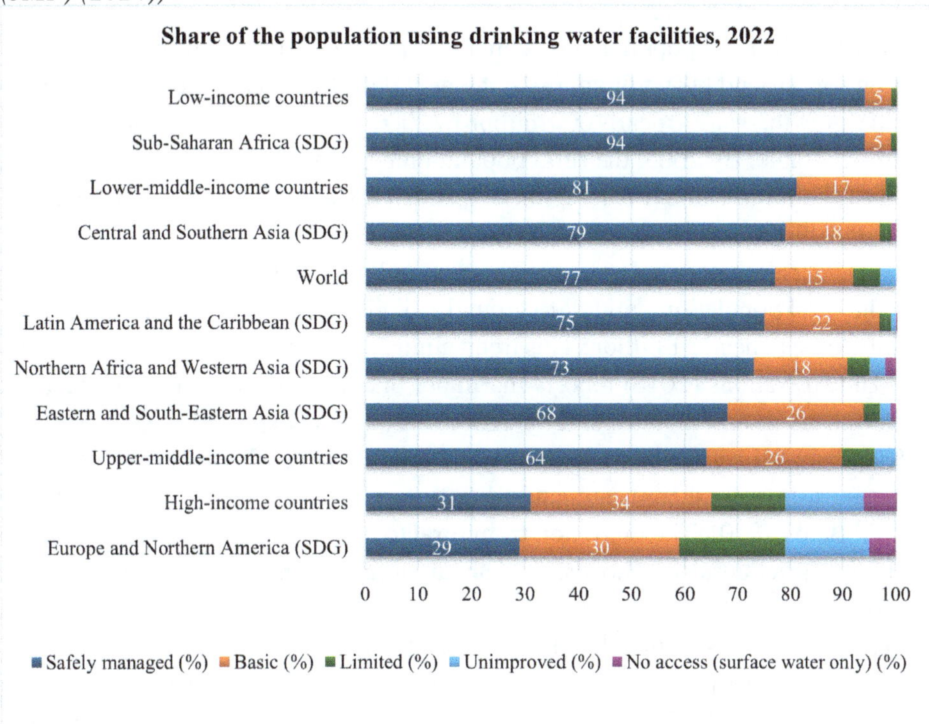

The chart illustrates the global disparities in access to drinking water facilities in 2022. High-income countries and Europe/Northern America report the highest share of safely managed water services (94%), while Sub-Saharan Africa and low-income countries face the greatest challenges, with only 31% and 29% safely managed access, respectively. Instead, these regions show higher reliance on limited, unimproved, or surface water sources. Globally, 73% of people have safely managed water, leaving nearly one-third with inadequate access. The data underscores the urgent need for investment, governance reform, and innovative financing to achieve Sustainable Development Goal 6 by 2030.

EXPLANATION OF URBAN AND RURAL POPULATION TRENDS

The graph illustrates the historical and projected distribution of populations in urban and rural regions between 1990 and 2030. It highlights a significant demographic shift, where the rural population, which has historically dominated, is declining, while the urban population is steadily increasing. This trend reflects global patterns of urbanization driven by industrialization, economic opportunities, and socio-political changes.

In 1990, the rural population was approximately 2 billion, while the urban population was around 500 million. Over the next decade, urban regions experienced gradual growth, rising to nearly 600 million by 2000, while rural populations remained relatively stable. By 2005, the rural population began to decline slightly, dropping below 1.9 billion, while urban regions expanded further to around 650 million. By 2015, the shift became more pronounced, with the rural population decreasing to 1.7 billion and the urban population reaching 750 million. Projections for 2030 suggest that the rural population will fall to around 1.5 billion, while the urban population will increase to nearly 1 billion. This trend aligns with global urbanization patterns. The United Nations (2018) reports that more than half of the world's population already lives in urban areas, and by 2050, this share is expected to rise to nearly 68%. Urbanization is primarily fueled by the promise of employment opportunities, better infrastructure, healthcare, and education in cities compared to rural regions (World Bank, 2020). The migration from rural to urban regions is also linked to the decline of agricultural livelihoods, mechanization, and changing economic structures (Bloom et al., 2008).

The implications of this demographic shift are profound. On the positive side, urbanization can foster innovation, productivity, and improved access to services (UN-Habitat, 2022). Cities often act as economic engines, driving national and regional growth. However, rapid urbanization also poses challenges, including overcrowding, inadequate housing, traffic congestion, environmental degradation, and the strain on public services (United Nations, 2018). Governments face increasing pressure to ensure sustainable urban planning to accommodate the growing urban population without exacerbating inequality and environmental stress.

For rural regions, population decline can lead to labor shortages in agriculture, abandonment of farmland, and reduced economic activity. However, rural depopulation may also offer opportunities for environmental recovery and sustainable land management if properly managed (World Bank, 2020). Policymakers are increasingly recognizing the importance of balancing urban and rural development to ensure equitable growth and prevent excessive urban-rural divides.

In conclusion, the chart underscores a global demographic transformation, where urban areas are becoming increasingly dominant while rural regions face steady population decline. This transformation demands strategic policy responses to harness the benefits of urbanization while mitigating its challenges. Simultaneously, rural development policies must address depopulation by promoting sustainable livelihoods, digital inclusion, and infrastructure development. The ongoing trend of urban growth and rural decline is not merely a demographic reality but also a socio-economic and environmental issue that requires comprehensive planning to ensure sustainable development for future generations.

Figure 6. Urban and rural population trends of clean water and sanitation (1990–2030, projected (WHO/UNICEF Joint Monitoring Programme (improved sanitation facilities) and UNCTADstat)

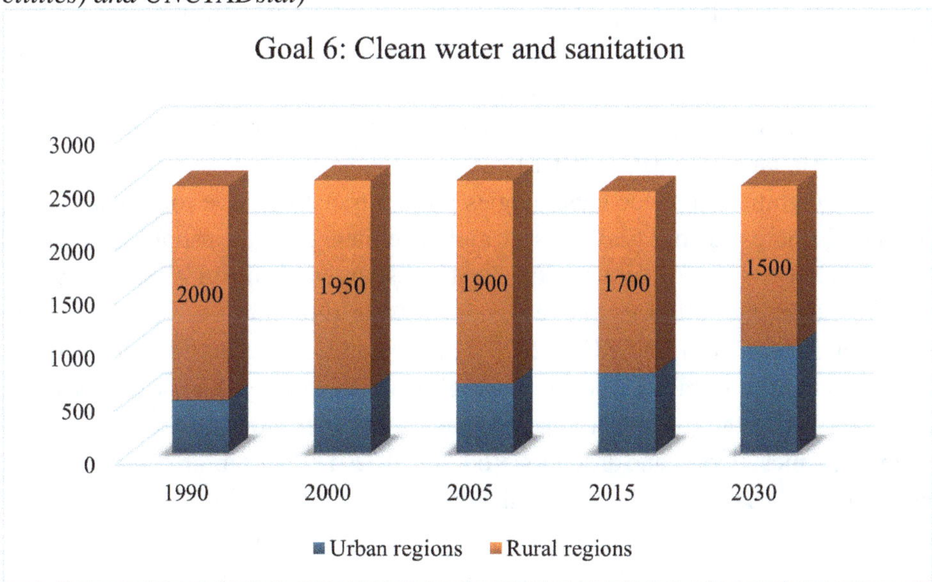

SDG 6 PROGRESS AND THE ROLE OF SPACE TECHNOLOGIES

Sustainable Development Goal (SDG) 6, which seeks to "ensure availability and sustainable management of water and sanitation for all," is a critical component of the United Nations 2030 Agenda for Sustainable Development. The graph presented illustrates the global progress on key SDG 6 indicators—safely managed drinking water, basic drinking water, safely managed sanitation, and basic sanitation—from 2000 through projections to 2030. The trends highlight steady progress across all four categories, with marked acceleration in the latter years, suggesting that global initiatives are producing measurable outcomes. However, the role of emerging technologies, particularly space-based systems, is becoming increasingly important in supporting monitoring, assessment, and decision-making to achieve universal access.

In 2000, safely managed drinking water access was significantly lower compared to the other indicators, hovering around 30%, whereas basic sanitation and drinking water were relatively higher, at about 60–80%. By 2015 and 2017, the data indicate notable improvements: safely managed sanitation increased significantly from around 55% to over 70%, while basic drinking water also showed consistent upward trends. Safely managed drinking water, though slower in growth, shows projections that align with other indicators by 2030, suggesting an ambitious acceleration trajectory. These projections align with global efforts, as reports by the WHO/UNICEF Joint Monitoring Programme show that billions of people have gained access to improved water, sanitation, and hygiene services in recent decades, yet progress must intensify to meet 2030 targets (WHO/UNICEF, 2021).

The integration of space technologies plays a transformative role in achieving SDG 6. Remote sensing, Earth observation satellites, and geospatial technologies provide vital data for monitoring freshwater resources, identifying water scarcity hotspots, tracking pollution sources, and assessing the impact of climate change on water availability (UNESCO, 2020). For example, satellite missions such as NASA's GRACE (Gravity Recovery and Climate Experiment) track groundwater depletion, while the European Space Agency's Sentinel satellites monitor surface water dynamics, floods, and droughts (Rodell et al., 2018). This information supports policymakers and international organizations in making evidence-based decisions to allocate resources and plan interventions more effectively.

Furthermore, space technologies enhance disaster risk management related to water, such as floods and droughts, which are intensifying under climate change. By integrating satellite-derived data with predictive models, governments can better prepare for and respond to water-related hazards, protecting vulnerable populations (United Nations Office for Outer Space Affairs, 2021). These tools also facilitate the monitoring of rural and remote areas where traditional survey methods are costly or impractical.

However, while technological progress is promising, challenges remain in ensuring equity. Space-based data must be accessible to all nations, particularly low-income countries that face the greatest challenges in achieving SDG 6. Capacity building, data sharing, and international cooperation are critical to ensure that advancements benefit everyone, not only technologically advanced nations.

In conclusion, the chart demonstrates positive progress toward achieving SDG 6, with projections suggesting that universal access to safe drinking water and sanitation is within reach by 2030 if current momentum continues. The role of space technologies is indispensable in supporting this goal, offering innovative solutions to monitor, manage, and sustain water resources globally. To maximize impact, policymakers must integrate space-based data into national planning frameworks while fostering global collaboration. This combined effort of human commitment and technological innovation offers the best pathway to achieving SDG 6 and ensuring water and sanitation for all.

Figure 7. SDG 6 progress and the role of space technologies (United Nations, 2020)

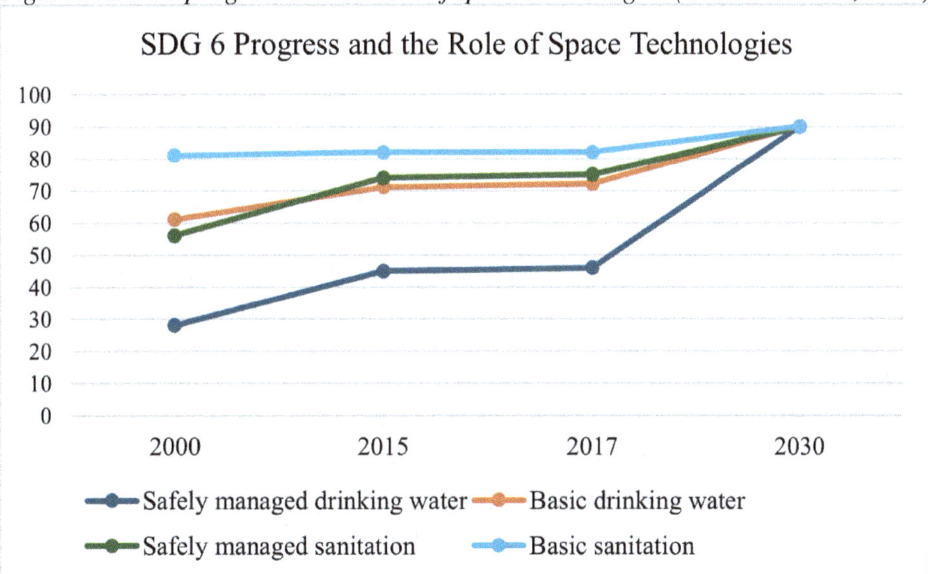

CONCLUSION

Achieving Sustainable Development Goal 6 requires integrating innovative financing with advanced technologies such as remote sensing and artificial intelligence. While financing gaps remain significant, technological tools enhance transparency, predictive capacity, and equitable governance in water management (Rodell et al., 2018; UNESCO, 2020). Remote sensing provides real-time data on water availability and risks, while AI enables smart forecasting and resource optimization (Alotaibi & Nassif, 2024). Together, they strengthen resilience, efficiency, and inclusivity in financing frameworks. By fostering global cooperation and equitable access, the combined application of finance, RS, and AI can accelerate progress toward universal clean water and sanitation.

Financing clean water and sanitation remains one of the most pressing challenges of sustainable development, demanding both financial innovation and systemic reforms. As outlined throughout this study, achieving Sustainable Development Goal 6 (SDG 6) universal access to clean water and sanitation by 2030 requires unprecedented levels of investment, governance reform, and multi-stakeholder collaboration. Despite progress in recent decades, billions of people still lack safe drinking water and adequate sanitation, underscoring the severity of the global financing gap (UNICEF & WHO, 2021). Estimates suggest that meeting global water and sanitation targets will require at least USD 1 trillion annually, far exceeding current investment flows (Organisation for Economic Co-operation and Development, 2023; World Bank, 2020).

The evidence indicates that reliance on traditional public financing is insufficient. While governments continue to provide over 90% of sectoral funding, this model cannot meet the accelerating needs driven by urbanization, climate change, and population growth (World Bank, 2020; Choudhury et al., 2023). Innovative financing strategies, including blended finance, microcredit, and public private partnerships (PPPs), provide practical avenues to mobilize private capital while mitigating investment risks (Organisation for Economic Co-operation and Development, 2018; Kolker et al., 2016). At the same time, approaches such as treating water as a new asset class demonstrate how financial markets can be leveraged for sustainable and climate-resilient water management (Elmahdi & Wang, 2022).

However, financial innovation alone will not guarantee success. The broader enabling environment comprising governance, regulation, and institutional capacity—plays an equally decisive role. Weak governance, corruption, and poor institutional frameworks often prevent resources from reaching the communities most in need (UNDP, 2019). Therefore, financing strategies must be paired with robust accountability systems, transparent pricing models, and strong institutional capacity to ensure equitable outcomes (Pories et al., 2019). Community engagement and attitudinal change also remain central, particularly in rural and peri-urban sanitation, where demand generation and behavioral shifts are necessary for sustainable adoption (Nigam, 1996).

The SDG 6 Global Acceleration Framework provides a blueprint for addressing these challenges, emphasizing five accelerators finance, data, capacity, innovation, and governance—that must work in tandem to achieve transformative progress (UN-Water, 2020). The integration of emerging technologies, such as space-based monitoring systems, further illustrates how innovation can complement financing by enhancing data-driven decision-making and resource management (Rodell et al., 2018; UNESCO, 2020). Yet, technological advancements must be accompanied by equitable access to ensure benefits reach vulnerable populations rather than exacerbating inequalities.

Ultimately, achieving SDG 6 is not merely a question of infrastructure investment but of systemic resilience and inclusivity. Every dollar invested in water and sanitation yields fourfold returns in improved health, productivity, and reduced healthcare costs (World Bank, 2020). This makes water financing not just a moral obligation but also a sound economic strategy that contributes to poverty reduction, gender equality, and long-term economic growth. Closing the financing gap will require a paradigm shift moving from fragmented, short-term solutions toward integrated, long-term strategies that align financial instruments with governance reforms and community participation.

If the global community can foster effective collaboration between governments, private investors, development agencies, and local communities, the vision of clean water and sanitation for all is achievable. By mobilizing diverse financing mechanisms, strengthening governance systems, and leveraging technological innovation, the pathway toward SDG 6 becomes not only feasible but also transformative delivering lasting benefits for health, equity, and sustainable development.

Social Implications

The findings of this paper highlight profound social implications in addressing the global financing gap for clean water and sanitation. Lack of adequate access disproportionately affects vulnerable groups, particularly women, children, and marginalized communities, who bear the brunt of poor health, lost educational opportunities, and economic exclusion (UNICEF & WHO, 2021). Financing models that expand equitable access not only improve public health outcomes by reducing waterborne diseases but also enhance social equity and gender equality. For example, investments in sanitation infrastructure can reduce the burden on women and girls who often travel long distances for water, thereby improving school attendance and participation in the workforce. Moreover, clean water access is foundational to breaking cycles of poverty, fostering community resilience, and ensuring dignity in daily life. Thus, innovative financing and governance reforms are not solely economic interventions but transformative tools for building inclusive, just, and healthier societies (Organisation for Economic Co-operation and Development, 2023; World Bank, 2020).

Practical Implication

The chapter outlines key themes in financing water, which can guide countries in developing effective financial strategies to meet their water and sanitation goals as part of the Sustainable Development Goals (SDGs). It highlights new initiatives that are emerging to support countries in their efforts, suggesting a collaborative approach to enhance investment and financing mechanisms for water-related projects (Organisation for Economic Co-operation and Development, 2022). Sustainable financing strategies for water supply and sanitation (WATSAN) must adopt a holistic approach that considers the socioeconomic and long-term environmental perspectives, rather than merely focusing on cost estimation and funding sources. This implies that stakeholders need to integrate various aspects of water management, including extraction, use, and re-use, to ensure effective resource management. In rural and peri-urban sanitation, the emphasis should be on fostering attitudinal change and creating effective demand for improved access to sanitation services. This suggests that financing strategies should not only address hardware and software components but also prioritize community engagement and education to enhance the demand for better sanitation solutions, Nigam, A. (1996).

The paper emphasizes the need for a comprehensive framework for allocating risks in water supply and sanitation (WSS) sector investments, which involves establishing enforceable commercial agreements between public and private parties. This framework aims to address the practical limits of what governments can achieve within their existing environments and necessitates significant policy changes regarding the roles and responsibilities of key stakeholders. It highlights the importance of empowering municipalities to act as effective guarantors for financing, which requires them to possess sufficient creditworthiness and legal authority. This empowerment is crucial for enabling local governments to secure financing for WSS projects, particularly in contexts where central government support is limited or where local utilities are facing operational cash losses (Baietti & Raymond, 2005). The paper highlights the urgent need for major investments in infrastructure and governance of water provisioning to ensure public health and increase resilience for transmissible diseases and virus outbreaks, emphasizing the importance of clean water and sanitation for human well-being and economic development. It underscores the challenges faced in achieving the targets of SDG-6, such as financial constraints, capacity building issues, and governance gaps, stressing the critical role of finance, capacity development, and effective governance in overcoming these challenges and ensuring sustainable water management for all (Sage, 2022). The paper highlights the urgent need to scale up financing for water security, indicating that current investment levels must be tripled to achieve universal and equitable access to safe drinking water by 2030. This suggests that policymakers and stakeholders must prioritize and increase funding allocations for water-related projects. It emphasizes the importance of better cooperation among various entities to address the financing challenge for water security, indicating that collaborative efforts can lead to more effective solutions and increased financial flows necessary to meet the estimated costs of over US$1 trillion per year for achieving Sustainable Development Goal 6 ("Financing," 2023). The paper emphasizes the need for a paradigm shift in defining, developing, and financing water assets, which involves creating an enabling investment environment that supports water resource conservation and integrated water management approaches. This shift is crucial for increasing the resilience of the water sector to climate change and addressing the projected 40% shortfall in water availability by 2030, particularly in developing countries. It proposes an innovative financial model that treats water as a new asset class, which aims to overcome financial market barriers and ensure the affordability and bankability of water reuse and sanitation investments. This model encourages collaboration between municipalities and the private sector, facilitating the scaling up of water reuse, sanitation, and desalination projects while addressing the interconnected challenges of water security and climate resilience (Elmahdi & Wang, 2022).

REFERENCES

Ahmed, A., Khan, M., & Li, Y. (2023). Remote sensing and artificial intelligence for sustainable water resource management: Opportunities and challenges. *Sustainability, 15*(12), 10045.

Alotaibi, R., & Nassif, A. B. (2024). AI-driven remote sensing for sustainable water resource management: A comprehensive review. *Environmental Science and Pollution Research International, 31*(12), 15321–15340. DOI: 10.1007/s11356-023-29152-2

Baietti, A., & Raymond, P. (2005). *Financing water supply and sanitation investments: Utilizing risk mitigation instruments to bridge the financing gap*. World Bank. https://openknowledge.worldbank.org/bitstream/handle/10986/17235/320280WSS1Investments.pdf

Bloom, D. E., Canning, D., & Fink, G. (2008). Urbanization and the wealth of nations. *Science, 319*(5864), 772–775. DOI: 10.1126/science.1153057 PMID: 18258907

Boretti, A., & Rosa, L. (2019). Reassessing the projections of the World Water Development Report. *Sustainability, 11*(3), 760. DOI: 10.3390/su11030760

Choudhury, D., & Bhattacharjee, A. (2023). Environmental communication as a mediator in the relationship between environmental concern and attitude towards green services in a sequential mediation model: Evidence from restaurant industry. *Purushartha, 16*(1). journals.smsvaranasi.comDOI: 10.21844/16202116105

Csala, J., & Mwangi, J. W. (2019). *Nature-Based Solutions for Water Security and the Role of Education for Enhancing the Implementation of the 2030 Agenda.* In *Building Sustainability Through Environmental Education* (pp. chapter 3). IGI Global. DOI: DOI: 10.4018/978-1-5225-7727-0.ch003

Elmahdi, A., & Wang, L. (2022). Water asset transition through treating water as a new asset class for paradigm shift for climate–water resilience. *Climate (Basel), 10*(12), 191. DOI: 10.3390/cli10120191

Fernández-Pacheco, D., & Andreu, J. (Eds.). (2023b). *Special issue: Use of remote sensing technologies for water resources management. Water*. MDPI. https://www.mdpi.com/journal/water/special_issues/7AFF236H6J

Financing. (2023). The United Nations world water development report (pp. 162–171). United Nations. DOI: 10.18356/9789210026208c020

Ghobadi, F., & Kang, D. (2023). F.; Kang, D. Application of Machine Learning in Water Resources Management: A Systematic Literature Review. *Water (Basel), 15*(4), 620. DOI: 10.3390/w15040620

Ghobadi, F., & Kang, D. (2023). Application of machine learning in water resources management: A systematic literature review. *Water (Basel), 15*(4), 620. DOI: 10.3390/w15040620

Hutton, G., & Varughese, M. (2016). *The costs of meeting the 2030 sustainable development goal targets on drinking water, sanitation, and hygiene*. World Bank., DOI: 10.1596/K8543

Joint Monitoring Programme. (2015). *Water supply and sanitation*. World Health Organization & UNICEF. https://www.wssinfo.org/data-estimates/tables/

Kolker, J., Kingdom, B., & Trémolet, S. (2016). *Financing options for the 2030 water agenda*. World Bank. https://openknowledge.worldbank.org/handle/10986/25399

Machete, I., & Marques, R. (2021). Financing the water and sanitation sectors: A hybrid literature review. *Infrastructures, 6*(1), 9. DOI: 10.3390/infrastructures6010009

Marley, G., Bazaanah, P., & Oppong, P. (2022). *Examining the Role of NGOs in Community Water and Sanitation Improvement: A Case of the Tunayilli Community in the Northern Region, Ghana.* In *Handbook of Research on Resource Management and the Struggle for Water Sustainability in Africa* (pp. chapter 2). IGI Global. DOI: DOI: 10.4018/978-1-7998-8809-3.ch002

Mohanty, A., Ghosh, A., & Roy, S. (2024). Calling for better ecological values: Integrating Indigenous Knowledge System with sustainable policies. *Purushartha, 17*(1), 39–50. Advance online publication. DOI: 10.21844/16202117103

Mundonde, J., & Makoni, P. L. (2024). Framework model for financing sustainable water and sanitation infrastructure in Zimbabwe. *Water (Basel), 16*(12), 1691. DOI: 10.3390/w16121691

Nigam, A. (1996). *Sustainable financing of WATSAN*. Loughborough University Repository. https://repository.lboro.ac.uk/articles/conference_contribution/Sustainable_financing_of_WATSAN/9588650

Nigam, A. (1996). *Financing of water supply and sanitation investments in developing countries*. UNDP–World Bank Water and Sanitation Program.

Organisation for Economic Co-operation and Development. (2018). *Making blended finance work for the sustainable development goals*. OECD Publishing. https://www.oecd.org/dac/making-blended-finance-work.htm

Organisation for Economic Co-operation and Development. (2022). *Accelerating action on financing water: An agenda for the future*. OECD Studies on Water., DOI: 10.1787/6bf7264b-

Organisation for Economic Co-operation and Development. (2023). *Financing a water secure future*. OECD Publishing. https://www.oecd.org/en/publications/financing-a-water-secure-future_a2ecb261-en.html

Pories, L., Fonseca, C., & Delmon, V. (2019). *Mobilising finance for WASH: Getting the foundations right*. WaterAid. https://washmatters.wateraid.org/publications/mobilising-finance-for-wash-getting-the-foundations-right

Rodell, M., Famiglietti, J. S., Wiese, D. N., Reager, J. T., Beaudoing, H. K., Landerer, F. W., & Lo, M.-H. (2018). Emerging trends in global freshwater availability. *Nature*, *557*(7707), 651–659. DOI: 10.1038/s41586-018-0123-1 PMID: 29769728

Sage, N. (2022). SDG-6 clean water and sanitation (pp. 289–304). *Springer eBooks*. DOI: 10.1007/978-3-031-07127-0_8

UN-Habitat. (2022). *World cities report 2022: Envisaging the future of cities*. United Nations Human Settlements Programme. https://unhabitat.org/

UN-Water. (2020). *The SDG 6 global acceleration framework*. United Nations. https://www.unwater.org/publications/sdg-6-global-acceleration-framework

UNESCO. (2020). *The United Nations world water development report 2020: Water and climate change*. UNESCO Publishing. https://unesdoc.unesco.org/ark:/48223/pf0000372985

UNICEF, & World Health Organization. (2021). *State of the world's drinking water 2021*. UNICEF & WHO. https://www.unicef.org/reports/state-worlds-drinking-water

United Nations. (2015). *Millennium development goals and beyond 2015*. United Nations. https://www.un.org/millenniumgoals//environ.shtml

United Nations. (2018). *World urbanization prospects: The 2018 revision*. Department of Economic and Social Affairs. https://population.un.org/wup/

United Nations Development Programme (UNDP). (2019). *Human development report 2019: Beyond income, beyond averages, beyond today*. UNDP. https://hdr.undp.org/

United Nations Office for Outer Space Affairs. (2021). *Space solutions for the SDGs*. UNOOSA. https://www.unoosa.org/oosa/en/ourwork/space4sdgs/index.html

Willems, R. C., & MacDonald, S. A. (2016). *Access to Safe Drinking Water: A Low-Income Sustainable Water Purification System.* In *Impact of Water Pollution on Human Health and Environmental Sustainability* (pp. chapter 13). IGI Global. DOI: DOI: 10.4018/978-1-4666-9559-7.ch013

World Bank. (2020). *Urban development overview.* The World Bank. https://www.worldbank.org/en/topic/urbandevelopment/overview

World Health Organization (WHO), & UNICEF. (2021). *Progress on household drinking water, sanitation and hygiene 2000–2020: Five years into the SDGs.* WHO & UNICEF. https://washdata.org/

Zhou, Y., & Chen, X. (Eds.). (2023a). *Special issue: Applications of remote sensing and machine learning in water resources management. Water.* MDPI. https://www.mdpi.com/journal/water/special_issues/0MX8979401

Compilation of References

Abbasi, L., Hedayati, K., & Ghanbari, D. (2021). Magnetic properties and kinetic roughening study of prepared polyaniline: Lead ferrite, cobalt ferrite and nickel ferrite nanocomposites electrodeposited thin films. *Journal of Materials Science Materials in Electronics*, *32*(11), 14477–14493. DOI: 10.1007/s10854-021-06006-1

Abdel-Aal, S. K., & Abdel-Rahman, A. S. (2020). Graphene influence on the structure, magnetic, and optical properties of rare-earth perovskite. *Journal of Nanoparticle Research*, *22*(9), 267. DOI: 10.1007/s11051-020-05001-7

Abd-Elhamid, A. I., El Morsy, S. M., Soliman, H. M. A., & El-Sayed, M. E. A. (2022). Superior adsorption performance of citrate modified graphene oxide as nanomaterial for removal of organic and inorganic pollutants from aqueous solution. *Scientific Reports*, *12*(1), 1863. DOI: 10.1038/s41598-022-13111-6 PMID: 35654871

Abdel-Shafy, H. I., Mansour, M. S. M., & Al-Sulaiman, A. M. (2019). Anaerobic/aerobic integration via UASB/enhanced aeration for greywater treatment and unrestricted reuse. *Water Practice & Technology*, *14*(4), 837–850. DOI: 10.2166/wpt.2019.065

Abdullah, Md., Obayedullah, M., Shariful Islam Shuvo, Md., Abul Khair, Md., Hossain, D. and Nahidul Islam, Md. (2025). A review on multifunctional applications of nanoparticles: Analyzing their multi-physical properties. Results in Surfaces and Interfaces, [online] 21, p.100635. doi:.DOI: 10.1016/j.rsurfi.2025.100635

Achak, M., Ouazzani, N., Yaacoubi, A., & Mandi, L. (2008). Caractérisation des margines issues d'une huilerie moderne et essais de leur traitement par coagulation-floculation par la chaux et le sulfate d'aluminium. *Revue des sciences de l'eau / Journal of Water Science*, *21*(1), 53–67. DOI: 10.7202/017930ar

Adamiec, P. (1998). *Caractérisation d'une cendre volante silico-alumineuse et sa reconstruction: Corrélation avec les caractéristiques géologiques, gîtologiques et minéralogiques du charbon* [These de doctorat, Lille 1]. https://theses.fr/1998LIL10023

Adamiec, P., Benezet, J. C., & Benhassaine, A. (2011). *Relation entre une cendre volante silico-alumineuse et son charbon* (arXiv:1111.6847). arXiv. https://doi.org//arXiv.1111.6847DOI: 10.48550

Ademe, E. (2001). *Les boues d'épuration municipales et leur utilisation en agriculture*. Dossier Documentaire.

Ahmaruzzaman, M. (2010). A review on the utilization of fly ash. *Progress in Energy and Combustion Science, 36*(3), 327–363. DOI: 10.1016/j.pecs.2009.11.003

Ahmed, A., Ghosh, P. K., Hasan, M., & Rahman, A. (2020). Surface and groundwater quality assessment and identification of hydrochemical characteristics of a southwestern coastal area of Bangladesh. *Environmental Monitoring and Assessment, 192*(4), 258. DOI: 10.1007/s10661-020-8227-0 PMID: 32239354

Ahmed, A., Khan, M., & Li, Y. (2023). Remote sensing and artificial intelligence for sustainable water resource management: Opportunities and challenges. *Sustainability, 15*(12), 10045.

Alam, M. M., Mustehasan., Akram, U., Ahmad, M., & Ahmed, N. Z. (2025). Nanotechnology in the Unani system of medicine: A comprehensive review and hypothesis. *Advances in Integrative Medicine, 12*(4), 100564. DOI: 10.1016/j.aimed.2025.100564

Alam, S. K., Li, P., Rahman, M., Fida, M., & Elumalai, V. (2025). Key factors affecting groundwater nitrate levels in the Yinchuan Region, Northwest China: Research using the eXtreme Gradient Boosting (XGBoost) model with the SHapley Additive exPlanations (SHAP) method. *Environmental Pollution, 364*, 125336. DOI: 10.1016/j.envpol.2024.125336 PMID: 39566709

Alazab, A. A., & Saleh, T. A. (2023). Underwater superoleophobic cellulose/acrylamide-modified magnetic polyurethane foam for efficient oil/water separation. *Materials Chemistry and Physics, 302*, 127609.

Alazaiza, M. Y. D., Albahnasawi, A., Ali, G. A. M., Bashir, M. J. K., Copty, N. K., Amr, S. S. A., Abushammala, M. F. M., & Al Maskari, T. (2021). Recent Advances of Nanoremediation Technologies for Soil and Groundwater Remediation: A Review. *Water (Basel), 13*(16), 2186. DOI: 10.3390/w13162186

Al-Gheethi, A. A. S., Noman, E. A., Mohamed, R. M. S. R., Talip, B. A., Abdullah, A. H., & Kassim, A. H. M. (2019). Reuse of greywater for irrigation purpose. In Mohamed, R. M. S. R., Al-Gheethi, A. A. S., & Kassim, A. H. M. (Eds.), *Management of greywater in developing countries* (pp. 73–87). Springer. DOI: 10.1007/978-3-319-90269-2_4

Alharbi, SE, Zarewa, MU, Saleh, TA, 2025., Integrated adsorption–electrocoagulation for dyes removal using solid petcoke-derived activated carbon coupled with predictive machine learning models, Diamond and Related Materials, 159, Part B, 112825

Ali, N., Said, A., Ali, F., Raziq, F., Ali, Z., Bilal, M., Reinert, L., Begum, T., & Iqbal, H. M. N. (2020). Photocatalytic degradation of Congo Red dye from aqueous environment using cobalt ferrite nanostructures: Development, characterization, and photocatalytic performance. *Water, Air, and Soil Pollution, 231*(2), 50. DOI: 10.1007/s11270-020-4410-8

Alkenani, A., & Saleh, T. A. (2022). Synthesis of amine-modified graphene integrated membrane as protocols for simultaneous rejection of hydrocarbons, metal ions, and salts from water, Journal of Molecular Liquids, 367. *Part B, 120291*. Advance online publication. DOI: 10.1016/j.molliq.2022.120291

Alkhadra, M.A., Su, X., Suss, M.E., Tian, H., Guyes, E.N., Shocron, A.N., Conforti, K.M., de Souza, J.P., Kim, N., Tedesco, M., Khoiruddin, K., Wenten, I.G., Santiago, J.G., Hatton, T.A. and Bazant, M.Z. (2022). Electrochemical Methods for Water Purification, Ion Separations, and Energy Conversion. Chemical Reviews, [online] 122(16), pp.13547–13635. doi:.DOI: 10.1021/acs.chemrev.1c00396

Allen, A. (2001). Containment landfills: The myth of sustainability. *Engineering Geology, 60*(1–4), 3–19. DOI: 10.1016/S0013-7952(00)00084-3

Al-Marri, J. S., Abouedwan, A. B., Ahmad, M. I., & Bensalah, N. (2023). Electrocoagulation using aluminum electrodes as a sustainable and economic method for the removal of kinetic hydrate inhibitor (polyvinyl pyrrolidone) from produced wastewaters. *Frontiers in Water, 5*, 1305347. Advance online publication. DOI: 10.3389/frwa.2023.1305347

Al-Mefleh, N. K., Othman, Y. A., Tadros, M. J., Al-Assaf, A., & Talozi, S. (2021). An assessment of treated greywater reuse in irrigation on growth and protein content of *Prosopis* and *Albizia*. *Horticulturae, 7*(3), 38. DOI: 10.3390/horticulturae7030038

Alotaibi, R., & Nassif, A. B. (2024). AI-driven remote sensing for sustainable water resource management: A comprehensive review. *Environmental Science and Pollution Research International, 31*(12), 15321–15340. DOI: 10.1007/s11356-023-29152-2

Alsaba, M. T., Al Dushaishi, M. F., & Abbas, A. K. (2020). A comprehensive review of nanoparticles applications in the oil and gas industry. *Journal of Petroleum Exploration and Production Technology*, *10*(4), 1389–1399. Advance online publication. DOI: 10.1007/s13202-019-00825-z

Altammar, K. A. (2023). A review on nanoparticles: Characteristics, synthesis, applications, and challenges. [PubMed]. *Frontiers in Microbiology*, *14*(1155622), 1155622. Advance online publication. DOI: 10.3389/fmicb.2023.1155622

Ameer, S., Abdelrahman, E.A., Batool, F., Batool, A., Akhtar, T., Alhamzani, A.G. and Abou-Krisha, M.M. (2025). Bimetallic nanoparticles loaded on waste derived carbon quantum dots: A green approach for degradation of azo dye. International Biodeterioration & Biodegradation, [online] 205, p.106184. doi:.DOI: 10.1016/j.ibiod.2025.106184

Amiri, M. J., Bahrami, M., Badkouby, M., & Kalavrouziotis, I. (2019). Greywater treatment using single and combined adsorbents for landscape irrigation. *Environmental Processes*, *6*(1–3), 43–63. DOI: 10.1007/s40710-019-00362-1

Ansari, K., & Shrikhande, A. N. (2019). Feasibility on grey water treatment by electrocoagulation process: A review. *International Journal of Emerging Technology*, *10*(1), 92–99.

Apul, O. G., & Karanfil, T. (2015). Adsorption of synthetic organic contaminants by carbon nanotubes: A critical review. *Water Research*, *68*, 34–55. DOI: 10.1016/j.watres.2014.09.032 PMID: 25462715

Arias Arias, F., Guevara, M., Tene, T., Angamarca, P., Molina, R., Valarezo, A., Salguero, O., Vacacela Gomez, C., Arias, M., & Caputi, L. S. (2020). The Adsorption of Methylene Blue on Eco-Friendly Reduced Graphene Oxide. *Nanomaterials (Basel, Switzerland)*, *10*(4), 681. DOI: 10.3390/nano10040681 PMID: 32260330

Ariho, A., Aja, L., Muhammad, T., & Mohammad, L. (2024). Water pollution: causes, impacts and current efforts to address the issues of water pollution along river Meizimera-kihihi, Kanugu District,Uganda. F1000Research, [online] 13, pp.1298–1298. doi:DOI: 10.12688/f1000research.155162.1

Ariho, A., Aja, L., Muhammad, T., & Mohammad, L. (2024). Water pollution: causes, impacts and current efforts to address the issues of water pollution along river... *F1000Research*, [online] 13, p.1298. DOI: 10.12688/f1000research.155162.2

Aroor, G., Khan, M.A., Shetty, A.R., Rai, R., Ganesha, H. and Navada, M.K. (2025). From chemistry to performance: How nano additives are transforming bio-lubricants for enhanced tribological applications. Journal of Molecular Liquids, [online] 425, p.127242. doi:.DOI: 10.1016/j.molliq.2025.127242

Arora, R. (2019). Adsorption of Heavy Metals–A Review. *Materials Today: Proceedings, 18*, 4745–4750. DOI: 10.1016/j.matpr.2019.07.462

Arslan-Alaton, I., Kabdaşlı, I., Hanbaba, D., & Kuybu, E. (2008). Electrocoagulation of a real reactive dyebath effluent using aluminum and stainless steel electrodes. *Journal of Hazardous Materials, 150*(1), 166–173. DOI: 10.1016/j.jhazmat.2007.09.032 PMID: 17945416

Asadollah, S. B. H. S., Sharafati, A., Motta, D., & Yaseen, Z. M. (2021). River water quality index prediction and uncertainty analysis: A comparative study of machine learning models. *Journal of Environmental Chemical Engineering, 9*(1), 104599. DOI: 10.1016/j.jece.2020.104599

Asghar, N., Hussain, A., Nguyen, D. A., Ali, S., Hussain, I., Junejo, A., & Ali, A. (2024). Advancement in nanomaterials for environmental pollutants remediation: A systematic review on bibliometrics analysis, material types, synthesis pathways, and related mechanisms. *Journal of Nanobiotechnology, 22*(1), 26. Advance online publication. DOI: 10.1186/s12951-023-02151-3 PMID: 38200605

Asif, Z., Chen, Z., An, C. and Dong, J. (2022). Environmental Impacts and Challenges Associated with Oil Spills on Shorelines. Journal of Marine Science and Engineering, [online] 10(6), p.762. doi:.DOI: 10.3390/jmse10060762

Astruc, D. (2020). Introduction: Nanoparticles in Catalysis. [PubMed]. *Chemical Reviews, 120*(2), 461–463. DOI: 10.1021/acs.chemrev.8b00696

Aydin, D., Gübbük, İ. H., & Ersöz, M. (2024). Recent advances and applications of nanostructured membranes in water purification. *Turkish Journal of Chemistry, 48*(1), 1–20.

Azrour, M., Mabrouki, J., Fattah, G., Guezzaz, A., & Aziz, F. (2022). Machine learning algorithms for efficient water quality prediction. *Modeling Earth Systems and Environment, 8*(2), 2793–2801. DOI: 10.1007/s40808-021-01266-6

Azzaoui, K., Barboucha, B., Hammouti, B., & Touzani, R. (2022). Nanotechnology: History and Various Applications, a Mini Review. PRSM. [online] doi:DOI: 10.34874/prsm.ehei-jst-vol2iss1.36531

Baietti, A., & Raymond, P. (2005). *Financing water supply and sanitation investments: Utilizing risk mitigation instruments to bridge the financing gap*. World Bank. https://openknowledge.worldbank.org/bitstream/handle/10986/17235/320280WSS1Investments.pdf

Balamurugan, J., Kasthuri, N., Sudarsan, J. S., Basi Reddy, A., Sakhare, D. T., & Sampathkumar, V. (2024). An efficient wastewater collection model for groundwater resource protection in smart cities. *Groundwater for Sustainable Development*, *25*, 101091. DOI: 10.1016/j.gsd.2024.101091

Balusamy, S. R., Joshi, A. S., Perumalsamy, H., Mijakovic, I., & Singh, P.Sri Renukadevi Balusamy. (2023). Advancing sustainable agriculture: A critical review of smart and eco-friendly nanomaterial applications. *Journal of Nanobiotechnology*, *21*(1), 372. Advance online publication. DOI: 10.1186/s12951-023-02135-3 PMID: 37821961

Barhoum, A., García-Betancourt, M. L., Jeevanandam, J., Hussien, E. A., Mekkawy, S. A., Mostafa, M., Omran, M. M. S., Abdalla, M., & Bechelany, M. (2022). Review on Natural, Incidental, Bioinspired, and Engineered Nanomaterials: History, Definitions, Classifications, Synthesis, Properties, Market, Toxicities, Risks, and Regulations. *Nanomaterials (Basel, Switzerland)*, *12*(2), 177. DOI: 10.3390/nano12020177 PMID: 35055196

Bayda, S., Adeel, M., Tuccinardi, T., Cordani, M. and Rizzolio, F. (2019). The History of Nanoscience and Nanotechnology: From Chemical–Physical Applications to Nanomedicine. Molecules, [online] 25(1), p.112. doi:.DOI: 10.3390/molecules25010112

Belkhiri, L., Tiri, A., & Mouni, L. (2020). Spatial distribution of the groundwater quality using kriging and Co-kriging interpolations. *Groundwater for Sustainable Development*, *11*, 100473. DOI: 10.1016/j.gsd.2020.100473

Bellarbi, A., Graich, A., Monkade, M., Zradba, A., & Laamyem, A. (2017). Valorization of bottom ash generated by a thermal power plant in the treatment of cardboard manufacturing wastewater. *Global J. Eng. Sci. Res. Manage*, *4*(10), 11–22.

Bellier, N., Baipaywad, P., Ryu, N., Lee, J. Y., & Park, H. (2022). Recent biomedical advancements in graphene oxide- and reduced graphene oxide-based nanocomposite nanocarriers. *Biomaterials Research*, *26*(1), 65. Advance online publication. DOI: 10.1186/s40824-022-00313-2 PMID: 36435846

Bhattacharya, A. K., Naiya, T. K., Mandal, S. N., & Das, S. K. (2008). Adsorption, kinetics and equilibrium studies on removal of Cr(VI) from aqueous solutions using different low-cost adsorbents. *Chemical Engineering Journal, 137*(3), 529–541. DOI: 10.1016/j.cej.2007.05.021

Bloom, D. E., Canning, D., & Fink, G. (2008). Urbanization and the wealth of nations. *Science, 319*(5864), 772–775. DOI: 10.1126/science.1153057 PMID: 18258907

Bohli, N. (2009). *Etude physique de la formation de films à base de polymères conducteurs et applications en micro-éléctronique* [These de doctorat, Bordeaux 1]. https://theses.fr/2009BOR13961

Boretti, A., & Rosa, L. (2019). Reassessing the projections of the World Water Development Report. *Sustainability, 11*(3), 760. DOI: 10.3390/su11030760

Bui, D. T., Khosravi, K., Tiefenbacher, J., Nguyen, H., & Kazakis, N. (2020). Improving prediction of water quality indices using novel hybrid machine-learning algorithms. *The Science of the Total Environment, 721*, 137612. DOI: 10.1016/j.scitotenv.2020.137612 PMID: 32169637

Burlec, A. F., Corciova, A., Boev, M., Batir-Marin, D., Mircea, C., Cioanca, O., Danila, G., Danila, M., Bucur, A. F., & Hancianu, M. (2023). Current Overview of Metal Nanoparticles' Synthesis, Characterization, and Biomedical Applications, with a Focus on Silver and Gold Nanoparticles. [PubMed]. *Pharmaceuticals (Basel, Switzerland), 16*(10), 1410–1410. DOI: 10.3390/ph16101410

Cai, X., Wu, L., Li, Y., Lei, S., Xu, J., Lyu, H., Li, J., Wang, H., Dong, X., Zhu, Y. and Wang, G. (2023). Remote sensing identification of urban water pollution source types using hyperspectral data. Journal of Hazardous Materials, [online] 459, p.132080. doi:.DOI: 10.1016/j.jhazmat.2023.132080

Campbell, H. W., & Crescuolo, P. J. (1982). The use of rheology for sludge characterization. *Water Science and Technology, 14*(6–7), 475–489. DOI: 10.2166/wst.1982.0120

Cao, Y., Malmali, M., Qian, X., & Wickramasinghe, S. R. (2022). Continuous electrocoagulation-membrane distillation unit for treating hydraulic fracturing produced water. *Journal of Water Process Engineering, 50*, 103219. DOI: 10.1016/j.jwpe.2022.103219

Casey, T. J., & Casey, T. J. (1997). *Unit treatment processes in water and wastewater engineering*. Wiley Chichester. http://www.aquavarra.ie/Utpbkpdf/chap0.pdf

Chahid, E. G., Moznine, R. E., Zradba, A., Dani, A., Melouky, A. E., Idrissi, L., Cherkaoui, O., Choukri, E., Mezzane, D., & Belboukhari, A. (2013, November-December). Analysis of relaxation processes and low frequency dispersion in waste water. *Journal of Optoelectronics and Advanced Materials*, *15*, 1209–1216.

Chandran, D. G., Loganathan, M., & Biswas, R. (2023). A review on adsorption of heavy metals from wastewater using carbon nanotube and graphene-based nanomaterials. *Environmental Science and Pollution Research International*, *30*(51), 110010–110046. DOI: 10.1007/s11356-023-30192-6 PMID: 37804379

Chatterjee, A. K. (2011). Indian Fly Ashes: Their Characteristics and Potential for Mechanochemical Activation for Enhanced Usability. *Journal of Materials in Civil Engineering*, *23*(6), 783–788. DOI: 10.1061/(ASCE)MT.1943-5533.0000279

Chaturvedi, S. and Dave, P.N. (2019). Water Purification Using Nanotechnology an Emerging Opportunities. Chemical Methodologies, [online] 3(1), pp.115–144. doi:.DOI: 10.22034/chemm.2018.143461.1069

Chen, M., Blanc, D., Gautier, M., Mehu, J., & Gourdon, R. (2013). Environmental and technical assessments of the potential utilization of sewage sludge ashes (SSAs) as secondary raw materials in construction. *Waste Management (New York, N.Y.)*, *33*(5), 1268–1275. DOI: 10.1016/j.wasman.2013.01.004 PMID: 23415711

Chen, Y., Song, L., Liu, Y., Yang, L., & Li, D. (2020). A review of the artificial neural network models for water quality prediction. *Applied Sciences (Basel, Switzerland)*, *10*(17), 5776. DOI: 10.3390/app10175776

Cheong, S., Watt, J. D., & Tilley, R. D. (2010). Shape control of platinum and palladium nanoparticles for catalysis. *Nanoscale*, *2*(10), 2045. DOI: 10.1039/c0nr00276c PMID: 20694209

Choudhury, D., & Bhattacharjee, A. (2023). Environmental communication as a mediator in the relationship between environmental concern and attitude towards green services in a sequential mediation model: Evidence from restaurant industry. *Purushartha, 16*(1). journals.smsvaranasi.comDOI: 10.21844/16202116105

Chukeaw, T., Fang, K., Xiong, Q., Gu, Y., He, C., Wang, K. and Liu, G. (2025). Synergistic integration of electrocoagulation and membrane pre-concentration: A critical review on advanced strategies for organic recovery and fouling mitigation in municipal wastewater treatment. Separation and Purification Technology, [online] 382, p.135994. doi:.DOI: 10.1016/j.seppur.2025.135994

Colin, F., & Gazbar, S. (1995). Distribution of water in sludges in relation to their mechanical dewatering. *Water Research*, *29*(8), 2000–2005. DOI: 10.1016/0043-1354(94)00274-B

Csala, J., & Mwangi, J. W. (2019). *Nature-Based Solutions for Water Security and the Role of Education for Enhancing the Implementation of the 2030 Agenda.* In *Building Sustainability Through Environmental Education* (pp. chapter 3). IGI Global. DOI: DOI: 10.4018/978-1-5225-7727-0.ch003

Daneshyar, A., Ghaedi, M., Sabzehmeidani, M. M., & Daneshyar, A. (2017). H2S adsorption onto Cu-Zn–Ni nanoparticles loaded activated carbon and Ni-Co nanoparticles loaded γ-Al2O3: Optimization and adsorption isotherms. *Journal of Colloid and Interface Science*, *490*, 553–561. DOI: 10.1016/j.jcis.2016.11.068 PMID: 27923140

Dang, V., Thamilselvan, A., Khedulkar, A., Lin, J. T., Adorna, J. A., Wang, Y., Pandit, B., Huynh, T., & Doong, R. (2023). S-scheme N-doped carbon dots anchored g-C3N4/Fe2O3 shell/core composite for photoelectrocatalytic trimethoprim degradation and water splitting. 320, pp.121928–121928. DOI: 10.1016/j.apcatb.2022.121928

Das, A. K., & Chen, L. (2024). A Review on Electrochemical Advanced Oxidation Treatment of Dairy Wastewater. *Environments (Basel, Switzerland)*, *11*(6), 124–124. DOI: 10.3390/environments11060124

Dehghani Dastjerdi, O., Shokrollahi, H., & Mirshekari, S. (2023). A review of synthesis, characterization, and magnetic properties of soft spinel ferrites. *Inorganic Chemistry Communications*, *153*, 110797. DOI: 10.1016/j.inoche.2023.110797

Del Prado-Audelo, M. L., García Kerdan, I., Escutia-Guadarrama, L., Reyna-González, J. M., Magaña, J. J., & Leyva-Gómez, G. (2021). Nanoremediation: Nanomaterials and Nanotechnologies for Environmental Cleanup. *Frontiers in Environmental Science*, *9*, 793765. Advance online publication. DOI: 10.3389/fenvs.2021.793765

Diamadopoulos, E., Ioannidis, S., & Sakellaropoulos, G. P. (1993). As(V) removal from aqueous solutions by fly ash. *Water Research*, *27*(12), 1773–1777. DOI: 10.1016/0043-1354(93)90116-Y

Dieudé-Fauvel, E., & Dentel, S. K. (2011). *Sludge conditioning: Impact of Polymers on Floc structure.*

Djamel-Eddine, M. (2015). *Utilisation des énergies renouvelables dans le dessalement: Cas de l'Algérie* [Thesis]. https://e-biblio.univ-mosta.dz/handle/123456789/951

Drexler, K. E. (1981). Molecular engineering: An approach to the development of general capabilities for molecular manipulation. Proceedings of the National Academy of Sciences, [online] 78(9), pp.5275–5278. doi:DOI: 10.1073/pnas.78.9.5275

Dube, A., Malode, S.J., Alshehri, M.A. and Shetti, N.P. (2025). Electrochemical water treatment: Review of different approaches. Journal of Environmental Management, [online] 373, p.123911. doi:.DOI: 10.1016/j.jenvman.2024.123911

Dubey, S. S., Singh, A. K., & Tripathi, U. N. (2022). Water Quality - A Review. International Journal of Chemical Sciences, [online] 20(3), pp.1–7. Available at: https://www.tsijournals.com/abstract/water-quality--a-review-15778.html [Accessed 11 Nov. 2025].

Dukuly, L. P., Gupta, M., Ghani, S., & Akram, W. (2025). PCA-integrated machine learning framework for predicting rock fragmentation in blasting operations. *Multiscale and Multidisciplinary Modeling. Experiments and Design*, *8*(9), 409. DOI: 10.1007/s41939-025-01008-6

Dutta, B., Basu, J. K., & DasGupta, S. (2003). Removal of Cresol from Aqueous Solution Using Fly Ash as Adsorbent: Experiments and Modeling. *Separation Science and Technology*, *38*(6), 1345–1360. DOI: 10.1081/SS-120018813

El Alami, S. E. M., Monkade, M., & Lahlou, K. (2011). Valorisation d'un mélange de cendres d'une centrale thermique et de scories d'aciérie dans les chaussées routières. *Environnement, Ingénierie & Développement*. https://eid.episciences.org/7863

El Bilali, A., Taleb, A., & Brouziyne, Y. (2021). Groundwater quality forecasting using machine learning algorithms for irrigation purposes. *Agricultural Water Management*, *245*, 106625. DOI: 10.1016/j.agwat.2020.106625

El Hafidi, E. M., Chahid, E. G., Mortadi, A., & Laasri, S. (2024). Study on a new solar-powered desalination system to alleviate water scarcity using impedance spectroscopy. *Materials Today: Proceedings*. Advance online publication. DOI: 10.1016/j.matpr.2024.02.038

El Hafidi, E. M., Mortadi, A., Chahid, E. G., & Laasri, S. (2023). Monitoring of domestic wastewater treatment via infiltration percolation using impedance spectroscopy. *Environmental Technology & Innovation*, *32*, 103421. DOI: 10.1016/j.eti.2023.103421

El Hamdouni, Y., El Hajjaji, S., Szabó, T., Trif, L., Felhősi, I., Abbi, K., Labjar, N., Harmouche, L., & Shaban, A.Youssra El Hamdouni. (2022). Biomass valorization of walnut shell into biochar as a resource for electrochemical simultaneous detection of heavy metal ions in water and soil samples: Preparation, characterization, and applications. *Arabian Journal of Chemistry, 15*(11), 104252–104252. DOI: 10.1016/j.arabjc.2022.104252

El-Kalliny, A.S., Abdel-Wahed, M.S., El-Zahhar, A.A., Hamza, I.A. and Gad-Allah, T.A. (2023). Nanomaterials: a review of emerging contaminants with potential health or environmental impact. Discover Nano, [online] 18(1), p.68. doi:.DOI: 10.1186/s11671-023-03787-8

El-kordy, A., Elgamouz, A., Khalil, A.K.A., Tijani, N., Kawde, A.-N. and Laoui, T. (2025). Development and electrochemical evaluation of Bismuth/Clay-based membranes for heavy metal rejection from wastewater. Surfaces and Interfaces, [online] 72, p.106953. doi:.DOI: 10.1016/j.surfin.2025.106953

Elmahdi, A., & Wang, L. (2022). Water asset transition through treating water as a new asset class for paradigm shift for climate–water resilience. *Climate (Basel), 10*(12), 191. DOI: 10.3390/cli10120191

EPA. U. (2000). Wastewater technology fact sheet chemical precipitation. *Office of Water and Watersheds, Washington DC: United States Environmental Protection Agency. Doi: EPA.*

Fadillah, G., Hidayat, R., Saputra, A., Wasisto, H.S., Rani, R., Saeki, K., Toda, K. and Ohira, S.-I. (2025). Advanced electrodialysis techniques for analytical separation: A comprehensive review. Analytica Chimica Acta, [online] 1380, p.344637. doi:.DOI: 10.1016/j.aca.2025.344637

Fadillah, G., Alarifi, N. T. S., Suryawan, I., & Saleh, T. A. (2024). Advances in designed reactors for water treatment process: A review highlighting the designs and performance. *Journal of Water Process Engineering*, *63*, 105417. DOI: 10.1016/j.jwpe.2024.105417

Fan, K., Li, Y., Elsworth, D., Dong, M., Yin, C., Li, Y., & Chen, Z. (2018). Three stages of methane adsorption capacity affected by moisture content. *Fuel, 231*, 352–360. DOI: 10.1016/j.fuel.2018.05.120

Fayet, T., Pellissier, C., Hjiaj, M., Abriak, N. E., & Klein, C. (1999). Mechanical and physical properties of fly ashes. *Journal de Chimie Physique et de Physico-Chimie Biologique, 96.* https://www.osti.gov/etdeweb/biblio/348688

Feculak, M., Loureiro, S., White, J.C., Xing, B., Wu, K.C.-W. ., Sheteiwy, M.S., Gao, Y., Oleszczuk, P. and Jośko, I. (2025). Engineered nanoparticle transformations: Rethinking toxicity in water. Nano Today, [online] 65, p.102804. doi:.</unknown>DOI: 10.1016/j.nantod.2025.102804

Fernández-Pacheco, D., & Andreu, J. (Eds.). (2023b). *Special issue: Use of remote sensing technologies for water resources management. Water.* MDPI. https://www.mdpi.com/journal/water/special_issues/7AFF236H6J

Financing. (2023). The United Nations world water development report (pp. 162–171). United Nations. DOI: 10.18356/9789210026208c020

Foorginezhad, S., Zerafat, M.M., Ismail, A.F. and Goh, P.S. (2025). Emerging membrane technologies for sustainable water treatment: a review on recent advances. Environmental Science: Advances, [online] 4(4), pp.530–570. doi:.DOI: 10.1039/D4VA00378K

Fouda-Mbanga, B.G., Nyoni, B., Mnyango, J.I., Phiri, C., Hlangothi, S.P., Miracle, A., Boluwatife, A., Celestina, O., Voukeng, S.N.K., Tsague, C.F., Nyamen, L.D., Mokolokolo, P.P., Mofokeng, J.P. and Ngobeni, P. (2025). The Nexuses of Micro(nano)plastics at Environmental Interfaces: Challenges in Wastewater Treatment and Sustainable Water Management. Journal of Environmental Chemical Engineering, [online] p.119952.

Fu, M., Jiao, Q., & Zhao, Y. (2013). In situ fabrication and characterization of cobalt ferrite nanorods/graphene composites. *Materials Characterization, 86*, 303–315. DOI: 10.1016/j.matchar.2013.10.019

Gankanda, A., Cwiertny, D. M., & Grassian, V. H. (2016). Role of Atmospheric CO_2 and H_2O Adsorption on ZnO and CuO Nanoparticle Aging: Formation of New Surface Phases and the Impact on Nanoparticle Dissolution. *Journal of physical chemistry. C. The Journal of Physical Chemistry. C, Nanomaterials and Interfaces, 120*(34), 19195–19203. DOI: 10.1021/acs.jpcc.6b05931

Gazzaz, N. M., Yusoff, M. K., Aris, A. Z., Juahir, H., & Ramli, M. F. (2012). Artificial neural network modeling of the water quality index for Kinta River (Malaysia) using water quality variables as predictors. *Marine Pollution Bulletin, 64*(11), 2409–2420. DOI: 10.1016/j.marpolbul.2012.08.005 PMID: 22925610

Ghaffour, N., Bundschuh, J., Mahmoudi, H., & Goosen, M. F. A. (2015). Renewable energy-driven desalination technologies: A comprehensive review on challenges and potential applications of integrated systems. *Desalination, 356*, 94–114. DOI: 10.1016/j.desal.2014.10.024

Ghasemi, Z., Seif, A., Ahmadi, T. S., Zargar, B., Rashidi, F., & Rouzbahani, G. M. (2011). Thermodynamic and kinetic studies for the adsorption of Hg(II) by nano-TiO2 from aqueous solution. *Advanced Powder Technology*, *23*(2), 148–156. DOI: 10.1016/j.apt.2011.01.004

Ghawi, A. H. (2019). Development of the greywater domestic treatment unit for irrigation of the garden in rural areas. *Journal of Ecological Engineering*, *20*(3), 201–208. DOI: 10.12911/22998993/99732

Ghobadi, F., & Kang, D. (2023). F.; Kang, D. Application of Machine Learning in Water Resources Management: A Systematic Literature Review. *Water (Basel)*, *15*(4), 620. DOI: 10.3390/w15040620

Goodwin, J. A. S., & Forster, C. F. (1985). A further examination into the composition of activated sludge surfaces in relation to their settlement characteristics. *Water Research*, *19*(4), 527–533. DOI: 10.1016/0043-1354(85)90045-4

Graich, A., Mghaiouini, R., Bellarbi, A., Khaidar, M., Laamyem, A., Monkade, M., & Zradba, A. (2021). Use of a cascade system based on marine sand and bottom ash in the treatment of cardboard manufacturing wastewater by infiltration percolation. *Physica Scripta*, *96*(12), 125711. DOI: 10.1088/1402-4896/ac2659

Guo, M., Li, M., Fu, H., Zhang, Y., Chen, T., Tang, H., Zhang, T., & Li, H. (2023). Quantitative analysis of polycyclic aromatic hydrocarbons (PAHs) in water by surface-enhanced Raman spectroscopy (SERS) combined with Random Forest. *Spectrochimica Acta. Part A: Molecular and Biomolecular Spectroscopy*, *287*, 122057. DOI: 10.1016/j.saa.2022.122057 PMID: 36332395

Han, C., Kim, H., & Yang, H. (2020). Quantum Dots and Applications. *Materials (Basel)*, *13*(4), 897. DOI: 10.3390/ma13040897 PMID: 32085387

Hashemi, S. M., Hasani, S., Ardakani, K. J., & Davar, F. (2019). The effect of simultaneous addition of ethylene glycol and agarose on the structural and magnetic properties of $CoFe_2O_4$ nanoparticles prepared by the sol–gel auto-combustion method. *Journal of Magnetism and Magnetic Materials*, *492*, 165714. DOI: 10.1016/j.jmmm.2019.165714

Hedayati, K., Azarakhsh, S., Saffari, J., & Ghanbari, D. (2017). Magnetic and photocatalyst $CoFe_2O_4$–CdS nanocomposites: Simple preparation of Ni-, Co-, Zn- or Ag-doped CdS nanoparticles. *Journal of Materials Science Materials in Electronics*, *28*(7), 5472–5481. DOI: 10.1007/s10854-016-6209-3

Heiligtag, F. J., & Niederberger, M. (2013). The fascinating world of nanoparticle research. *Materials Today*, *16*(7-8), 262–271. DOI: 10.1016/j.mattod.2013.07.004

Herman, A. P., Yusup, S., Shahbaz, M., & Patrick, D. O. (2016). Bottom ash characterization and its catalytic potential in biomass gasification. *Procedia Engineering*, *148*, 432–436. DOI: 10.1016/j.proeng.2016.06.447

Hummers, W. S.Jr, & Offeman, R. E. (1958). Preparation of graphitic oxide. *Journal of the American Chemical Society*, *80*(6), 1339. DOI: 10.1021/ja01539a017

Hutton, G., & Varughese, M. (2016). *The costs of meeting the 2030 sustainable development goal targets on drinking water, sanitation, and hygiene*. World Bank., DOI: 10.1596/K8543

Idris SADLI. (n.d.). *MODELISATION PAR IMPEDANCE D'UNE PILE A COMBUSTIBLE PEM POUR UTILISATION EN ELECTRONIQUE DE PUISSANCE.*

Iijima, S. (1991). Helical microtubules of graphitic carbon. *Nature*, *354*(6348), 56–58. DOI: 10.1038/354056a0

Iijima, S. (1991). Helical microtubules of graphitic carbon. *Nature*, *354*, 56–58.

Inyinbor, A. A., Bello, O. S., Oluyori, A. P., Inyinbor, H. E., & Fadiji, A. E. (2019). Wastewater conservation and reuse in quality vegetable cultivation: Overview, challenges, and future prospects. *Food Control*, *98*, 489–500. DOI: 10.1016/j.foodcont.2018.12.008

Iqbal, A., Cevik, E., Mustafa, A., Qahtan, T. F., Zeeshan, M., & Bozkurt, A. (2024). Emerging developments in polymeric nanocomposite membrane-based filtration for water purification: A concise overview of toxic metal removal. *Chemical Engineering Journal*, *481*, 148760.

Irshad, M., Ain, Q., Zaman, M., Aslam, M. Z., Kousar, N., Asim, M., Rafique, M., Siraj, K., Tabish, A. N., Usman, M., Hassan Farooq, M., Assiri, M. A., & Imran, M. (2022). Photocatalysis and perovskite oxide-based materials: A remedy for a clean and sustainable future. *RSC Advances*, *12*(12), 7009–7039. DOI: 10.1039/D1RA08185C PMID: 35424711

Ito, A., Sato, T., & Goto, T. (2014). Transparent anatase and rutile TiO2 films grown by laser chemical vapor deposition. *Thin Solid Films*, *551*, 37–41.

Jahan, N., Roy, H., & Reaz, A. H. Shasawti Arshi, Rahman, E., Firoz, S.H. and Islam, M.S. (2022). A comparative study on sorption behavior of graphene oxide and reduced graphene oxide towards methylene blue. *Case Studies in Chemical and Environmental Engineering*, 6, pp.100239–100239. DOI: 10.1016/j.cscee.2022.100239

Jamil, R. (2019). Reduction in fresh water consumption by grey water reuse for flushing and irrigation: A case study of a multistorey hotel building. In *Proceedings of the 1st Conference on Sustainable Civil Engineering.*

Janani, T., Sudarsan, J. S., & Prasanna, K. (2019). Grey water recycling with corn cob as an adsorbent. *AIP Conference Proceedings, 2112*(1), 020181. DOI: 10.1063/1.5112366

Janoš, P., Buchtová, H., & Rŷznarová, M. (2003). Sorption of dyes from aqueous solutions onto fly ash. *Water Research, 37*(20), 4938–4944. DOI: 10.1016/j.watres.2003.08.011 PMID: 14604640

Jeanmarie, D. L. (1977). Surface Raman spectroelectrochemistry Part I. Heterocyclic, aromatic, and aliphatic amines adsorbed on the anodized silver electrode. *Journal of Electroanalytical Chemistry, 84*(1), 1–20. DOI: 10.1016/0368-1874(77)80399-7

Jiang, M., Althomali, R. H., Ansari, S. A., Abdu, E., & Gupta, J. (2023). Advances in preparation, biomedical, and pharmaceutical applications of chitosan-based gold, silver, and magnetic nanoparticles: A review. [PubMed]. *International Journal of Biological Macromolecules, 251*, 126390–126390. DOI: 10.1016/j.ijbiomac.2023.126390

Ji, J., Lu, Z., Lei, Y., & Turner, C. H. (2018). Mechanistic insights into the direct propylene epoxidation using Au nanoparticles dispersed on TiO2/SiO2. *Chemical Engineering Science, 191*, 169–182. DOI: 10.1016/j.ces.2018.06.064

Joint Monitoring Programme. (2015). *Water supply and sanitation.* World Health Organization & UNICEF. https://www.wssinfo.org/data-estimates/tables/

Joudeh, N. and Linke, D. (2022). Nanoparticle classification, physicochemical properties, characterization, and applications: a comprehensive review for biologists. Journal of Nanobiotechnology, [online] 20(1). doi:.DOI: 10.1186/s12951-022-01477-8

Kao, P.-C., Tzeng, J.-H., & Huang, T.-L. (2000). Removal of chlorophenols from aqueous solution by fly ash. *Journal of Hazardous Materials, 76*(2), 237–249. DOI: 10.1016/S0304-3894(00)00201-6 PMID: 10936536

Karandish, F. (2019). Applying grey water footprint assessment to achieve environmental sustainability within a nation under intensive agriculture: A high-resolution assessment for common agrochemicals and crops. *Environmental Earth Sciences, 78*(6), 1–22. DOI: 10.1007/s12665-019-8199-y

Kayalvizhi, S., Jiavana, K. F. K., Suganthi, K., & Malarvizhi, S. (2023). Prediction of ground water quality in western regions of Tamil Nadu using deep auto encoders. *Urban Climate, 49*, 101458. DOI: 10.1016/j.uclim.2023.101458

Keiding, K., Wybrandt, L., & Nielsen, P. H. (2001). Remember the water—A comment on EPS colligative properties. *Water Science and Technology*, *43*(6), 17–23. DOI: 10.2166/wst.2001.0330 PMID: 11381964

Khan, M. S. I., Islam, N., Uddin, J., Islam, S., & Nasir, M. K. (2022). Water quality prediction and classification based on principal component regression and gradient boosting classifier approach. *Journal of King Saud University. Computer and Information Sciences*, *34*(8), 4773–4781. DOI: 10.1016/j.jksuci.2021.06.003

Khorshidi, S., Sabbaghi, S., & Saboori, R. (2020). Removal of hydrogen sulfide from water-based drilling fluid using titania/carbon nanotubes nano-hybrid: Optimization, kinetics, and isotherms. *Desalination and Water Treatment*, *200*, 154–166. DOI: 10.5004/dwt.2020.26089

Khurshid, R., Ali, F., Qureshi, M. T., Afzal, A., & Ali, Z. (2019). Polyol-mediated coprecipitation and aminosilane grafting of superparamagnetic, spinel $ZnFe_2O_4$ nanoparticles for room-temperature ethanol sensors. *Journal of the Electrochemical Society*, *166*(4), B258–B265. DOI: 10.1149/2.1021904jes

Kjeldsen, P., Barlaz, M. A., Rooker, A. P., Baun, A., Ledin, A., & Christensen, T. H. (2002). Present and Long-Term Composition of MSW Landfill Leachate: A Review. *Critical Reviews in Environmental Science and Technology*, *32*(4), 297–336. DOI: 10.1080/10643380290813462

Klimonda, A., & Kowalska, I. (2021). Sequential process: Membrane filtration and ion exchange as an effective method for water solution purification containing cationic surfactants. *Desalination and Water Treatment*, *214*, 232–241. DOI: 10.5004/dwt.2021.26663

Kolker, J., Kingdom, B., & Trémolet, S. (2016). *Financing options for the 2030 water agenda*. World Bank. https://openknowledge.worldbank.org/handle/10986/25399

Kouadri, S., Elbeltagi, A., Islam, A. R. M. T., & Kateb, S. (2021). Performance of machine learning methods in predicting water quality index based on irregular data set: Application on Illizi region (Algerian southeast). *Applied Water Science*, *11*(12), 190. DOI: 10.1007/s13201-021-01528-9

Kouadri, S., Pande, C. B., Panneerselvam, B., Moharir, K. N., & Elbeltagi, A. (2022). Prediction of irrigation groundwater quality parameters using ANN, LSTM, and MLR models. *Environmental Science and Pollution Research International*, *29*(14), 21067–21091. DOI: 10.1007/s11356-021-17084-3 PMID: 34748181

Krishnan, D., Raidongia, R., Shao, J., & Huang, J. (2014). Graphene Oxide Assisted Hydrothermal Carbonization of Carbon Hydrates. *ACS Nano*, *8*, 449–457.

Kudo, A., & Miseki, Y. (2009). Heterogeneous photocatalyst materials for water splitting. *Chemical Society Reviews, 38*(1), 253–278. DOI: 10.1039/B800489G PMID: 19088977

Kulkarni, S., Kaware, J., & Mumbai, N. (2014). Regeneration and Recovery in Adsorption- a Review. *International Journal of Innovative Science. Engineering & Technology*, *1*(8), 61–64.

Kumar, R., Rose, P. K., Sharma, P. K., Lamba, J., Kumar, M., & Bhattacharya, P. (2025). Micro(nano)plastic and per- and polyfluoroalkyl substances in soil/sediment–water ecosystems: Sources, transport, interactions, and challenges. *Current Opinion in Chemical Engineering*, *48*, 101125. DOI: 10.1016/j.coche.2025.101125

Laaffat, J., Aziz, F., Ouazzani, N., & Mandi, L. (2019). Biotechnological approach of greywater treatment and reuse for landscape irrigation in small communities. *Saudi Journal of Biological Sciences*, *26*(1), 83–90. DOI: 10.1016/j.sjbs.2017.01.006 PMID: 30622410

Léonard, A. (2003). *Etude du séchage convectif de boues de station d'épuration-Suivi de la texture par microtomographie à rayons X.* https://orbi.uliege.be/handle/2268/39648

Liang, C., Wei, D., Zhang, S., Ren, Q., Shi, J. and Liu, L. (2021). Removal of antibiotic resistance genes from swine wastewater by membrane filtration treatment. Ecotoxicology and Environmental Safety, [online] 210, p.111885. doi:.DOI: 10.1016/j.ecoenv.2020.111885

Li, G., Wu, L., Li, F., Xu, P., Zhang, D., & Li, H. (2013b). Photoelectrocatalytic degradation of organic pollutants via a CdS quantum dots enhanced TiO2 nanotube array electrode under visible light irradiation. *Nanoscale*, *5*(5), 2118. DOI: 10.1039/c3nr34253k PMID: 23381869

Lin, L., Yang, H., & Xu, X. (2022). Effects of Water Pollution on Human Health and Disease Heterogeneity: a Review. *Frontiers in Environmental Science*, [online] 10. Available at: https://www.frontiersin.org/journals/environmental-science/articles/10.3389/fenvs.2022.880246/full

Li, P., Qian, H., & Wu, J. (2019). Groundwater quality prediction using machine learning approaches in an arid region of China. *The Science of the Total Environment*, *653*, 1311–1321.

Liu, Z., Wang, Q., Huang, X., & Qian, X. (2022). Surface Functionalization of Graphene Oxide with Hyperbranched Polyamide-Amine and Microcrystalline Cellulose for Efficient Adsorption of Heavy Metal Ions. *ACS Omega*, *7*(13), 10944–10945. https://pubs.acs.org/journal/acsodf. DOI: 10.1021/acsomega.1c06647 PMID: 35415369

Li, X., Qin, Y., Picraux, S. T., & Guo, Z.-X. (2011). Noncovalent assembly of carbon nanotube-inorganic hybrids. *Journal of Materials Chemistry*, *21*, 7527–7547.

Luo, X., Liu, Y., Wang, X., Li, Y., & Zhang, L. (2022). Removal of methylene blue by crosslinked egg white protein/graphene oxide bionanocomposite aerogels. *Nanomaterials (Basel, Switzerland)*, *12*(3), 487. PMID: 35159833

Lu, Y., Yousaf, M., Akhtar, M. N., Noor, A., Akbar, M., Shah, M. A. K. Y., Yan, S., & Wang, F. (2022). Effect of Gd and Co contents on the microstructural, magneto-optical and electrical characteristics of cobalt ferrite ($CoFe_2O_4$) nanoparticles. *Ceramics International*, *48*(2), 2782–2792. DOI: 10.1016/j.ceramint.2021.10.067

Machete, I., & Marques, R. (2021). Financing the water and sanitation sectors: A hybrid literature review. *Infrastructures*, *6*(1), 9. DOI: 10.3390/infrastructures6010009

Magdy, G., Aboelkassim, E., Abd Elhaleem, S. M., & Belal, F. (2024). A comprehensive review on silver nanoparticles: Synthesis approaches, characterization techniques, and recent pharmaceutical, environmental, and antimicrobial applications. *Microchemical Journal*, *196*, 109615–109615. DOI: 10.1016/j.microc.2023.109615

Mahboubeh, H., Fatemeh, Z., Zahra, J. R., Ali, A., & Zohreh, A. (2014). Synthesis of cobalt ferrite ($CoFe_2O_4$) nanoparticles using combustion, coprecipitation, and precipitation methods: A comparison study of size, structural, and magnetic properties. *Journal of Magnetism and Magnetic Materials*, *371*, 43–48. DOI: 10.1016/j.jmmm.2014.06.059

Mao, Y., Zhao, Y. and Cotterill, S. (2023). Examining Current and Future Applications of Electrocoagulation in Wastewater Treatment. Water, [online] 15(8), p.1455. doi:.DOI: 10.3390/w15081455

Marley, G., Bazaanah, P., & Oppong, P. (2022). *Examining the Role of NGOs in Community Water and Sanitation Improvement: A Case of the Tunayilli Community in the Northern Region, Ghana.* In *Handbook of Research on Resource Management and the Struggle for Water Sustainability in Africa* (pp. chapter 2). IGI Global. DOI: DOI: 10.4018/978-1-7998-8809-3.ch002

Marttinen, S. K., Kettunen, R. H., & Rintala, J. A. (2003). Occurrence and removal of organic pollutants in sewages and landfill leachates. *The Science of the Total Environment, 301*(1–3), 1–12. DOI: 10.1016/S0048-9697(02)00302-9 PMID: 12493180

Menchaca-Nal, S., Jativa-Herrera, J. A., Moscoso-Londoño, O., Pampillo, L. G., Martínez-García, R., Knobel, M., & Londoño-Calderón, C. L. (2023). Composite magnetic properties of cobalt ferrite nanoparticles embedded in bacterial nanocellulose of different porosity levels. *Materials Chemistry and Physics, 303*, 127753. DOI: 10.1016/j.matchemphys.2023.127798

Méthodes électrochimiques—Mesures d'impédances: Dossier complet | Techniques de l'Ingénieur. (n.d.). Retrieved July 19, 2024, from https://www.techniques-ingenieur.fr/base-documentaire/materiaux-th11/essais-et-expertise-en-corrosion-et-vieillissement-42499210/methodes-electrochimiques-p2210/

Mghaiouini, R., Graich, A., Elaoud, A., Garmim, T., Belghiti, M. E., Benzbiria, N., Hozayn, M., Monkade, M., & El Bouari, A. (2020). Formulation and physico-mechanical characterization of an eco-mortar composite based on bottom ash and magnetized water. *Indian Journal of Science and Technology, 13*(10), 1172–1187. DOI: 10.17485/ijst/2020/v13i10/149889

Mininni, G., Sbrilli, A., Guerriero, E., & Rotatori, M. (2004). Dioxins and furans formation in pilot incineration tests of sewage sludge spiked with organic chlorine. *Chemosphere, 54*(9), 1337–1350. DOI: 10.1016/S0045-6535(03)00252-2 PMID: 14659427

Misman, N. A., & Sharif, M. F. Jalal, A. and Nur Hafizah Azizan (2023). Water pollution and the assessment of water quality parameters: a review. DESALINATION AND WATER TREATMENT, 294, pp.79–88. doi:DOI: 10.5004/dwt.2023.29433

Mogashane, T. M., Maree, J. P., & Mokoena, L. (2024). Adsorption of polycyclic aromatic hydrocarbons from wastewater using iron oxide nanomaterials recovered from acid mine water: A review. *Minerals (Basel), 14*(8), 826.

Mohamed, R. M. S. R., Al-Gheethi, A. A. S., & Kassim, A. H. M. (Eds.). (2019). *Management of greywater in developing countries: Alternative practices, treatment and potential for reuse and recycling*. Springer International Publishing. DOI: 10.1007/978-3-319-90269-2

Mohammed, M. I., Ismael, M. K., & Gönen, M. (2020). Synthesis of Chitosan-Silica Nanocomposite for Removal of Methyl Orange from Water: Composite Characterization and Adsorption Performance. *IOP Conference Series. Materials Science and Engineering, 745*(1), 012084. DOI: 10.1088/1757-899X/745/1/012084

Mohan, D., Singh, K. P., Singh, G., & Kumar, K. (2002). Removal of Dyes from Wastewater Using Flyash, a Low-Cost Adsorbent. *Industrial & Engineering Chemistry Research, 41*(15), 3688–3695. DOI: 10.1021/ie010667+

Mohanty, A., Ghosh, A., & Roy, S. (2024). Calling for better ecological values: Integrating Indigenous Knowledge System with sustainable policies. *Purushartha, 17*(1), 39–50. Advance online publication. DOI: 10.21844/16202117103

Molinié, P. (2006). *Comment modéliser un isolant en électrostatique? Approches classiques et modernes d'un problème complexe.* 27. https://centralesupelec.hal.science/hal-00239414

Morsy, A., Hatem, N., Ewais, H.A., Mohamed, S., Morsy, A., Fathi, H., Tarek, A. and Nassef, E. (2025). High-performance self-cleaning cellulose acetate membranes reinforced with nanochitosan for heavy metal removal. Environmental Research, [online] 288, p.123197. doi:.DOI: 10.1016/j.envres.2025.123197

Mortadi, A., Chahid, E. G., Elmelouky, A., Chahbi, M. E. L., Ghyati, N., Zaim, S., Cherkaoui, O., & El Moznine, R. (2020). Complex electrical conductivity as a new technique to monitor the coagulation-flocculation processes in the wastewater treatment of the textile Industry. *Water Resources and Industry, 24*, 100130. DOI: 10.1016/j.wri.2020.100130

Mortadi, A., Chahid, E. G., Nasrellah, H., Cherkaoui, O., & El Moznine, R. (2019). Dielectric and electric properties as a tool to investigate the coagulation mechanism during sludge treatment. *Environmental Technology, 40*(1), 72–85. DOI: 10.1080/09593330.2017.1378927 PMID: 28891396

Mortadi, A., Mghaiouini, R., Elmelouky, A., Chahid, E., Hairch, Y., Saifaoui, D., Monkade, M., Cherkaoui, O., & El Moznine, R. (2022). New approach to investigate and to monitor the coagulation process during wastewater treatment. *Materials Today: Proceedings, 66*, 325–328. DOI: 10.1016/j.matpr.2022.05.424

Mundonde, J., & Makoni, P. L. (2024). Framework model for financing sustainable water and sanitation infrastructure in Zimbabwe. *Water (Basel), 16*(12), 1691. DOI: 10.3390/w16121691

Nabila Chaabani, Ncib, S., Lasâad Dammak, Larchet, C., Wided Bouguerra and Elimame Elaloui (2025). High-Efficiency Removal of Copper and Nickel via Donnan Dialysis Using Fujifilm Cation-Exchange Membranes: Process Optimization Through Response Surface Methodology. Membranes, [online] 15(12), pp.346–346. doi:.DOI: 10.3390/membranes15120346

Nammari, D. R., Hogland, W., Marques, M., Nimmermark, S., & Moutavtchi, V. (2004). Emissions from a controlled fire in municipal solid waste bales. *Waste Management (New York, N.Y.)*, *24*(1), 9–18. DOI: 10.1016/j.wasman.2003.08.003 PMID: 14672722

Nandhini, G., Vignesh, D., Shobana, M. K., Pazhanivel, T., Kavita, S., & Balaji, P. (2025). Graphene-decorated magnetic cobalt ferrite for effective UV-accelerated photocatalytic degradation of methylene blue: Experimental and theoretical insights by DFT. *Emergent Materials*, *8*(1), 29–44. DOI: 10.1007/s42247-024-00799-7

Nel, N., & Jacobs, H. E. (2019). Investigation into untreated greywater reuse practices by suburban households under the threat of intermittent water supply. *Journal of Water, Sanitation, and Hygiene for Development : a Journal of the International Water Association*, *9*(4), 627–634. DOI: 10.2166/washdev.2019.055

Nigam, A. (1996). *Financing of water supply and sanitation investments in developing countries*. UNDP–World Bank Water and Sanitation Program.

Nigam, A. (1996). *Sustainable financing of WATSAN*. Loughborough University Repository. https://repository.lboro.ac.uk/articles/conference_contribution/Sustainable_financing_of_WATSAN/9588650

Norén, A., Modin, O., Rauch, S., Fedje, K.K., Björklund, S., Thoutam, A.K., Markocsan, N., Andersson-Sköld, Y. and Strömvall, A.-M. (2025). Enhanced photoelectrocatalysis for oxidation of organic pollutants and metal recovery from polluted water and sediments. Journal of Hazardous Materials, [online] 500, p.140375. doi:. DOI: 10.1016/j.jhazmat.2025.140375

Nuengmatcha, P., Mahachai, R., & Chanthai, S. (2014). Thermodynamic and kinetic study of the intrinsic adsorption capacity of graphene oxide for malachite green removal from aqueous solution. *Oriental Journal of Chemistry*, *30*(4), 1463–1474. DOI: 10.13005/ojc/300403

Obernberger, I., & Supancic, K. (2009). Possibilities of ash utilisation from biomass combustion plants. *Proceedings of the 17th European Biomass Conference & Exhibition, 29*. https://www.best-research.eu/files/publications/pdf/Paper-Obernberger-ash-utilisation-2009.pdf

Olabode, O. F., & Comte, J. C. (2025). Long-term groundwater security in the fast-growing coastal megacity of Lagos, Nigeria: Insights from distributed recharge simulations. *Hydrological Sciences Journal*, *70*(11), 1813–1833. Advance online publication. DOI: 10.1080/02626667.2025.2505171

Olawade, D. B., Wada, O. Z., Egbewole, B. I., Fapohunda, O., Ige, A. O., Usman, S. O., & Ajisafe, O. (2024). Metal and metal oxide nanomaterials for heavy metal remediation: Novel approaches for selective, regenerative, and scalable water treatment. *Frontiers in Nanotechnology*, *6*, 1466721. Advance online publication. DOI: 10.3389/fnano.2024.1466721

Organisation for Economic Co-operation and Development. (2018). *Making blended finance work for the sustainable development goals*. OECD Publishing. https://www.oecd.org/dac/making-blended-finance-work.htm

Organisation for Economic Co-operation and Development. (2022). *Accelerating action on financing water: An agenda for the future*. OECD Studies on Water., DOI: 10.1787/6bf7264b-

Organisation for Economic Co-operation and Development. (2023). *Financing a water secure future*. OECD Publishing. https://www.oecd.org/en/publications/financing-a-water-secure-future_a2ecb261-en.html

Organization, W. H. (2022). Guidelines for drinking-water quality*: Fourth edition incorporating the first and second addenda*. World Health Organization.

Othman, N. H., Alias, N. H., Fuzil, N. S., Marpani, F., Shahruddin, M. Z., Chew, C. M., & Ismail, A. F. (2021). A review on the use of membrane technology systems in developing countries. *Membranes*, *12*(1), 30.

Ouyang, W., Teng, F., He, J., & Fang, X. (2019). Enhancing the Photoelectric Performance of Photodetectors Based on Metal Oxide Semiconductors by Charge-Carrier Engineering. *Advanced Functional Materials*, *29*(9), 1807672–1807672. DOI: 10.1002/adfm.201807672

Pandey, A. K., Reji Kumar, R., B, K., Laghari, I. A., Samykano, M., Kothari, R., Abusorrah, A. M., Sharma, K., & Tyagi, V. V. (2021). Utilization of solar energy for wastewater treatment: Challenges and progressive research trends. *Journal of Environmental Management*, *297*, 113300. DOI: 10.1016/j.jenvman.2021.113300 PMID: 34293672

Park, S., & Ruoff, R. S. (2009). Chemical methods for the production of graphenes. *Nature Nanotechnology*, *4*, 217–224.

Patel, S. K., Shukla, S. C., Natarajan, B. R., Asaithambi, P., Dwivedi, H. K., Sharma, A., & Prajapati, A. K. (2025). State of the art review for industrial wastewater treatment by electrocoagulation process: Mechanism, cost and sludge analysis. *Desalination and Water Treatment*, *321*, 100915.

Pories, L., Fonseca, C., & Delmon, V. (2019). *Mobilising finance for WASH: Getting the foundations right.* WaterAid. https://washmatters.wateraid.org/publications/mobilising-finance-for-wash-getting-the-foundations-right

Prévot, H. (2000). La récupération de l'énergie issue du traitement des déchets. *Rapport Du Conseil Général Des Mines Au Ministère de l'Economie, Des Finances et de l'Industrie, Paris.* http://emericmenard.free.fr/dossierenergies/Documents/Rapport-%20R%E9cup%E9ration%20d'energie%20issue%20des%20d%E9chet%20juillet%20200.pdf

Puri, C., & Sumana, G. (2018). Highly effective adsorption of crystal violet dye from contaminated water using graphene oxide intercalated montmorillonite nanocomposite. *Applied Clay Science, 166*, 102–112. DOI: 10.1016/j.clay.2018.09.012

Puri, N., Gupta, A., & Mishra, A. (2021). Recent advances on nano-adsorbents and nanomembranes for the remediation of water. *Journal of Cleaner Production, 322*, 129051. DOI: 10.1016/j.jclepro.2021.129051

Qi, Y., Li, D., Zhang, S., Li, F., & Hua, T. (2023). Electrochemical filtration for drinking water purification: A review on membrane materials, mechanisms and roles. [online]. Journal of Environmental Sciences (China). Advance online publication. doi: PubMedDOI: 10.1016/j.jes.2023.06.033

Raicy, M. C., Aju, C. D., & Achu, A. L. (2026). Groundwater quality and risk in the Ganga River Basin: An integrated machine learning appraisal. *Environmental Geochemistry and Health, 48*(2), 72. DOI: 10.1007/s10653-025-02964-1 PMID: 41483424

Raju, G. G. (2016). *Dielectrics in Electric Fields: Tables, Atoms, and Molecules* (2nd ed.). CRC Press., DOI: 10.1201/9781315373270

Rakotoarisoa, Z., & Tiruta-Barna, L. (n.d.). *Étude des écoulements à travers des couches de cendres volantes compactées saturées et non saturées en eau.* Retrieved July 19, 2024, from https://scholar.google.com/scholar?hl=en&as_sdt=0%2C5&q=%29.+%C3%89tude+des+%C3%A9coulements+%C3%A0+travers+des+couches+de+cendres+volantes+compact%C3%A9es+satur%C3%A9es+et+non+satur%C3%A9es+en+eau.+Environnement%2C+Ing%C3%A9nierie+%26+D%C3%A9veloppement.&btnG=

Rathod, S. M., Khandelwal, P. S., & Rajuskar, R. S. (2019). Influence of gadolinium doped in nickel nano ferrite on magnetic and optical properties prepared by sol–gel technique. *Materials Today: Proceedings, 15*, 560–565. DOI: 10.1016/j.matpr.2019.04.121

Ravindra, A. V., Chandrika, M., Rajesh, C., Kollu, P., Ju, S., & Ramarao, S. D. (2019). Simple synthesis, structural and optical properties of cobalt ferrite nanoparticles. *The European Physical Journal Plus, 134*(6), 1–9. DOI: 10.1140/epjp/i2019-12690-2

Raynaud, M. (2010). *Couplage de caractérisations mécanique et physico-chimique en vue d'analyser les limites de la déshydratation des boues résiduaires urbaines* [PhD Thesis, Doctorat Génie des Procédés, Université de Pau et des Pays de l'Adour]. https://hal.science/tel-02594092

Reihanifar, M., Takallou, A., Taheri, M., Gholizadeh Lonbar, A., Ahmadi, M., & Sharifi, A. (2024). Nanotechnology advancements in groundwater remediation: A comprehensive analysis of current research and future prospects. *Groundwater for Sustainable Development, 27*, 101330. DOI: 10.1016/j.gsd.2024.101330

Rezaee, R. (2024). The Potential of Salinity Gradient Energy Using Reverse Electrodialysis to Generate Electricity for Seawater Desalination Plants, an Example from Western Australia. *Clean Energy and Sustainability, 2*(2), 10006. DOI: 10.35534/ces.2024.10006

Rezania, S., Darajeh, N., Rupani, P. F., Mojiri, A., Kamyab, H., & Taghavijeloudar, N. (2024). Recent Advances in the Adsorption of Different Pollutants from Wastewater Using Carbon-Based and Metal-Oxide Nanoparticles. *Applied Sciences (Basel, Switzerland), 14*(24), 11492–11492. DOI: 10.3390/app142411492

Ricou, P., Lécuyer, I., & Le Cloirec, P. (1999). Removal of Cu2+, Zn2+ and Pb2+ by adsorption onto fly ash and fly ash/lime mixing. *Water Science and Technology, 39*(10), 239–247. DOI: 10.2166/wst.1999.0663

Rodell, M., Famiglietti, J. S., Wiese, D. N., Reager, J. T., Beaudoing, H. K., Landerer, F. W., & Lo, M.-H. (2018). Emerging trends in global freshwater availability. *Nature, 557*(7707), 651–659. DOI: 10.1038/s41586-018-0123-1 PMID: 29769728

Rovel, J.-M. (2010). *Dessalement de l'eau de mer*. Editions TI. https://perso.ens-lyon.fr/timothee.audinet-de-pieuchon/fichiers/LC4_TI_Dessalement_eau_de_mer.pdf

Sadaf, S., Roy, H., Fariha, A., Rahman, T.U., Tasnim, N., Jahan, N., Sokan-Adeaga, A.A., Safwat, S.M. and Islam, M.S. (2024). Electrocoagulation-based wastewater treatment process and significance of anode materials for the overall improvement of the process: A critical review. Journal of Water Process Engineering, [online] 62, p.105409. doi:.DOI: 10.1016/j.jwpe.2024.105409

Sage, N. (2022). SDG-6 clean water and sanitation (pp. 289–304). *Springer eBooks*. DOI: 10.1007/978-3-031-07127-0_8

Sakthivel, S., Neppolian, B., Shankar, M. V., Arabindoo, B., Palanichamy, M., & Murugesan, V. (2003). Solar photocatalytic degradation of azo dye: Comparison of photocatalytic efficiency of ZnO and TiO_2. *Solar Energy Materials and Solar Cells*, *77*(1), 65–82. DOI: 10.1016/S0927-0248(02)00255-6

Saleem, H., Zaidi, S. J., Ismail, A. F., & Goh, P. S. (2022). Advances of nanomaterials for air pollution remediation and their impacts on the environment. [PubMed]. *Chemosphere*, *287*, 132083. DOI: 10.1016/j.chemosphere.2021.132083

Saleh, T. A. (2024). Materials, nanomaterials, nanocomposites, and methods used for the treatment and removal of hazardous pollutants from wastewater: Treatment technologies for water recycling and sustainability. *Nano-Structures & Nano-Objects*, *39*, 101231.

Saleh, T. A., & Al-Daous, M. A. (2016). *Graphene science handbook: Fabrication methods*. CRC Press., DOI: 10.1201/b19606-39

Saleh, T. A., Muhammad, A. M., Tawabini, B., & Ali, S. A. (2016). Aminomethylphosphonate chelating ligand and octadecyl alkyl chain in a resin for simultaneous removal of Co(II) ions and organic contaminants. *Journal of Chemical & Engineering Data*, *61*(9), 3377–3385. DOI: 10.1021/acs.jced.6b00475

Saqr, A. M., Nasr, M., Fujii, M., Yoshimura, C., & Ibrahim, M. G. (2024). Delineating suitable zones for solar-based groundwater exploitation using multi-criteria analysis: A techno-economic assessment for meeting sustainable development goals (SDGs). *Groundwater for Sustainable Development*, *25*, 101087. DOI: 10.1016/j.gsd.2024.101087

Sarmah, S., Patra, K. P., Maji, P. K., Ravi, S., & Bora, T. (2023). A comparative study on the structural, magnetic and dielectric properties of magnesium substituted cobalt ferrites. *Ceramics International*, *49*(1), 1444–1463. DOI: 10.1016/j.ceramint.2022.09.126

Sharifianjazi, F., Moradi, M., Parvin, N., Nemati, A., Jafari Rad, A., Sheysi, N., Abouchenari, A., Mohammadi, A., Karbasi, S., Ahmadi, Z., Esmaeilkhanian, A., Irani, M., Pakseresht, A., Sahmani, S., & Shahedi Asl, M. (2020). Magnetic $CoFe_2O_4$ nanoparticles doped with metal ions: A review. *Ceramics International*, *46*(11), 18391–18412. DOI: 10.1016/j.ceramint.2020.04.202

Shobana, M. K., Nandhini, G., Kavita, S., Kumar, V. S., & Pazhanivel, T. (2022). Photocatalytic and magnetic properties of Mg substituted cobalt ferrite. *Materials Science and Engineering B*, *286*, 116030. DOI: 10.1016/j.mseb.2022.116030

Singha, S., Pasupuleti, S., Singha, S. S., Singh, R., & Kumar, S. (2021). Prediction of groundwater quality using efficient machine learning technique. *Chemosphere, 276*, 130265. DOI: 10.1016/j.chemosphere.2021.130265 PMID: 34088106

Singh, R., Arora, V., Tandon, R. P., Mansingh, A., & Chandra, S. (1999). Dielectric spectroscopy of doped polyaniline. *Synthetic Metals, 104*(2), 137–144. DOI: 10.1016/S0379-6779(99)00043-0

Singh, R., Upreti, P., Allemailem, K. S., Almatroudi, A., Rahmani, A. H., & Albalawi, G. M. (2022). Geospatial assessment of ground water quality and associated health problems in the Western Region of India. *Water (Basel), 14*(3), 296. DOI: 10.3390/w14030296

Smith, J. K., & Vesilind, P. A. (1995). Dilatometric measurement of bound water in wastewater sludge. *Water Research, 29*(12), 2621–2626. DOI: 10.1016/0043-1354(95)00144-A

Sonia, S., Sheetal, , Kumari, H., Sharma, S., Chahal, S., Duhan, S., Kumar, A., & Kumar, P. (2025). Magnetic cobalt ferrite–tin oxide nanocomposites for efficient photocatalytic dye degradation. *Water, Air, and Soil Pollution, 236*(9), 602. DOI: 10.1007/s11270-025-08238-x

Sorkin, V., & Zhang, Y. W. (2011). Graphene-based pressure nano-sensors. *Journal of Molecular Modeling, 17*(11), 2825–2830. DOI: 10.1007/s00894-011-0972-0 PMID: 21287214

Subudhi, S., Pati, A. K., Bose, S., Sahoo, S., Pattanaik, A., Acharya, B. M., & Thakur, R. R. (2025). Prediction of groundwater quality assessment by integrating boosted learning with DE optimizer. *Scientific Reports, 16*(1), 618. Advance online publication. DOI: 10.1038/s41598-025-30122-1 PMID: 41326700

Sun, H. (2019). Grand Challenges in Environmental Nanotechnology. *Frontiers in Nanotechnology, 1*, 2. Advance online publication. DOI: 10.3389/fnano.2019.00002

Sutherland, I. W. (2001). Exopolysaccharides in biofilms, flocs and related structures. *Water Science and Technology, 43*(6), 77–86. DOI: 10.2166/wst.2001.0345 PMID: 11381975

Su, Z., Wu, J., He, X., & Elumalai, V. (2020). Temporal changes of groundwater quality within the groundwater depression cone and prediction of confined groundwater salinity using Grey Markov model in Yinchuan area of northwest China. *Exposure and Health, 12*(3), 447–468. DOI: 10.1007/s12403-020-00355-8

Sylvestre, A. (2006). *Apport de la spectroscopie diélectrique basse fréquence dans l'analyse de matériaux isolants à forte permittivité* [Thesis, Université Joseph-Fourier - Grenoble I]. https://theses.hal.science/tel-00651493

Toshima, N. (2013). Metal nanoparticles for energy conversion. *Pure and Applied Chemistry, 85*(2), 437–451. DOI: 10.1351/PAC-CON-12-08-17

Tripathy, D. B., & Gupta, A.Divya Bajpai Tripathy and Gupta. (2023). Nanomembranes-Affiliated Water Remediation: Chronology, Properties, Classification, Challenges and Future Prospects. [PubMed]. *Membranes, 13*(8), 713–713. DOI: 10.3390/membranes13080713

Ullah, A., Shahzada, K., Khan, S. W., & Starov, V. (2020). Purification of produced water using oscillatory membrane filtration. *Desalination, 491*, 114428. DOI: 10.1016/j.desal.2020.114428

UNESCO. (2020). *The United Nations world water development report 2020: Water and climate change*. UNESCO Publishing. https://unesdoc.unesco.org/ark:/48223/pf0000372985

UN-Habitat. (2022). *World cities report 2022: Envisaging the future of cities*. United Nations Human Settlements Programme. https://unhabitat.org/

UNICEF, & World Health Organization. (2021). *State of the world's drinking water 2021*. UNICEF & WHO. https://www.unicef.org/reports/state-worlds-drinking-water

United Nations Development Programme (UNDP). (2019). *Human development report 2019: Beyond income, beyond averages, beyond today*. UNDP. https://hdr.undp.org/

United Nations Office for Outer Space Affairs. (2021). *Space solutions for the SDGs*. UNOOSA. https://www.unoosa.org/oosa/en/ourwork/space4sdgs/index.html

United Nations. (2015). *Millennium development goals and beyond 2015*. United Nations. https://www.un.org/millenniumgoals//environ.shtml

United Nations. (2018). *World urbanization prospects: The 2018 revision*. Department of Economic and Social Affairs. https://population.un.org/wup/

UN-Water. (2020). *The SDG 6 global acceleration framework*. United Nations. https://www.unwater.org/publications/sdg-6-global-acceleration-framework

USMAN, M. (2015). Studies of Ferroelectric and Multiferroic Behavior in [Ba (Zr, Ti) O3] 1-y:[CoFe2O4] y System (Doctoral dissertation, Quaid-i-Azam University, Islamabad). (n.d.).

Valjarević, A. (2025). Oil field impacts on Venezuela's rivers and water stress with environmental challenges. Journal of South American Earth Sciences, [online] 164, p.105675. doi:.DOI: 10.1016/j.jsames.2025.105675

Vijayaraghavan, T., Suriyaraj, S. P., Selvakumar, R., Venkateswaran, R., & Ashok, A. (2016). Rapid and efficient visible light photocatalytic dye degradation using AFe_2O_4 (A = Ba, Ca and Sr) complex oxides. *Materials Science and Engineering B*, *210*, 43–50. DOI: 10.1016/j.mseb.2016.04.005

Vuppaladadiyam, A. K., Merayo, N., Prinsen, P., Luque, R., Blanco, A., & Zhao, M. (2019). A review on greywater reuse: Quality, risks, barriers and global scenarios. *Reviews in Environmental Science and Biotechnology*, *18*(1), 77–99. DOI: 10.1007/s11157-018-9487-9

Wang, C., Zhou, J., Chu, L., Zhang, M., Xu, C., Liu, J., & Li, S. (2024). Diethylenetriamine-functionalized reduced graphene oxide having more amino groups for methylene blue removal. *RSC Advances*, *14*(5), 3280–3288. DOI: 10.1039/D3RA08767K PMID: 38249680

Wang, D., Sun, W., Xu, Y., Tang, H., & Gregory, J. (2004). Speciation stability of inorganic polymer flocculant–PACl. *Colloids and Surfaces. A, Physicochemical and Engineering Aspects*, *243*(1), 1–10. DOI: 10.1016/j.colsurfa.2004.04.073

Wang, L., Chen, Y., Han, X., Zhang, J., & Zhao, L. (2024). Diethylenetriamine-functionalized reduced graphene oxide having more amino groups for methylene blue removal. *Journal of Water Process Engineering*, *59*, 103652. PMID: 38249680

Warsinger, D. M., Mistry, K. H., Nayar, K. G., Chung, H. W., & Lienhard, V. J. H. (2015). Entropy generation of desalination powered by variable temperature waste heat. *Entropy (Basel, Switzerland)*, *17*(11), 7530–7566. DOI: 10.3390/e17117530

Willems, R. C., & MacDonald, S. A. (2016). *Access to Safe Drinking Water: A Low-Income Sustainable Water Purification System.* In *Impact of Water Pollution on Human Health and Environmental Sustainability* (pp. chapter 13). IGI Global. DOI: DOI: 10.4018/978-1-4666-9559-7.ch013

Willner, M. R., & Vikesland, P. J. (2018). Nanomaterial enabled sensors for environmental contaminants. *Journal of Nanobiotechnology*, *16*(1), 95. Advance online publication. DOI: 10.1186/s12951-018-0419-1 PMID: 30466465

World Bank. (2020). *Urban development overview*. The World Bank. https://www.worldbank.org/en/topic/urbandevelopment/overview

World Health Organization (WHO), & UNICEF. (2021). *Progress on household drinking water, sanitation and hygiene 2000–2020: Five years into the SDGs*. WHO & UNICEF. https://washdata.org/

Wu, B. (2019). Membrane-based technology in greywater reclamation: A review. *The Science of the Total Environment*, *656*, 184–200. DOI: 10.1016/j.scitotenv.2018.11.347 PMID: 30504020

Xu, P., Zeng, G. M., Huang, D. L., Lai, C., Zhao, M. H., Wei, Z., Li, N. J., Huang, C., & Xie, G. X. (2012). Adsorption of Pb(II) by iron oxide nanoparticles immobilized Phanerochaete chrysosporium: Equilibrium, kinetic, thermodynamic and mechanisms analysis. *Chemical Engineering Journal*, *203*, 423–431. DOI: 10.1016/j.cej.2012.07.048

Yagub, M. T., Sen, T. K., Afroze, S., & Ang, H. M. (2014). Dye and its removal from aqueous solution by adsorption: A review. *Advances in Colloid and Interface Science*, *209*, 172–184. DOI: 10.1016/j.cis.2014.04.002 PMID: 24780401

Yusof, N. F., Raffi, A. A., Yahaya, N. Z. S., Abas, K. H., Othman, M. H. D., Jaafar, J., & Rahman, M. A. (2023). Surface Modification of UiO-66 on Hollow Fibre Membrane for Membrane Distillation. [PubMed]. *Membranes*, *13*(3), 253. DOI: 10.3390/membranes13030253

Yusuf, A., Almotairy, A. R. Z., Henidi, H., Alshehri, O. Y., & Aldughaim, M. S. (2023). Nanoparticles as drug delivery systems: A review of the implication of nanoparticles' physicochemical properties on responses in biological systems. *Polymers*, *15*(7), 1596. DOI: 10.3390/polym15071596 PMID: 37050210

Yuvaraj, N., Praghash, K., Logeshwaran, J., Peter, G., & Stonier, A. A. (2023). An artificial intelligence based sustainable approaches—IoT systems for smart cities. In *AI Models for Blockchain-Based Intelligent Networks in IoT Systems: Concepts, Methodologies, Tools, and Applications* (pp. 105–120). Springer International Publishing., DOI: 10.1007/978-3-031-31952-5_5

Yuvaraj, N., & Suresh Ghana Dhas, C. (2020). High-performance link-based cluster ensemble approach for categorical data clustering. *The Journal of Supercomputing*, *76*(6), 4556–4579. DOI: 10.1007/s11227-018-2526-z

Zaharieva, K. L., Milenova, K. I., Rives, V., Trujillano, R., Cherkezova-Zheleva, Z. P., Eliyas, A. E., & Mitov, I. G. (2015). Mixed cobalt–copper ferrite-type materials: Synthesis and photocatalytic efficiency in degradation of Reactive Black 5 dye under UV-light irradiation. *Izvestiia po Himiia*, *47*, 105–111.

Zahid, M., Rashid, A., Akram, S., Rehan, Z.A. and Razzaq, W. (2018). A Comprehensive Review on Polymeric Nano-Composite Membranes for Water Treatment. Journal of Membrane Science & Technology, [online] 08(01). doi:.DOI: 10.4172/2155-9589.1000179

Zhang, J., Li, G., Yuan, X., Li, P.-P., Yu, Y., Yang, W., & Zhao, S. (2023). Reduction of Ultrafiltration Membrane Fouling by the Pretreatment Removal of Emerging Pollutants: A Review. [PubMed]. *Membranes*, *13*(1), 77–77. DOI: 10.3390/membranes13010077

Zhi, W., Feng, D., Tsai, W. P., Sterle, G., Harpold, A., Shen, C., & Li, L. (2021). From hydrometeorology to river water quality: Can a deep learning model predict dissolved oxygen at the continental scale? *Environmental Science & Technology*, *55*(4), 2357–2368. DOI: 10.1021/acs.est.0c06783 PMID: 33533608

Zhou, Y., & Chen, X. (Eds.). (2023a). *Special issue: Applications of remote sensing and machine learning in water resources management. Water*. MDPI. https://www.mdpi.com/journal/water/special_issues/0MX8979401

Zhou, P., Xu, Y., & Cui, T. (2025). *A New Photoelectrocatalytical Water Purification System for Simultaneous Removal of Organic Pollutants and Heavy Metal Ions in Water*. Research., DOI: 10.34133/research.0654

Zhu, M., Wang, J., Yang, X., Zhang, Y., Zhang, L., Ren, H., Wu, B., & Ye, L. (2022). A review of the application of machine learning in water quality evaluation. *Eco-Environment & Health*, *1*(2), 107–116. DOI: 10.1016/j.eehl.2022.06.001 PMID: 38075524

About the Contributors

Tawfik A. Saleh is currently a faculty member at King Fahd University of Petroleum and Minerals. He has published several papers as well as proceeding in well-known journals. He also published several books and book chapters and reviews. His research work focused largely in the area of materials technology; synthesis, characterization and applications of inorganic-organic nanocomposites, adsorption and separation. He is in the editorial board of several international journals.

***

El Ghaouti Chahid is a Professor and Ph.D. holder at the Polydisciplinary Faculty of Sidi Bennour, Chouaïb Doukkali University in Morocco, specializing in materials science, solid-state physics, and materials physics.

El Mokhtar El Hafidi is a researcher at the Laboratory of Engineering Sciences for Energy (LabSIPE) at Chouaïb Doukkali University, specializing in environmental technologies, materials science, desalination and water treatment, electrochemistry, and perovskite solar cells. His current project focuses on optimizing water treatment processes using impedance spectroscopy.

Mohammad Irfan is a Professor at the School of Business and Commerce, Chandigarh University, Lucknow (Unnao), India. He earned his Ph.D. from the Central University of Haryana and holds an MBA in Finance, an M.Com in Accounts and Law, and an MA in Economics. A UGC-JRF/SRF/NET qualifier in Management and Commerce, he also holds certifications from NSE (NCFM) and BSE. With more than sixteen years of teaching and research experience, Dr. Irfan's expertise includes Security Analysis and Portfolio Management (SAPM), Artificial Intelligence, Machine Learning, Blockchain, Cryptocurrency, Financial Engineering, FinTech, Green Finance, and Alternative Finance. He has authored over 109 Scopus-indexed research papers in reputed journals such as The Journal of Economic Cooperation and Development (Q2), International Journal of Business Excellence, Montenegrin Journal of Economics (Q2), Cogent Business & Management (Q2), and Journal of Islamic Monetary Economics and Finance. His work has received 700+ citations with an H-index of 15. He has also published 12 books with Springer and IGI Global and is listed among the World's Top 2% Scientists.

Neetha Delphin Mary is an Assistant Professor in the Department of Civil Engineering at Tagore Institute of Engineering and Technology, Deviyakurichi, Salem District, Tamil Nadu, India. With a strong academic background and a commitment to advancing the field of civil engineering, she focuses on structural analysis, sustainable construction practices, and environmental impact assessments. Passionate about research and education, she actively engages in academic development and student mentoring. She can be reached at neethak81@gmail.com.

Said Laasri is a Full Professor at the National School of Applied Sciences of El Jadida, Chouaïb Doukkali University, Morocco. With over 20 years of experience, he specializes in materials science and energy, with expertise in functional materials, thermal processes, energy efficiency, and green hydrogen technologies. He is also Vice Dean of the Polydisciplinary Faculty of Sidi Bennour and a leading contributor to academic innovation, research valorization, and sustainable energy projects

M. Vairavel, M.E., Ph.D., PDF, is an Associate Professor in the Department of Mechanical Engineering at Dhanalakshmi Srinivasan College of Engineering, Coimbatore, Tamil Nadu, India. He serves as the Chief Editor of Kalaivani Publications. He holds two Ph.Ds.—the first (2017) from the Indian School of Science, Gorgand, and the second (2021) in welding from Vels Institute of Technology—and is currently pursuing postdoctoral research. With 123 journal articles and 19 honors, his expertise spans robotics, AI, neural networks, RSM, and solar desalination. From 2010 to 2014, he worked as a steel metal designer at CRAFTSMAN AUTOMATION PVT LTD. He later served as Assistant Professor at various institutions, including Annapoorna Engineering College. He is also active in publishing, having edited and published over 52 books. His contributions include 9 Indian and 3 international patents in solar energy, Rs 1.4 crore in research funding, and extensive CFD-based design projects. He continues to mentor students and professionals in academic and industrial sectors.

Bondala Ramakrishna is an Assistant Professor in the Faculty of Civil Engineering at Annapoorana Engineering College (Autonomous), Salem, Tamil Nadu, India. She brings academic excellence and practical insight into areas such as geotechnical engineering, construction materials, and environmental sustainability.

Dedicated to academic advancement and student development, she actively contributes to teaching, research, and institutional initiatives.

Index

www.ingramcontent.com/pod-product-compliance
Lightning Source LLC
LaVergne TN
LVHW060002031026
849161LV00025B/1274